Precalculus

Precalculus

Sandra DeGuzman and Johanna Halsey

Contents

FUNCTIONS

1.1 Functions and Function Notation

A jetliner changes altitude as its distance from the starting point of a flight increases. The weight of a growing child increases with time. In each case, one quantity depends on another. There is a relationship between the two quantities that we can describe, analyze, and use to make predictions. In this section, we will analyze such relationships.

Determining Whether a Relation Represents a Function

A relation is a set of ordered pairs. The set of the first components of each ordered pair is called the **domain** and the set of the second components of each ordered pair is called the **range**. Consider the following set of ordered pairs. The first numbers in each pair are the first five natural numbers. The second number in each pair is twice that of the first.

$$\{(1,\ 2),\ (2,\ 4),\ (3,\ 6),\ (4,\ 8),\ (5,\ 10)\}$$

The domain is $\{1,\ 2,\ 3,\ 4,\ 5\}$. The range is $\{2,\ 4,\ 6,\ 8,\ 10\}$.

Note that each value in the domain is also known as an **input** value, or independent variable, and is often labeled with the lowercase letter x. Each value in the range is also known as an **output** value, or dependent variable, and is often labeled lowercase letter y.

A function f is a relation that assigns a single value in the range to each value in the domain. In other words, no x-values are repeated. For our example that relates the first five natural numbers to numbers double their values, this relation is a function because each element in the domain, $\{1,\ 2,\ 3,\ 4,\ 5\}$, is paired with exactly one element in the range, $\{2,\ 4,\ 6,\ 8,\ 10\}$.

Now let's consider the set of ordered pairs that relates the terms "even" and "odd" to the first five natural numbers. It would appear as

$$\{(\text{odd},\ 1),\ (\text{even},\ 2),\ (\text{odd},\ 3),\ (\text{even},\ 4),\ (\text{odd},\ 5)\}$$

Notice that each element in the domain, $\{\text{even},\ \text{odd}\}$ is *not* paired with exactly one element in the range, $\{1,\ 2,\ 3,\ 4,\ 5\}$. For example, the term "odd" corresponds to three values from the range, $\{1,\ 3,\ 5\}$ and the term "even" corresponds to two values from the range, $\{2,\ 4\}$. This violates the definition of a function, so this relation is not a function.

Figure 1 compares relations that are functions and not functions.

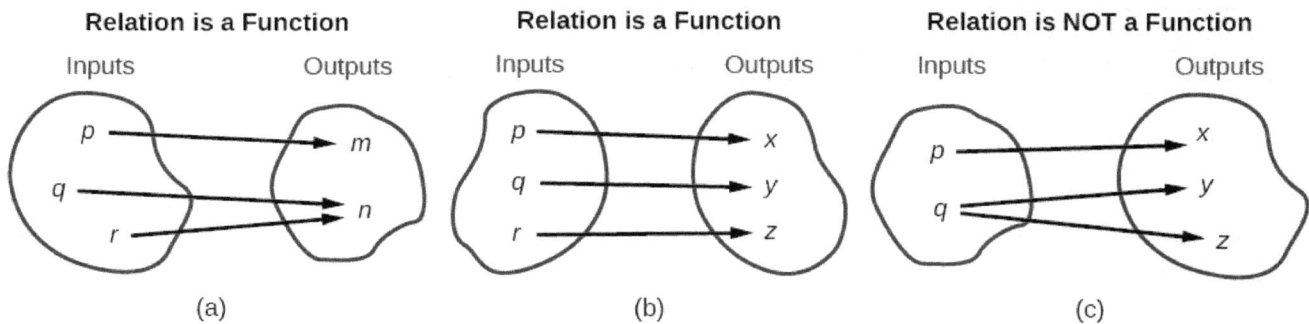

Figure 1: (a) This relationship is a function because each input is associated with a single output. Note that input q and r both give output n. (b) This relationship is also a function. In this case, each input is associated with a single output. (c) This relationship is not a function because input q is associated with two different outputs.

DEFINITION

A *function* is a relation in which each possible input value leads to exactly one output value. We say "the output is a function of the input."

The **input values** make up the **domain**, and the **output values** make up the **range**.

How To

Given a relationship between two quantities, determine whether the relationship is a function.

1. Identify the input values.
2. Identify the output values.
3. If each input value leads to only one output value, classify the relationship as a function. If any input value leads to two or more outputs, do not classify the relationship as a function.

EXAMPLE 1: DETERMINING IF MENU PRICE LISTS ARE FUNCTIONS

The coffee shop menu, shown in Figure 2 consists of items and their prices.

    a. Is price a function of the item?

    b. Is the item a function of the price?

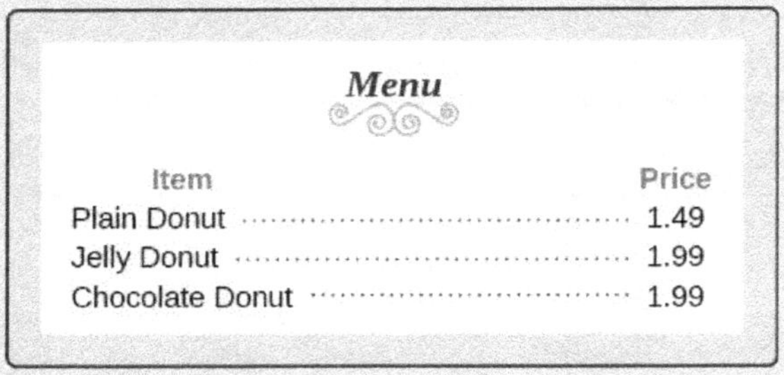

Menu

Item	Price
Plain Donut	1.49
Jelly Donut	1.99
Chocolate Donut	1.99

Figure 2

Answer

    a. Let's begin by considering the input as the items on the menu. The output values are then the prices. See Figure 3.

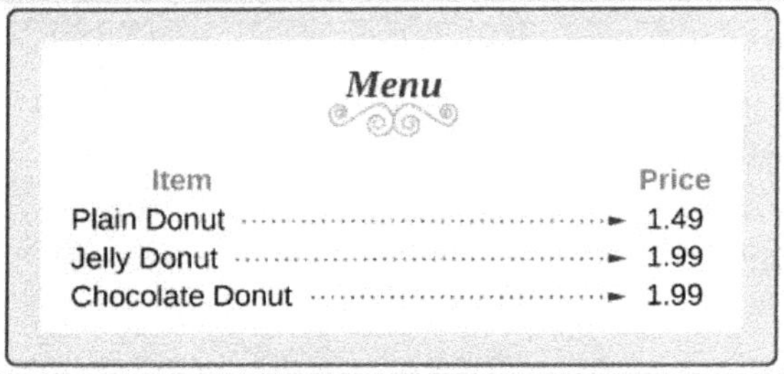

Menu

Item	Price
Plain Donut	► 1.49
Jelly Donut	► 1.99
Chocolate Donut	► 1.99

Figure 3

Each item on the menu has only one price, so the price is a function of the item.

    b. Two items on the menu have the same price. If we consider the prices to be the input values and the items to be the output, then the same input value could have more than one output associated with it. See Figure 4. Therefore, the item is a not a function of price.

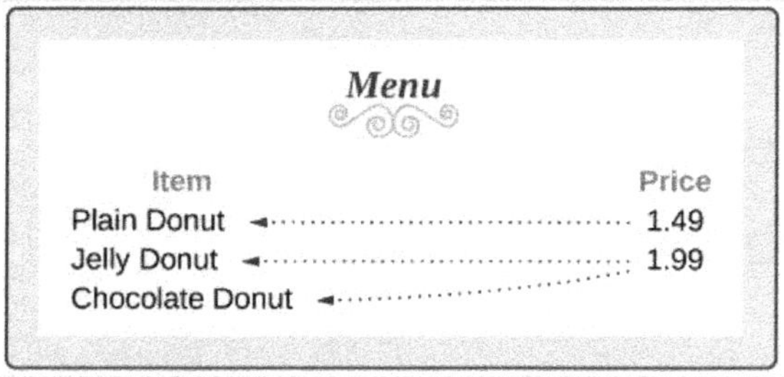

Menu

Item	Price
Plain Donut ◄	1.49
Jelly Donut ◄	1.99
Chocolate Donut ◄	

Figure 4

EXAMPLE 2: DETERMINING IF CLASS GRADE RULES ARE FUNCTIONS

In a particular math class, the overall percent grade corresponds to a grade point average. Is grade point average a function of the percent grade? Is the percent grade a function of the grade point average? Table 1 shows a possible rule for assigning grade points.

Table 1

Percent grade	0–56	57–61	62–66	67–71	72–77	78–86	87–91	92–100
Grade point average	0.0	1.0	1.5	2.0	2.5	3.0	3.5	4.0

Answer

For any percent grade earned, there is an associated grade point average, so the grade point average is a function of the percent grade. In other words, if we input the percent grade, the output is a specific grade point average.

In the grading system given, there is a range of percent grades that correspond to the same grade point average. For example, students who receive a grade point average of 3.0 could have a variety of percent grades ranging from 78 all the way to 86. Thus, percent grade is not a function of grade point average.

TRY IT #1

Table 2[1] lists the five greatest baseball players of all time in order of rank.

Table 2

Player	Rank
Babe Ruth	1
Willie Mays	2
Ty Cobb	3
Walter Johnson	4
Hank Aaron	5

a. Is the rank a function of the player name?
b. Is the player name a function of the rank?

Answer

a. Yes. The input would be the player name, and the output would be the rank. Each player is mapped to exactly one rank. This meets the definition of function.

---

1. http://www.baseball-almanac.com/legendary/lisn100.shtml. Accessed 3/24/2014

b. Yes. The input would be the rank, and the output would be the player name. Each rank is mapped to exactly one name, so this meets the definition of function. However, if two players had been tied for, say, 4th place, then the name would not have been a function of rank.

Using Function Notation

Once we determine that a relationship is a function, we need to display and define the functional relationships so that we can understand and use them, and sometimes also so that we can program them into computers. There are various ways of representing functions. A standard function notation is one representation that facilitates working with functions.

To represent "height is a function of age," we start by identifying the descriptive variables h for height and a for age. The letters f, g, and h are often used to represent functions just as we use x, y, and z to represent numbers and A, B, and C to represent sets.

h is f of a We name the function f; height is a function of age.

$h = f(a)$ We use parentheses to indicate the function input.

$f(a)$ We name the function f; the expression is read as "f of a."

Remember, we can use any letter to name the function; the notation $f(a)$ shows us that height, h, depends on age, a. The value a must be put into the function f to get the height. The parentheses indicate that age is input into the function; they do not indicate multiplication.

We can also give an algebraic expression as the input to a function. For example $f(a + b)$ means "first add a and b, and the result is the input for the function *f*." The operations must be performed in this order to obtain the correct result.

DEFINITION

function notation: The notation $y = f(x)$ defines a function named f. This is read as "y is a function of x." The letter x represents the input value, or independent variable. The letter y, or $f(x)$, represents the output value, or dependent variable.

EXAMPLE 3: USING FUNCTION NOTATION FOR DAYS IN A MONTH

Use function notation to represent a function whose input is the name of a month and output is the number of days in that month. Assume that the domain does not include leap years.

Answer

The number of days in a month is a function of the name of the month, so if we name the function f, we write $\text{days} = f(\text{month})$ or $d = f(m)$. The name of the month is the input to a "rule" that associates a specific number (the output) with each input.

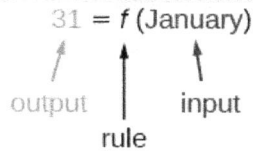

$$31 = f(\text{January})$$

output ↑ ↑ ↑ input

rule

Figure 5

For example, $f(\text{March}) = 31$, because March has 31 days. The notation $d = f(m)$ reminds us that the number of days, d (the output), is dependent on the name of the month, m (the input).

Analysis

Note that the inputs to a function do not have to be numbers; function inputs can be names of people, labels of geometric objects, or any other element that determines some kind of output. However, most of the functions we will work with in this book will have numbers or algebraic expressions as inputs and outputs.

EXAMPLE 4: INTERPRETING FUNCTION NOTATION

A function $N = f(y)$ gives the number of police officers, N, in a town in year y. What does $f(2005) = 300$ represent?

Answer

When we read $f(2005) = 300$, we see that the input year is 2005. The value for the output, the number of police officers, N, is 300. Remember, $N = f(y)$. The statement $f(2005) = 300$ tells us that in the year 2005 there were 300 police officers in the town.

TRY IT #2

Use function notation to express the weight of a pig in pounds as a function of its age in days d.

Answer

$w = f(d)$ since the input would be days and the output would be weight.

Q&A

Instead of a notation such as $y = f(x)$, could we use the same symbol for the output as for the function, such as $y = y(x)$, meaning "y is a function of x?"

Yes, this is often done, especially in applied subjects that use higher math, such as physics and engineering. However, in exploring math itself we like to maintain a distinction between a function such as f, which is a rule or procedure, and the output y we get by applying f to a particular input x. This is why we usually use notation such as $y = f(x), P = W(d)$, and so on.

Representing Functions Using Tables

A common method of representing functions is in the form of a table. The table rows or columns display the corresponding input and output values. In some cases, these values represent all we know about the relationship; other times, the table provides a few select examples from a more complete relationship.

Table 3 lists the input number of each month (January = 1, February = 2, and so on) and the output value of the number of days in that month. This information represents all we know about the months and days for a given year (that is not a leap year). Note that, in this table, we define a days-in-a-month function f where $D = f(m)$ identifies months by an integer rather than by name.

Table 3

Month number, m (input)	1	2	3	4	5	6	7	8	9	10	11	12
Days in month, D (output)	31	28	31	30	31	30	31	31	30	31	30	31

Table 4 defines a function $Q = g(n)$. Remember, this notation tells us that g is the name of the function that takes the input n and gives the output Q.

Table 4

n	1	2	3	4	5
Q	8	6	7	6	8

Table 5 displays the age of children in years and their corresponding heights. This table displays just some of the data available for the heights and ages of children. We can see right away that this table does not represent a function because the same input value, 5 years, has two different output values, 40 in. and 42 in.

Table 5

Age in years, a (input)	5	5	6	7	8	9	10
Height in inches, h (output)	40	42	44	47	50	52	54

HOW TO

Given a table of input and output values, determine whether the table represents a function.

1. Identify the input and output values.

---

2. Check to see if each input value is paired with only one output value. If so, the table represents a function.

EXAMPLE 5: IDENTIFYING TABLES THAT REPRESENT FUNCTIONS

Which table, Table 6, Table 7, or Table 8, represents a function (if any)?

Table 6

Input	Output
2	1
5	3
8	6

Table 7

Input	Output
−3	5
0	1
4	5

Table 8

Input	Output
1	0
5	2
5	4

Answer

Table 6 and Table 7 define functions. In both, each input value corresponds to exactly one output value. Table 8 does not define a function because the input value of 5 corresponds to two different output values.

When a table represents a function, corresponding input and output values can also be specified using function notation.

The function represented by Table 6 can be represented by writing

$$f(2) = 1, \ f(5) = 3, \text{ and } f(8) = 6.$$

Similarly, the statements

$$g(-3) = 5, \ g(0) = 1, \text{ and } g(4) = 5$$

represent the function in Table 7.

Table 8 cannot be expressed in a similar way because it does not represent a function.

TRY IT #3

Does Table 9 represent a function?

Table 9

Input	Output
1	10
2	100
3	1000

Answer

Yes. Each input corresponds to exactly one output.

Finding Input and Output Values of a Function

When we know an input value and want to determine the corresponding output value for a function, we **evaluate** the function. Evaluating will always produce one result because each input value of a function corresponds to exactly one output value.

When we know an output value and want to determine the input values that would produce that output value, we set the output equal to the function's formula and **solve** for the input. Solving can produce more than one solution because different input values can produce the same output value.

Evaluation of Functions in Algebraic Forms

When we have a function in formula form, it is usually a simple matter to evaluate the function. For example, the function $f(x) = 5 - 3x^2$ can be evaluated by squaring the input value, multiplying by 3, and then subtracting the product from 5.

HOW TO

Given the formula for a function, evaluate.

1. Replace the input variable in the formula with the value provided.
2. Calculate the result.

EXAMPLE 6: EVALUATING FUNCTIONS AT SPECIFIC VALUES

Evaluate $f(x) = x^2 + 3x - 4$ at

a. 2
b. a
c. $a + h$
d. $\frac{f(a+h) - f(a)}{h}$

Answer

Replace the x in the function with each specified value.

a. Because the input value is a number, 2, we can use simple algebra to simplify.

$$f(2) = 2^2 + 3(2) - 4$$
$$= 4 + 6 - 4$$
$$= 6$$

b. In this case, the input value is a letter so we cannot simplify the answer any further.

$$f(a) = a^2 + 3a - 4$$

c. With an input value of $a + h$, we must use the distributive property.

$$f(a + h) = (a + h)^2 + 3(a + h) - 4$$
$$= a^2 + 2ah + h^2 + 3a + 3h - 4$$

d. In this case, we apply the input values to the function more than once, and then perform algebraic operations on the result. We already found that

$$f(a + h) = a^2 + 2ah + h^2 + 3a + 3h - 4$$

and we know that

$$f(a) = a^2 + 3a - 4.$$

Now we combine the results and simplify.

$$\frac{f(a+h) - f(a)}{h}$$

$$= \frac{(a^2 + 2ah + h^2 + 3h - 4) - (a^2 + 3a - 4)}{h}$$

$$= \frac{2ah + h^2 + 3h}{h}$$

$$= \frac{h(2a + h + 3)}{h} \qquad \text{Factor out h.}$$

$$= 2a + h + 3 \qquad \text{Simplify.}$$

In Example 6, you worked with the expression below:

$$\frac{f(a+h) - f(a)}{h}$$

This is called the difference quotient. In Calculus, we use the difference quotient to develop an important concept called the derivative. You should work to become very comfortable with the difference quotient. We will work with it more in later sections of this chapter.

EXAMPLE 7: EVALUATING FUNCTIONS

Given the function $h(p) = p^2 + 2p$, evaluate $h(4)$.

Answer

To evaluate $h(4)$, we substitute the value 4 for the input variable p in the given function.

$$h(p) = p^2 + 2p$$
$$h(4) = (4)^2 + 2(4)$$
$$= 16 + 8$$
$$= 24$$

Therefore, for an input of 4, we have an output of 24.

TRY IT #4

Given the function $g(m) = \sqrt{m-4}$, evaluate $g(5)$.

Answer

$$g(5) = 1$$

EXAMPLE 8: SOLVING FUNCTIONS

Given the function $h(p) = p^2 + 2p$, solve for $h(p) = 3$.

Answer

$$h(p) = 3$$
$$p^2 + 2p = 3 \qquad \text{Substitute the original function.}$$
$$p^2 + 2p - 3 = 0 \qquad \text{Subtract 3 from each side.}$$
$$(p+3)(p-1) = 0 \qquad \text{Factor.}$$

If $(p+3)(p-1) = 0$, either $(p+3) = 0$ or $(p-1) = 0$ (or both of them equal 0). We will set each factor equal to 0 and solve for p in each case.

$$(p+3) = 0, \quad p = -3$$
$$(p-1) = 0, \quad p = 1$$

This gives us two solutions. The output $h(p) = 3$ when the input is either $p = 1$ or $p = -3$. We can also verify by graphing as in Figure 6. The graph verifies that $h(1) = h(-3) = 3$ and $h(4) = 24$.

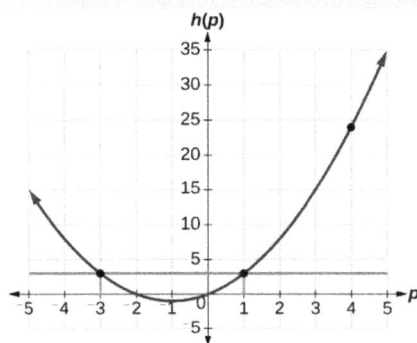

p	-3	-2	0	1	4
$h(p)$	3	0	0	3	24

Figure 6

TRY IT #5

Given the function $g(m) = \sqrt{m-4}$, solve $g(m) = 2$.

Answer

$$m = 8$$

Evaluating Functions Expressed in Formulas

Some functions are defined by mathematical rules or procedures expressed in equation form. f it is possible to express the function output with a formula involving the input quantity, then we can define a function in algebraic form. For example, the equation $2n + 6p = 12$ expresses a functional relationship between n and p. We can rewrite it to decide if p is a function of n.

HOW TO

Given a function in equation form, write its algebraic formula.

1. Solve the equation to isolate the output variable on one side of the equal sign, with the other side as an expression that involves *only* the input variable.
2. Use all the usual algebraic methods for solving equations, such as adding or subtracting the same quantity to or from both sides, or multiplying or dividing both sides of the equation by the same quantity.

EXAMPLE 9: FINDING AN EQUATION OF A FUNCTION

Express the relationship $2n + 6p = 12$ as a function $p = f(n)$, if possible.

Answer

To express the relationship in this form, we need to be able to write the relationship where p is a function of n, which means writing it as $p = [\text{expression involving } n]$.

$$2n + 6p = 12$$

$$6p = 12 - 2n \qquad \text{Subtract } 2n \text{ from both sides.}$$

$$p = \frac{12 - 2n}{6} \qquad \text{Divide both sides by 6 and simplify.}$$

$$p = \frac{12}{6} - \frac{2n}{6}$$

$$p = 2 - \frac{1}{3}n$$

Therefore, p as a function of n is written as

$$p = f(n) = 2 - \tfrac{1}{3}n.$$

Analysis

It is important to note that not every relationship expressed by an equation can also be expressed as a function with a formula.

EXAMPLE 10: EXPRESSING THE EQUATION OF A CIRCLE AS A FUNCTION

Does the equation $x^2 + y^2 = 1$ represent a function with x as input and y as output? If so, express the relationship as a function $y = f(x)$.

Answer

First we subtract x^2 from both sides.

$$y^2 = 1 - x^2$$

We now try to solve for y in this equation.

$$y = \sqrt{1 - x^2}$$
$$= +\sqrt{1 - x^2} \text{ and } -\sqrt{1 - x^2}$$

We get two outputs corresponding to the same input, so this relationship cannot be represented as a single function $y = f(x)$.

TRY IT #6

If $x - 8y^3 = 0$, express y as a function of x.

Answer

$$y = f(x) = \frac{\sqrt[3]{x}}{2}$$

Q&A

Are there relationships expressed by an equation that do represent a function but which still cannot be represented by an algebraic formula?

Yes, this can happen. For example, given the equation $x = y + 2^y$, if we want to express y as a function of x, there is no simple algebraic formula involving only x that equals y. However, each x does determine a unique value for y, and there are mathematical procedures by which y can be found to any desired accuracy. In this case, we say that the equation gives an implicit (implied) rule for y as a function of x, even though the formula cannot be written explicitly.

Evaluating a Function Given in Tabular Form

As we saw previously, we can represent functions in tables. Conversely, we can use information in tables to write functions, and we can evaluate functions using the tables. For example, how well do our pets recall the fond memories we share with them?

There is an urban legend that a goldfish has a memory of 3 seconds, but this is just a myth. Goldfish can remember up to 3 months, while the beta fish has a memory of up to 5 months. And while a puppy's memory span is no longer than 30 seconds, the adult dog can remember for 5 minutes. This is meager compared to a cat, whose memory span lasts for 16 hours.

The function that relates the type of pet to the duration of its memory span is more easily visualized with the use of a table. See Table 10.[2]

Table 10

Pet	Memory span in hours
Puppy	0.008
Adult dog	0.083
Cat	16
Goldfish	2160
Beta fish	3600

At times, evaluating a function in table form may be more useful than using equations. Here let us call the function P. The domain of the function is the type of pet and the range is a real number representing the number of hours the pet's memory span lasts. We can evaluate the function P at the input value of "goldfish." We would write $P\left(\text{goldfish}\right) = 2160$. Notice that, to evaluate the function in table form, we identify the input value and the corresponding output value from the pertinent row of the table. The tabular form for function P seems ideally suited to this function, more so than writing it in paragraph or function form.

HOW TO

Given a function represented by a table, identify specific output and input values.

1. Find the given input in the row (or column) of input values.
2. Identify the corresponding output value paired with that input value.
3. Find the given output values in the row (or column) of output values, noting every time that output value appears.
4. Identify the input value(s) corresponding to the given output value.

---

2. http://www.kgbanswers.com/how-long-is-a-dogs-memory-span/4221590. Accessed 3/24/2014.

EXAMPLE 11: EVALUATING AND SOLVING A TABULAR FUNCTION

Using Table 11,

    a. Evaluate $g\left(3\right)$.

    b. Solve $g\left(n\right)=6.$

Table 11

n	1	2	3	4	5
$g\left(n\right)$	8	6	7	6	8

Answer

    a. Evaluating $g\left(3\right)$ means determining the output value of the function g for the input value of $n=3.$ The table output value corresponding to $n=3$ is 7, so $g\left(3\right)=7.$

    b. Solving $g\left(n\right)=6$ means identifying the input values, $n,$ that produce an output value of 6. Table 11 shows two solutions: 2 and $4.$ When we input 2 into the function $g,$ our output is 6. When we input 4 into the function $g,$ our output is also 6.

TRY IT #7

Using Table 11, evaluate $g\left(1\right)$.

Answer

$$g\left(1\right)=8$$

Finding Function Values from a Graph

Evaluating a function using a graph also requires finding the corresponding output value for a given input value, only in this case, we find the output value by looking at the graph. Solving a function equation using a graph requires finding all instances of the given output value on the graph and observing the corresponding input value(s).

EXAMPLE 12: READING FUNCTION VALUES FROM A GRAPH

Given the graph in Figure 7,

    a. Evaluate $f\left(2\right)$.

    b. Solve $f\left(x\right)=4.$

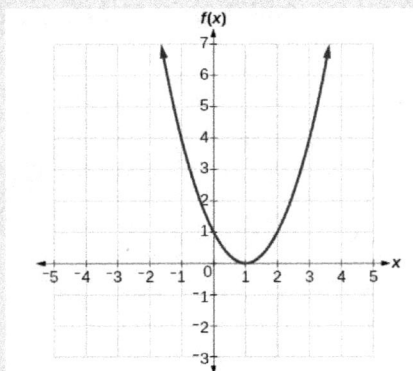

Figure 7

Answer

a. To evaluate $f(2)$, locate the point on the curve where $x = 2$, then read the *y*-coordinate of that point. The point has coordinates $(2, 1)$, so $f(2) = 1$. See Figure 8

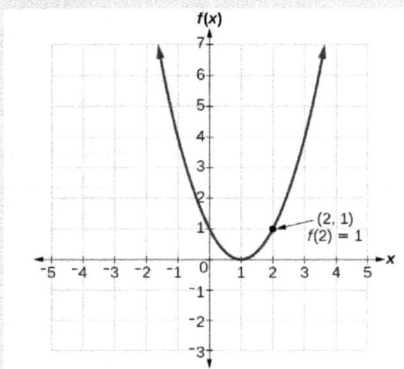

Figure 8

b. To solve $f(x) = 4$, we find the output value 4 on the vertical axis. Moving horizontally along the line $y = 4$, we locate two points of the curve with output value $4 : (-1, 4)$ and $(3, 4)$. These points represent the two solutions to $f(x) = 4 : -1$ or 3. This means $f(-1) = 4$ and $f(3) = 4$, or when the input is -1 or 3, the output is 4. See Figure 9.

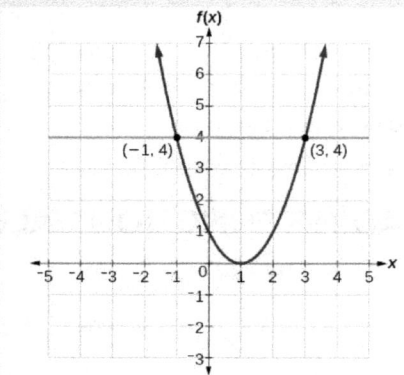

Figure 9

Determining Whether a Function is One-to-One

Some functions have a given output value that corresponds to two or more input values. For example, in the stock chart shown in Figure 10, the stock price was $1000 on five different dates, meaning that there were five different input values that all resulted in the same output value of $1000.

Figure 10

However, some functions have only one input value for each output value, as well as having only one output for each input. We call these functions one-to-one functions. As an example, consider a school that uses only letter grades and decimal equivalents, as listed in Table 12.

Table 12

Letter grade	Grade point average
A	4.0
B	3.0
C	2.0
D	1.0

This grading system represents a one-to-one function, because each letter input yields one particular grade point average output and each grade point average corresponds to one input letter.

To visualize this concept, let's look again at the two simple functions sketched in Figure 1**(a)** and Figure 1**(b)**. The function in part (a) shows a relationship that is not a one-to-one function because inputs q and r both give output n. The function in part (b) shows a relationship that is a one-to-one function because each input is associated with a single output.

DEFINITION

A *one-to-one function* is a function in which each output value corresponds to exactly one input value.

EXAMPLE 13: DETERMINING WHETHER A RELATIONSHIP IS A ONE-TO-ONE FUNCTION

Is the area of a circle a function of its radius? If yes, is the function one-to-one?

Answer

A circle of radius r has a unique area measure given by $A = \pi r^2$, so for any input, r, there is only one output, A. The area is a function of radius r.

If the function is one-to-one, the output value, the area, must correspond to a unique input value, the radius. Any area measure A is given by the formula $A = \pi r^2$. Because areas and radii are positive numbers, there is exactly one solution: $\sqrt{\dfrac{A}{\pi}}$. So the area of a circle is a one-to-one function of the circle's radius.

TRY IT #9

1. Is a balance a function of the bank account number?
2. Is a bank account number a function of the balance?
3. Is a balance a one-to-one function of the bank account number?

Answer

1. Yes, because each bank account has a single balance at any given time;
2. No, because several bank account numbers may have the same balance;
3. No, because the same output may correspond to more than one input.

TRY IT #10

Evaluate the following:

a. If each percent grade earned in a course translates to one letter grade, is the letter grade a function of the percent grade? Explain.

b. If so, is the function one-to-one? Explain.

Answer

a. Yes, the letter grade is a function of percent grade. Each input or percent grade is mapped to exactly one letter grade.

b. No, it is not one-to-one. Each letter grade must be associated with more than one input. There are 100 different percent numbers we could get but only about five possible letter grades, so there cannot be only one percent number that corresponds to each letter grade.

Using the Vertical Line Test

As we have seen in some examples above, we can represent a function using a graph. Graphs display a great many input-output pairs in a small space. The visual information they provide often makes relationships easier to understand. By convention, graphs are typically constructed with the input values along the horizontal axis and the output values along the vertical axis.

Very often graphs name the input value x and the output value y, and we say y is a function of x, or $y = f(x)$ when the function is named f. The graph of the function is the set of all points (x, y) in the plane that satisfies the equation $y = f(x)$. If the function is defined for only a few input values, then the graph of the function is only a few points, where the x-coordinate of each point is an input value and the y-coordinate of each point is the corresponding output value. For example, the black dots on the graph in Figure 11 tell us that $f(0) = 2$ and $f(6) = 1$. However, the set of all points (x, y) satisfying $y = f(x)$ is a curve. The curve shown includes $(0, 2)$ and $(6, 1)$ because the curve passes through those points.

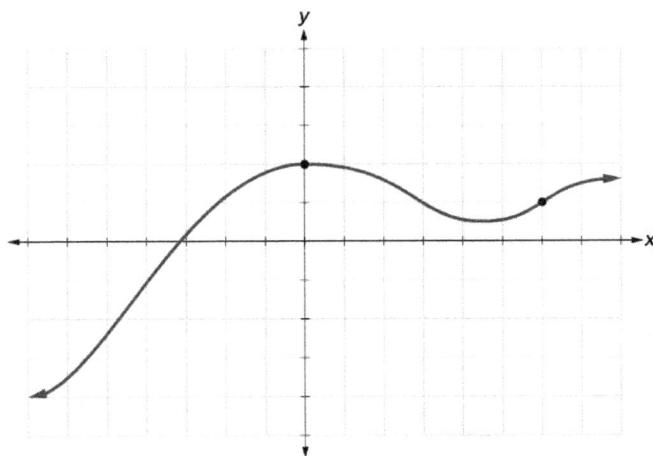

Figure 11

The **vertical line test** can be used to determine whether a graph represents a function. If we can draw any vertical line that intersects a graph more than once, then the graph does *not* define a function because a function can have only one output value for each input value. See Figure 12.

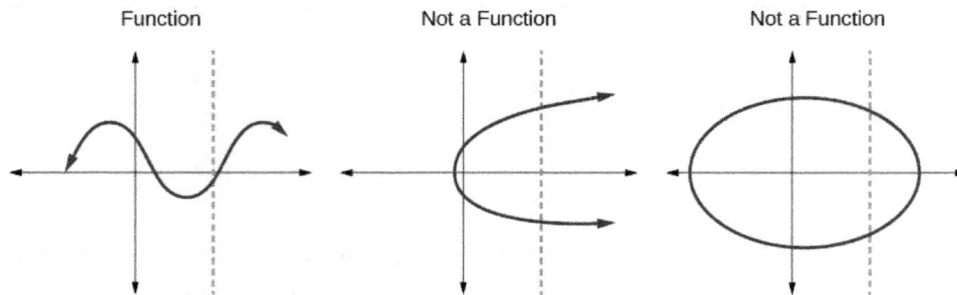

Function   Not a Function   Not a Function

Figure 12

The second and third graphs in Figure 12 are not functions because the input value represented by the dotted line has two output values in each case. The two output values are the y values where each dotted line intersects the solid line. This contradicts the definition of a function. Remember, a function is a rule where each input is mapped to **exactly** one output.

HOW TO

Given a graph, use the vertical line test to determine if the graph represents a function.

1. Inspect the graph to see if any vertical line drawn would intersect the curve more than once.
2. If there is any such line, determine that the graph does not represent a function.

EXAMPLE 14: APPLYING THE VERTICAL LINE TEST

Which of the graphs in Figure 13 represent(s) a function $y = f(x)$?

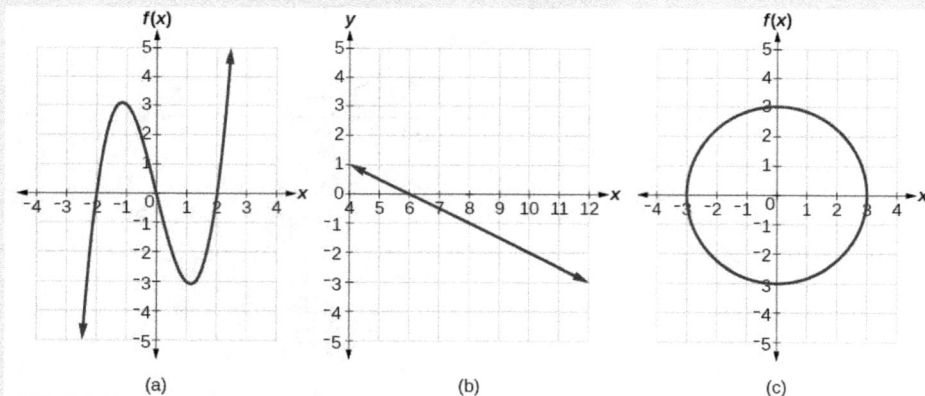

(a)   (b)   (c)

Figure 13

Answer

If any vertical line intersects a graph more than once, the relation represented by the graph is not a function. Notice that any vertical line would pass through only one point of the two graphs shown in parts (a) and (b) of Figure 13. From this we can conclude that these two graphs represent functions. The third graph does not represent a function because, at *x*-values between -3 and 3, a vertical line would intersect the graph at more than one point, as shown in Figure 14. This indicates that each of these inputs gets mapped to two different outputs. This contradicts the definition of a function, since we know a function maps each input to exactly one output.

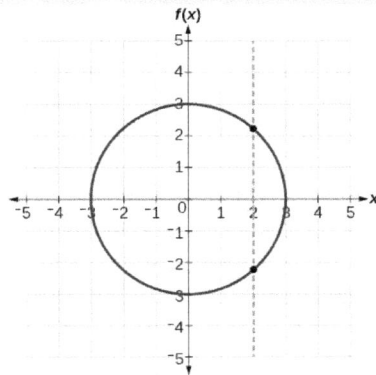

Figure 14

TRY IT #11

Does the graph in Figure 15 represent a function?

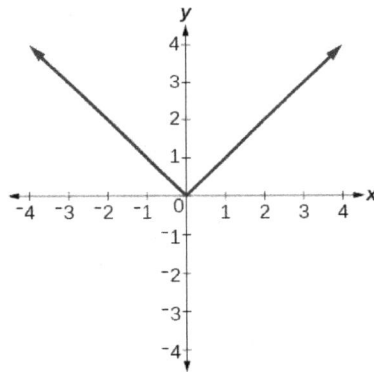

Figure 15

Answer

Yes. Any vertical line will only pass through the graph once. This confirms the definition of one-to-one, since each output corresponds to only one input.

Using the Horizontal Line Test

Once we have determined that a graph defines a function, an easy way to determine if it is a one-to-one function is to use the horizontal line test. Draw horizontal lines through the graph. If any horizontal line intersects the graph more than once, then the graph does not represent a one-to-one function. Each intersection along the horizontal line represents an x-value with the same output which contradicts the definition of one-to-one which states that each output value must be unique for the function to be one-to-one.

HOW TO

Given a graph of a function, use the horizontal line test to determine if the graph represents a one-to-one function.

1. Inspect the graph to see if any horizontal line drawn would intersect the curve more than once.
2. If there is any such line, determine that the function is not one-to-one.

EXAMPLE 15: APPLYING THE HORIZONTAL LINE TEST

Consider the functions shown in Figure 13 **(a)** and Figure 13**(b)**. Are either of the functions one-to-one?

Answer

The function in Figure 13**(a)** is not one-to-one. The horizontal line shown in Figure 16 intersects the graph of the function at two points. These two points have the same output value but different input values. Remember that the definition of a one-to-one function is that each output value corresponds to exactly one input value. Therefore, when a horizontal line intersects a graph at more than one point, we have a contradiction to this definition, and the function cannot be a one-to-one function.

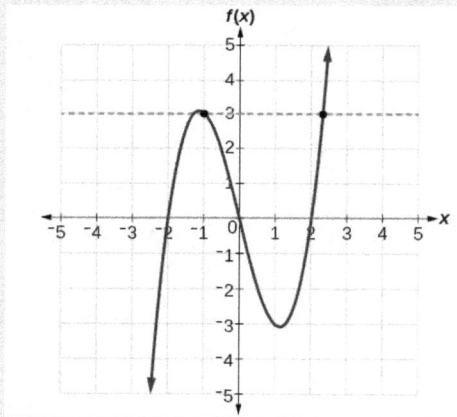

Figure 16

The function in Figure 13**(b)** is one-to-one. Any horizontal line will intersect a diagonal line at most once. This means that each output is associated with only one input value, satisfying the definition of a one-to-one function.

Is the graph shown in Figure 13 **(c)** one-to-one?

Answer

No, because it does not pass the horizontal line test. We can see that there are horizontal lines that would intersect the graph in more than one point, indicating that an output value has more than one input value that corresponds to it. This contradicts the definition of one-to-one.

Identifying Basic Toolkit Functions

In this text, we will be exploring functions including the shapes of their graphs, their unique characteristics, their algebraic formulas, and how to solve problems with them. When learning to read, we start with the alphabet. When learning to do arithmetic, we start with numbers. When working with functions, it is similarly helpful to have a base set of building-block elements. We call these our "toolkit functions," which form a set of basic named functions for which we know the graph, formula, and special properties. Some of these functions are programmed to individual buttons on many calculators. For these definitions we will use x as the input variable and $y = f(x)$ as the output variable.

We will see these toolkit functions, combinations of toolkit functions, their graphs, and their transformations frequently throughout this book. It will be very helpful if we can recognize these toolkit functions and their features quickly by name, formula, graph, and basic table properties. The graphs and sample table values are included with each function shown in Table 13.

Toolkit Functions

Table 13

Name	Function	Graph
Constant	$f(x) = c$, where c is a constant	

For the Graph column, the constant function graph is shown with the following table values:

x	f(x)
−2	2
0	2
2	2

Name	Function	Graph
Identity	$f(x) = x$	 	x	f(x)	 	---	---	 	−2	−2	 	0	0	 	2	2	
Absolute value	$f(x) =	x	$	 	x	f(x)	 	---	---	 	−2	2	 	0	0	 	2	2	
Quadratic	$f(x) = x^2$	 	x	f(x)	 |---|---| 	−2	4	 	−1	1	 	0	0	 	1	1	 	2	4	
Cubic	$f(x) = x^3$	 	x	f(x)	 |---|---| 	−1	−1	 	−0.5	−0.125	 	0	0	 	0.5	0.125	 	1	1	

Name	Function	Graph
Reciprocal	$f\left(x\right)=\frac{1}{x}$	

x	f(x)
−2	−0.5
−1	−1
−0.5	−2
0.5	2
1	1
2	0.5

Name	Function	Graph
Reciprocal squared	$f\left(x\right)=\frac{1}{x^2}$	

x	f(x)
−2	0.25
−1	1
−0.5	4
0.5	4
1	1
2	0.25

Name	Function	Graph
Square root	$f\left(x\right)=\sqrt{x}$	

x	f(x)
0	0
1	1
4	2

Name	Function	Graph
Cube root	$f\left(x\right)=\sqrt[3]{x}$	

x	f(x)
−1	−1
−0.125	−0.5
0	0
0.125	0.5
1	1

Access the following online resources for additional instruction and practice with functions.

- Determine if a Relation is a Function
  - https://youtu.be/zT69oxcMhPw

- Vertical Line Test
  - https://youtu.be/gO5WN9g1fJo

- Introduction to Functions
  - https://youtu.be/sW9-zBeQpCU

- Vertical Line Test on Graph
  - https://youtu.be/5Z8DaZPJLKY

- One-to-one Functions
  - https://youtu.be/QFOJmevha_Y

- Graphs as One-to-one Functions
  - https://youtu.be/tbSGdcSN8RE

Key Equations

Constant function	$f(x) = c$, where c is a constant
Identity function	$f(x) = x$
Absolute value function	$f(x) = |x|$
Quadratic function	$f(x) = x^2$
Cubic function	$f(x) = x^3$
Reciprocal function	$f(x) = \frac{1}{x}$
Reciprocal squared function	$f(x) = \frac{1}{x^2}$
Square root function	$f(x) = \sqrt{x}$
Cube root function	$f(x) = \sqrt[3]{x}$
Difference Quotient	$\frac{f(a+h)-f(a)}{h}$

KEY CONCEPTS

- A relation is a set of ordered pairs. A function is a specific type of relation in which each domain value, or input, leads to exactly one range value, or output.
- Function notation is a shorthand method for relating the input to the output in the form $y = f(x)$.
- In tabular form, a function can be represented by rows or columns that relate to input and output values.
- To evaluate a function, we determine an output value for a corresponding input value. Algebraic forms of a function can be evaluated by replacing the input variable with a given value.
- To solve for a specific function value, we determine the input values that yield the specific output value.

- An algebraic form of a function can be written from an equation.
- Input and output values of a function can be identified from a table.
- Relating input values to output values on a graph is another way to evaluate a function.
- A function is one-to-one if each output value corresponds to only one input value.
- A graph represents a function if any vertical line drawn on the graph intersects the graph at no more than one point.
- The graph of a one-to-one function passes the horizontal line test.

GLOSSARY

dependent variable

an output variable

domain

the set of all possible input values for a relation

function

a relation in which each input value yields a unique output value

horizontal line test

a method of testing whether a function is one-to-one by determining whether any horizontal line intersects the graph more than once

independent variable

an input variable

input

each object or value in a domain that relates to another object or value by a relationship known as a function

one-to-one function

a function for which each value of the output is associated with a unique input value

output

each object or value in the range that is produced when an input value is entered into a function

range

the set of output values that result from the input values in a relation

relation

a set of ordered pairs

vertical line test

a method of testing whether a graph represents a function by determining whether a vertical line intersects the graph no more than once

1.2 Domain and Range

If you're in the mood for a scary movie, you may want to check out one of the five most popular horror movies of all time—*I am Legend*, *Hannibal*, *The Ring*, *The Grudge*, and *The Conjuring*. Figure 1 shows the amount, in dollars, each of those movies grossed when they were released as well as the percent of ticket sales for horror movies in general by year. Notice that we can use the data to create a function of the amount each movie earned or the total ticket sales for all horror movies by year. In creating various functions using the data, we can identify different independent and dependent variables, and we can analyze the data and the functions to determine the domain and range. In this section, we will investigate methods for determining the domain and range of functions such as these.

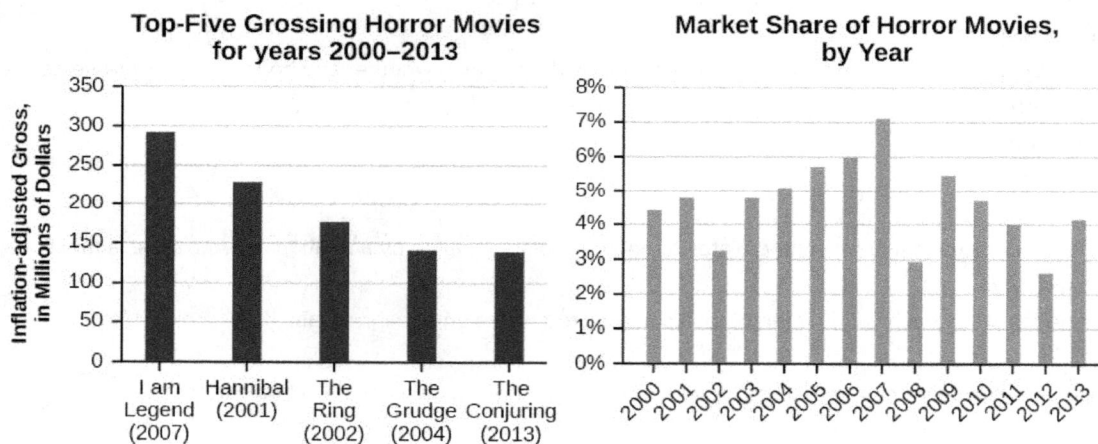

Top-Five Grossing Horror Movies for years 2000–2013

Market Share of Horror Movies, by Year

Figure 1

Based on data compiled by www.the-numbers.com.[1]

Finding the Domain of a Function Defined by an Equation

In Section 1.1, Functions and Function Notation, we were introduced to the concepts of domain and range. In this section, we will practice determining domains and ranges for specific functions. Keep in mind that, in determining domains and ranges, we need to consider what is physically possible or meaningful in real-world examples, such as tickets sales and year in the horror

---

1. The Numbers: Where Data and the Movie Business Meet. "Box Office History for Horror Movies." http://www.the-numbers.com/market/genre/Horror. Accessed 3/24/2014

movie example above. We also need to consider what is mathematically permitted. For example, we cannot include any input value that leads us to take an even root of a negative number if the domain and range consist of real numbers. Or in a function expressed as a formula, we cannot include any input value in the domain that would lead us to divide by 0.

We can visualize the domain as a "holding area" that contains "raw materials" for a "function machine" and the range as another "holding area" for the machine's products. See Figure 2.

For functions of real numbers, we can write the domain and range in interval notation, which uses values within brackets to describe a set of real numbers. In interval notation, we use a square bracket [when the set includes the endpoint and a parenthesis (to indicate that the endpoint is either not included or the interval is unbounded. For example, if a person has $100 to spend, he or she would need to express the interval that is more than 0 and less than or equal to 100 and write $(0, \ 100]$. We will discuss interval notation in greater detail later.

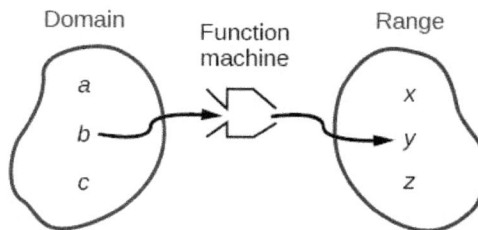

Figure 2

Let's turn our attention to finding the domain of a function whose equation is provided. Oftentimes, finding the domain of such functions involves remembering three different forms. First, if the function has no denominator or an even root, consider whether the domain could be all real numbers. Second, if there is a denominator in the function's equation, exclude values in the domain that force the denominator to be zero. Third, if there is an even root, exclude values that would make the radicand negative.

Before we begin, let us discuss the conventions of interval notation:

- The smallest value from the interval is written first, followed by a comma.
- The largest value in the interval is written second.
- Parentheses, (or), are used to signify that an endpoint is not included, called exclusive.
- Brackets, [or], are used to indicate that an endpoint is included, called inclusive.
- Use a symbol known as the union, ∪, to combine intervals if there are more than one.

See Figure 3 for a summary of interval notation.

Inequality	Interval Notation	Graph on Number Line	Description
$x > a$	(a, ∞)		x is greater than a
$x < a$	$(-\infty, a)$		x is less than a
$x \geq a$	$[a, \infty)$		x is greater than or equal to a
$x \leq a$	$(-\infty, a]$		x is less than or equal to a
$a < x < b$	(a, b)		x is strictly between a and b
$a \leq x < b$	$[a, b)$		x is between a and b, to include a
$a < x \leq b$	$(a, b]$		x is between a and b, to include b
$a \leq x \leq b$	$[a, b]$		x is between a and b, to include a and b

Figure 3

EXAMPLE 1: FINDING THE DOMAIN OF A FUNCTION AS A SET OF ORDERED PAIRS

Find the domain of the following function: $\{(2,\ 10), (3,\ 10), (4,\ 20), (5,\ 30), (6,\ 40)\}$.

Answer

First identify the input values. The input value is the first coordinate in an ordered pair. There are no restrictions, as the ordered pairs are simply listed. The domain is the set of the first coordinates of the ordered pairs.

$$\{2, 3, 4, 5, 6\}$$

We cannot use interval notation here, as we are not including all real numbers between 2 and 6.

TRY IT #1

Find the domain of the function:

$$\{(-5, 4), (0, 0), (5, -4), (10, -8), (15, -12)\}$$

Answer

$\{-5, \ 0, \ 5, \ 10, \ 15\}$ Again, we cannot use interval notation here.

HOW TO

Given a function written in equation form, find the domain.

1. Identify the input values.
2. Identify any restrictions on the input and exclude those values from the domain.
3. Write the domain in interval form, if possible.

EXAMPLE 2: FINDING THE DOMAIN OF A FUNCTION

Find the domain of the function $f(x) = x^2 - 1$.

Answer

The input value, shown by the variable x in the equation, is squared and then the result is lowered by one. Any real number may be squared and then be lowered by one, so there are no restrictions on the domain of this function. The domain is the set of real numbers.

In interval form, the domain of f is $(-\infty, \infty)$.

Notice that we use parenthesis if we are working with an infinite set of values in both directions from zero.

TRY IT #2

Find the domain of the function: $f(x) = 5 - x + x^3$.

Answer

$(-\infty, \infty)$. We know that any real number can be cubed, and have any other real numbers added or subtracted to it. This set of operations will produce a real number.

EXAMPLE 3: FINDING THE DOMAIN OF A FUNCTION INVOLVING A DENOMINATOR

Find the domain of the function $f(x) = \frac{x+1}{2-x}$.

Answer

When there is a denominator, we want to include only values of the input that do not force the denominator to be zero. So, we will set the denominator equal to 0 and solve for x.

$$2 - x = 0$$
$$-x = -2$$
$$x = 2$$

Now, we will exclude 2 from the domain. The answers are all real numbers where $x < 2$ or $x > 2$. We can use a symbol known as the union, $\cup$, to combine the two sets. In interval notation, we write the solution: $(-\infty, 2) \cup (2, \infty)$.

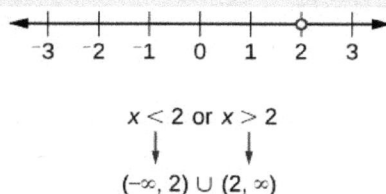

$x < 2$ or $x > 2$

$(-\infty, 2) \cup (2, \infty)$

Figure 4

In interval form, the domain of f is $(-\infty, 2) \cup (2, \infty)$.

TRY IT #3

Find the domain of the function: $f(x) = \frac{1+4x}{2x-1}$.

Answer

$\left(-\infty, \frac{1}{2}\right) \cup \left(\frac{1}{2}, \infty\right)$. We cannot let the denominator be equal to 0. This occurs when $x = \frac{1}{2}$.

HOW TO

Given a function written in equation form including an even root, find the domain.

1. Identify the input values.
2. Since there is an even root, exclude any real numbers that result in a negative number in the radicand. Set the radicand greater than or equal to zero and solve for x.
3. The solution(s) are the domain of the function. If possible, write the answer in interval form.

EXAMPLE 4: FINDING THE DOMAIN OF A FUNCTION WITH AN EVEN ROOT

Find the domain of the function $f(x) = \sqrt{7 - x}$.

Answer

When there is an even root in the formula, we exclude any real numbers that result in a negative number in the radicand.

Set the radicand greater than or equal to zero and solve for x.

$$7 - x \geq 0$$
$$-x \geq -7$$
$$x \leq 7$$

Now, we will exclude any number greater than 7 from the domain. The answers are all real numbers less than or equal to 7, or $\left(-\infty, 7\right]$.

TRY IT #4

Find the domain of the function $f(x) = \sqrt{5 + 2x}$.

Answer

$\left[-\frac{5}{2}, \infty\right)$. We know the radicand must be greater than or equal to 0. We find this occurs for values where the input is greater than or equal to $-\frac{5}{2}$.

Q&A

Can there be functions in which the domain and range do not intersect at all?

Yes. For example, the function $f\left(x\right) = -\frac{1}{\sqrt{x}}$ has the set of all positive real numbers as its domain but the set of all negative real numbers as its range. As a more extreme example, a function's inputs and outputs can be completely different categories (for example, names of weekdays as inputs and numbers as outputs, as on an attendance chart), in such cases the domain and range have no elements in common.

Using Notations to Specify Domain and Range

In the previous examples, we used inequalities and lists to describe the domain of functions. We can also use inequalities, or other statements that might define sets of values or data, to describe the behavior of the variable in set-builder notation. For example, $\left\{x \mid 10 \leq x < 30\right\}$ describes the behavior of x in set-builder notation. The braces $\left\{\right\}$ are read as "the set of," and the vertical bar | is read as "such that," so we would read $\left\{x \mid 10 \leq x < 30\right\}$ as "the set of x-values such that 10 is less than or equal to x, and x is less than 30."

Figure 5 compares inequality notation, set-builder notation, and interval notation.

	Inequality Notation	Set-builder Notation	Interval Notation
	$5 < h \le 10$	$\{h \mid 5 < h \le 10\}$	$(5, 10]$
	$5 \le h < 10$	$\{h \mid 5 \le h < 10\}$	$[5, 10)$
	$5 < h < 10$	$\{h \mid 5 < h < 10\}$	$(5, 10)$
	$h < 10$	$\{h \mid h < 10\}$	$(-\infty, 10)$
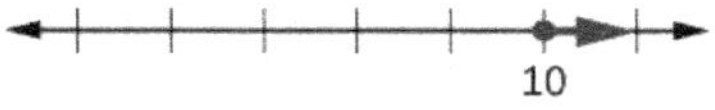	$h \ge 10$	$\{h \mid h \ge 10\}$	$[10, \infty)$
	All real numbers	$\mathbb{R}$	$(-\infty, \infty)$

Figure 5

To combine two intervals using inequality notation or set-builder notation, we use the word "or." As we saw in earlier examples, we use the union symbol, $\cup$, to combine two unconnected intervals. For example, the union of the sets $\{2, 3, 5\}$ and $\{4, 6\}$ is the set $\{2, 3, 4, 5, 6\}$. It is the set of all elements that belong to one *or* the other (or both) of the original two sets. For sets with a finite number of elements like these, the elements do not have to be listed in ascending order of numerical value. If the original two sets have some elements in common, those elements should be listed only once in the union set. For sets of real numbers on intervals, another example of a union is

$$\{x \mid |x| \ge 3\} = (-\infty, -3] \cup [3, \infty)$$

Set-Builder Notation and Interval Notation

DEFINITION

Set-builder notation is a method of specifying a set of elements that satisfy a certain condition. It takes the form $\{x \mid \text{statement about } x\}$ which is read as, "the set of all x such that the statement about x is true." For example,

$$\{x \mid 4 < x \le 12\}$$

Interval notation is a way of describing sets that include all real numbers between a lower limit that may or may not be included and an upper limit that may or may not be included. The endpoint values are listed between brackets or parentheses. A square bracket indicates inclusion in the set, and a parenthesis indicates exclusion from the set. For example, $(4, 12]$.

HOW TO

Given a line graph, describe the set of values using interval notation.

1. Identify the intervals to be included in the set by determining where the heavy line overlays the real line.
2. At the left end of each interval, use [with each end value to be included in the set (solid dot) or (for each excluded end value (open dot).
3. At the right end of each interval, use] with each end value to be included in the set (filled dot) or) for each excluded end value (open dot).
4. Use the union symbol, $\cup$, to combine all intervals into one set.

EXAMPLE 5: DESCRIBING SETS ON THE REAL-NUMBER LINE

Describe the intervals of values shown in Figure 6 using inequality notation, set-builder notation, and interval notation.

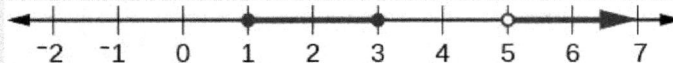

Figure 6

Answer

To describe the values, x, included in the intervals shown, we would say, "x is a real number greater than or equal to 1 and less than or equal to 3, or a real number greater than 5."

Inequality	$1 \leq x \leq 3$ or $x > 5$
Set-builder notation	$\{x \mid 1 \leq x \leq 3 \text{ or } x > 5\}$
Interval notation	$[1, 3] \cup (5, \infty)$

Remember that, when writing or reading interval notation, using a square bracket means the boundary is included in the set. Using a parenthesis means the boundary is not included in the set.

TRY IT #5

Given Figure 7, specify the graphed set in

a. words
b. set-builder notation
c. interval notation

Figure 7

Answer

a. Values that are less than or equal to –2, or values that are greater than or equal to –1 and less than 3;

b. $\{x | x \leq -2 \text{ or } -1 \leq x < 3\}$;

c. $(-\infty, -2] \cup [-1, 3)$

Finding Domain and Range from Graphs

Another way to identify the domain and range of functions is by using graphs. Because the domain refers to the set of possible input values, the domain of a graph consists of all the input values shown on the x-axis. The range is the set of possible output values, which are shown on the y-axis. Keep in mind that if the graph continues beyond the portion of the graph we can see, the domain and range may be greater than the visible values. This is indicated by the arrows in Figure 8.

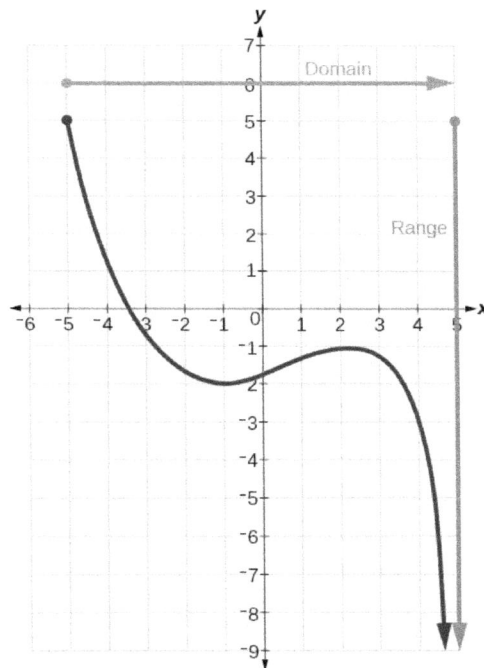

Figure 8

We can observe that the graph extends horizontally from -5 to the right without bound, so the domain is $[-5, \infty)$. The vertical extent of the graph is all range values 5 and below, so the range is $(-\infty, 5]$. Note that the domain and range are always written from lower to higher values, or from left to right for domain, and from the bottom of the graph to the top of the graph for range.

EXAMPLE 6: FINDING DOMAIN AND RANGE FROM A GRAPH

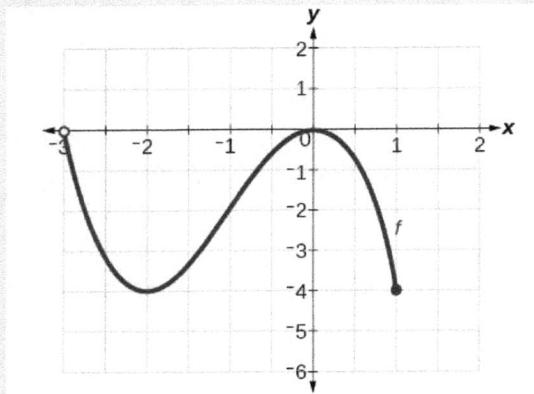

Figure 9

Find the domain and range of the function f whose graph is shown in Figure 9.

Answer

We can observe that the horizontal extent of the graph is –3 but not inclusive to 1 included, so the domain of f is $(-3, 1]$.

The vertical extent of the graph is 0 to –4, so the range is $[-4, 0]$. See Figure 10.

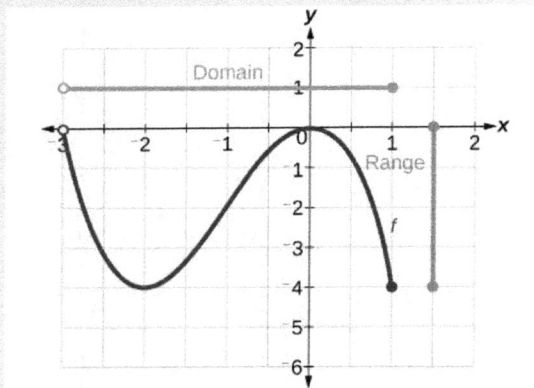

Figure 10

EXAMPLE 7: FINDING DOMAIN AND RANGE FROM A GRAPH OF OIL PRODUCTION

Find the domain and range of the function f whose graph is shown in Figure 11.

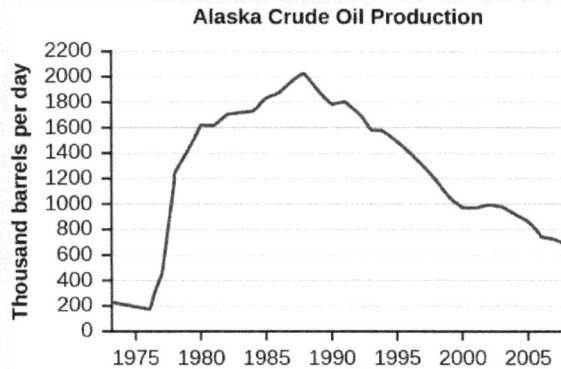

Alaska Crude Oil Production

Figure 11

(credit: modification of work by the U.S. Energy Information Administration)[2]

Answer

The input quantity along the horizontal axis is "years," which we represent with the variable t for time. The output quantity is "thousands of barrels of oil per day," which we represent with the variable b for barrels. The graph may continue to the left and right beyond what is viewed, but based on the portion of the graph that is visible, we can determine the domain as $1973 \leq t \leq 2008$ and the range as approximately $180 \leq b \leq 2010$.

In interval notation, the domain is [1973, 2008], and the range is about [180, 2010]. For the domain and the range, we approximate the smallest and largest values since they do not fall exactly on the grid lines.

TRY IT #6

Given Figure 12, identify the domain and range using interval notation.

---

2. http://www.eia.gov/dnav/pet/hist/LeafHandler.ashx?n=PET&s=MCRFPAK2&f=A.

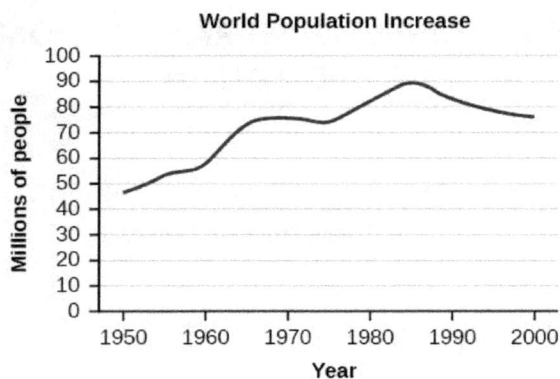

World Population Increase

Figure 12

Answer

domain = [1950,2002] range = [47,000,000,89,000,000]

Q&A

Can a function's domain and range be the same?

Yes. For example, the domain and range of the cube root function are both the set of all real numbers.

Finding Domains and Ranges of the Toolkit Functions

We will now return to our set of toolkit functions to determine the domain and range of each.

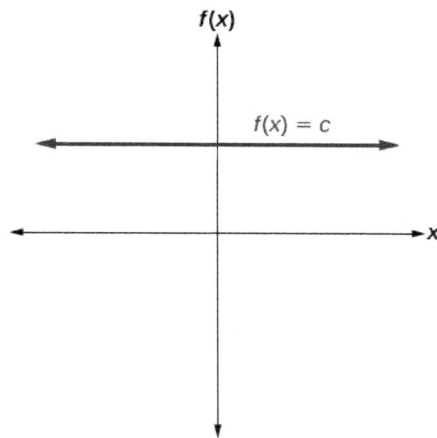

Domain: $(-\infty, \infty)$
Range: $[c, c]$

For the constant function $f\left(x\right) = c,$ the domain consists of all real numbers; there are no restrictions on the input. The only output value is the constant $c,$ so the range is the set $\left\{c\right\}$ that contains this single element. In interval notation, this is written as $\left[c, c\right]$, the interval that both begins and ends with $c.$

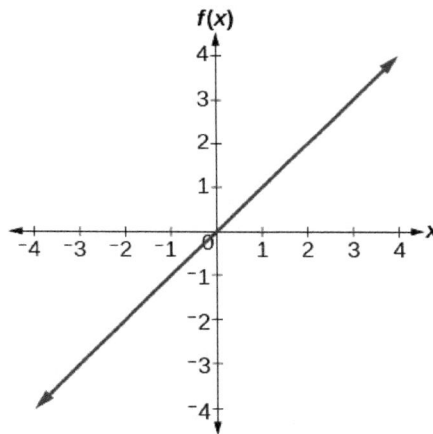

Domain: $(-\infty, \infty)$
Range: $(-\infty, \infty)$

For the identity function $f\left(x\right) = x,$ there is no restriction on $x.$ Both the domain and range are the set of all real numbers.

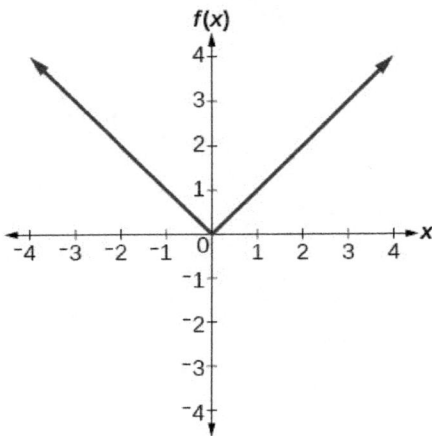

Domain: $(-\infty, \infty)$
Range: $[0, \infty)$

For the absolute value function $f\left(x\right) = \left|x\right|$, there is no restriction on x. However, because absolute value is defined as a distance from 0, the output can only be greater than or equal to 0.

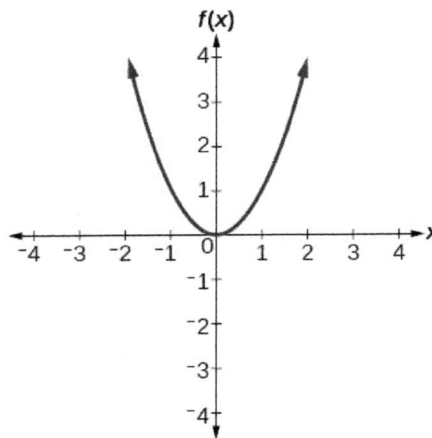

Domain: $(-\infty, \infty)$
Range: $[0, \infty)$

For the quadratic function $f\left(x\right) = x^2$, the domain is all real numbers since any real number can be squared and result in a real number. We can also see graphically that the horizontal extent of the graph is the whole real number line. We can also see vertically, the graph does not include any negative values for the range, so the range is only nonnegative real numbers.

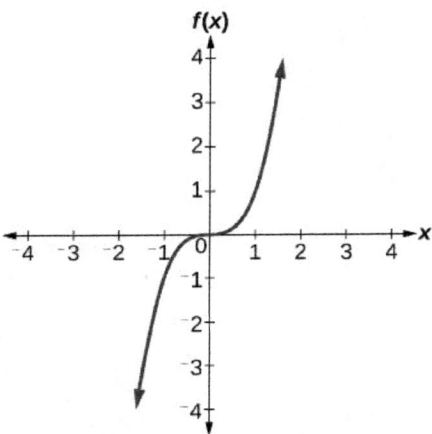

Domain: $(-\infty, \infty)$
Range: $(-\infty, \infty)$

For the cubic function $f\left(x\right) = x^3$, the domain is all real numbers because any real number raised to the third power will result in a real number. Graphically we can see that the horizontal extent of the graph is the whole real number line. The same applies to the vertical extent of the graph, so the domain and range include all real numbers.

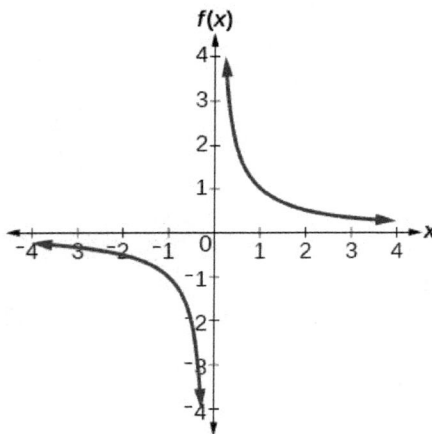

Domain: $(-\infty, 0) \cup (0, \infty)$
Range: $(-\infty, 0) \cup (0, \infty)$

For the reciprocal function $f\left(x\right) = \frac{1}{x}$, we cannot divide by 0, so we must exclude 0 from the domain. Further, 1 divided by any value can never be 0, so the range also will not include 0. The graph above indicates all other real numbers can be included in the range. In set-builder notation, we could also write $\left\{x\middle|\ x \neq 0\right\}$, the set of all real numbers that are not zero.

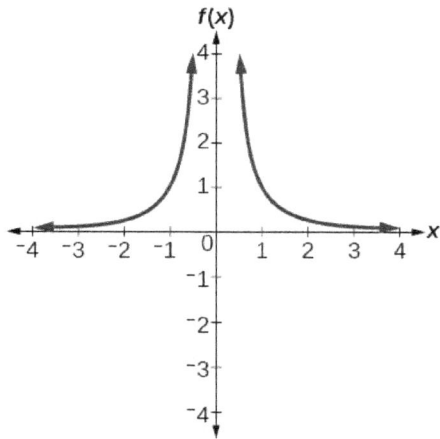

Domain: $(-\infty, 0) \cup (0, \infty)$
Range: $(0, \infty)$

For the reciprocal squared function $f\left(x\right) = \frac{1}{x^2}$, we cannot divide by 0, so we must exclude 0 from the domain. There is also no x that can give an output of 0, so 0 is excluded from the range as well. Note that the output of this function is always positive due to the square in the denominator, so the range includes only positive numbers.

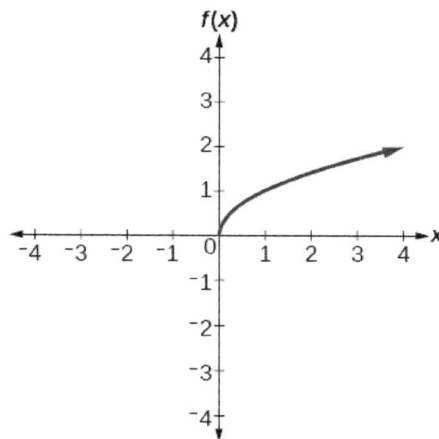

Domain: $[0, \infty)$
Range: $[0, \infty)$

For the square root function $f\left(x\right) = \sqrt{x}$, we cannot take the square root of a negative real number, so the domain must be 0 or greater. The range also excludes negative numbers because the square root of a positive number x is defined to be positive, even though the square of the negative number $-\sqrt{x}$ also gives us x.

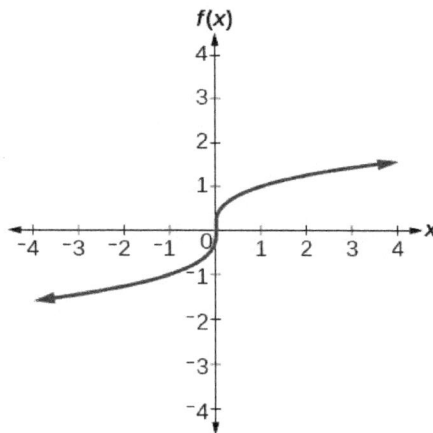

Domain: $(-\infty, \infty)$
Range: $(-\infty, \infty)$

For the cube root function $f\left(x\right) = \sqrt[3]{x}$, the domain and range include all real numbers. Note that there is no problem taking a cube root, or any odd-integer root, of a negative number, and the resulting output is negative (it is an odd function). We can see from the graph that we have not restrictions either horizontally or vertically.

HOW TO

Given the formula for a function, determine the domain and range.

1. Exclude from the domain any input values that result in division by zero.

---

2. Exclude from the domain any input values that have nonreal (or undefined) number outputs.

3. Use the valid input values to determine the range of the output values.

4. Look at the function graph and table values to confirm the actual function behavior.

EXAMPLE 8: FINDING THE DOMAIN AND RANGE USING TOOLKIT FUNCTIONS

Find the domain and range of $f(x) = 2x^3 - x$.

Answer

There are no restrictions on the domain, as any real number may be cubed and then subtracted from the result. The range can be found by graphing the function and viewing the vertical extent of the graph. The range of this type of function will be discussed in detail in chapter 4.

The domain is $(-\infty, \infty)$ and the range is also $(-\infty, \infty)$.

EXAMPLE 9: FINDING THE DOMAIN AND RANGE

Find the domain and range of $f(x) = \frac{2}{x+1}$.

Answer

We cannot evaluate the function at -1 because division by zero is undefined. The domain is $(-\infty, -1) \cup (-1, \infty)$. Because the function is never zero, we exclude 0 from the range. The range is $(-\infty, 0) \cup (0, \infty)$.

EXAMPLE 10: FINDING THE DOMAIN AND RANGE

Find the domain and range of $f(x) = 2\sqrt{x+4}$.

Answer

We cannot take the square root of a negative number, so the value inside the radical must be nonnegative.

$$x + 4 \geq 0 \text{ when } x \geq -4$$

The domain of $f(x)$ is $[-4, \infty)$.

We then find the range. We know that $f(-4) = 0$, and the function value increases as x increases without any upper limit. We conclude that the range of f is $[0, \infty)$.

Analysis

Figure 13 represents the function f.

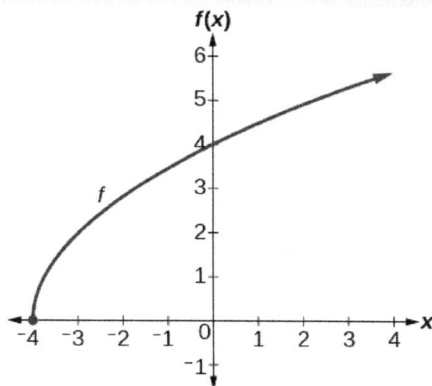

Figure 13

TRY IT #7

Find the domain and range of $f(x) = -\sqrt{2-x}$.

Answer

domain: $(-\infty, 2]$; range: $(-\infty, 0]$

EXAMPLE 11: FIND THE DOMAIN

Find the domain of the function $f(x) = \frac{1}{\sqrt{x-1}}$.

Answer

This function involves both a fraction and an even root so we need to consider how each of these properties will effect the domain. First we know that the expression under the radicand, $x-1$, needs to be greater than or equal to zero. However, if the expression equals zero, we will get a divide by zero. Therefore, because the square root is in the denominator, we solve

$$x - 1 > 0.$$

This simplifies to $x > 1$, so the domain is $(1, \infty)$.

Graphing Piecewise-Defined Functions (Optional)

Sometimes, we come across a function that requires more than one formula in order to obtain the given output. For example, in the toolkit functions, we introduced the absolute value function $f(x) = |x|$. With a domain of all real numbers and a range of values greater than or equal to 0, absolute value can be defined as the magnitude, or modulus, of a real number value regardless of sign. It is the distance from 0 on the number line. All of these definitions require the output to be greater than or equal to 0.

If we input 0, or a positive value, the output is the same as the input.

$$f(x) = x \text{ if } x \geq 0$$

If we input a negative value, the output is the opposite of the input.

$$f(x) = -x \text{ if } x < 0$$

Because this requires two different processes or pieces, the absolute value function is an example of a piecewise function. A piecewise function is a function in which more than one formula is used to define the output over different pieces of the domain.

We use piecewise functions to describe situations in which a rule or relationship changes as the input value crosses certain "boundaries." For example, we often encounter situations in business for which the cost per piece of a certain item is discounted once the number ordered exceeds a certain value. Tax brackets are another real-world example of piecewise functions. For example, consider a simple tax system in which incomes up to \$10,000 are taxed at 10%, and any additional income is taxed at 20%. The tax on a total income S would be $0.1S$ if $S \leq 10,000$ and $1000 + 0.2(S - 10,000)$ if $S > 10,000$.

DEFINITION

A **piecewise function** is a function in which more than one formula is used to define the output. Each formula has its own domain, and the domain of the function is the union of all these smaller domains. We notate this idea like this:

$$f(x) = \begin{cases} \text{formula 1 if } x \text{ is in domain 1} \\ \text{formula 2 if } x \text{ is in domain 2} \\ \text{formula 3 if } x \text{ is in domain 3} \end{cases}$$

In piecewise notation, the absolute value function is

$$|x| = \begin{cases} x \text{ if } x \geq 0 \\ -x \text{ if } x < 0 \end{cases}$$

HOW TO

Given a piecewise function, write the formula and identify the domain for each interval.

1. Identify the intervals for which different rules apply.
2. Determine formulas that describe how to calculate an output from an input in each interval.

---

3. Use braces and if-statements to write the function.

EXAMPLE 12: WRITING A PIECEWISE FUNCTION

A museum charges $5 per person for a guided tour with a group of 1 to 9 people or a fixed $50 fee for a group of 10 or more people. Write a function relating the number of people, n, to the cost, C.

Answer

Two different formulas will be needed. For n-values under 10, $C = 5n$. For values of n that are 10 or greater, $C = 50$.

$$C(n) = \begin{cases} 5n & \text{if} \quad 0 < n < 10 \\ 50 & \text{if} \quad n \geq 10 \end{cases}$$

Analysis

The function is represented in Figure 14. The graph is a diagonal line from $n = 0$ to $n = 10$ and a constant after that. In this example, the two formulas agree at the meeting point where $n = 10$, but not all piecewise functions have this property.

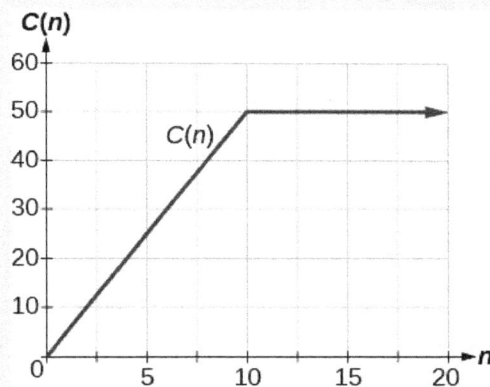

Figure 14

EXAMPLE 13: WORKING WITH A PIECEWISE FUNCTION

A cell phone company uses the function below to determine the cost, C, in dollars for g gigabytes of data transfer.

$$C(g) = \begin{cases} 25 & \text{if} \quad 0 < g < 2 \\ 25 + 10(g - 2) & \text{if} \quad g \geq 2 \end{cases}$$

Find the cost of using 1.5 gigabytes of data and the cost of using 4 gigabytes of data.

Answer

To find the cost of using 1.5 gigabytes of data, $C\left(1.5\right)$, we first look to see which part of the domain our input falls in. Because 1.5 is less than 2, we use the first formula.

$$C(1.5) = 25$$

To find the cost of using 4 gigabytes of data, $C\left(4\right)$, we see that our input of 4 is greater than 2, so we use the second formula.

$$C\left(4\right) = 25 + 10\left(4 - 2\right) = 45$$

Analysis

The function is represented in Figure 15. We can see where the function changes from a constant to a linear function with slope 10 at $g = 2.$ We plot the graphs for the different formulas on a common set of axes, making sure each formula is applied on its proper domain.

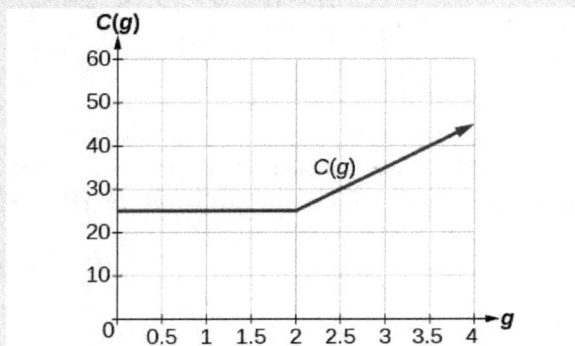

Figure 15

HOW TO

Given a piecewise function, sketch a graph.

1. Indicate on the x-axis the boundaries defined by the intervals on each piece of the domain.
2. For each piece of the domain, graph on that interval using the corresponding equation pertaining to that piece. Do not graph two functions over one interval because it would violate the criteria of a function.

EXAMPLE 14: GRAPHING A PIECEWISE FUNCTION

Sketch a graph of the function.

$$f(x) = \begin{cases} x^2 & \text{if} & x \le 1 \\ 3 & \text{if} & 1 < x \le 2 \\ x & \text{if} & x > 2 \end{cases}$$

Answer

Each of the component functions is from our library of toolkit functions, so we know their shapes. We can imagine graphing each function and then limiting the graph to the indicated domain. At the endpoints of the domain, we draw open circles to indicate where the endpoint is not included because of a less-than or greater-than inequality; we draw a closed circle where the endpoint is included because of a less-than-or-equal-to or greater-than-or-equal-to inequality.

Figure 16 shows the three components of the piecewise function graphed on separate coordinate systems.

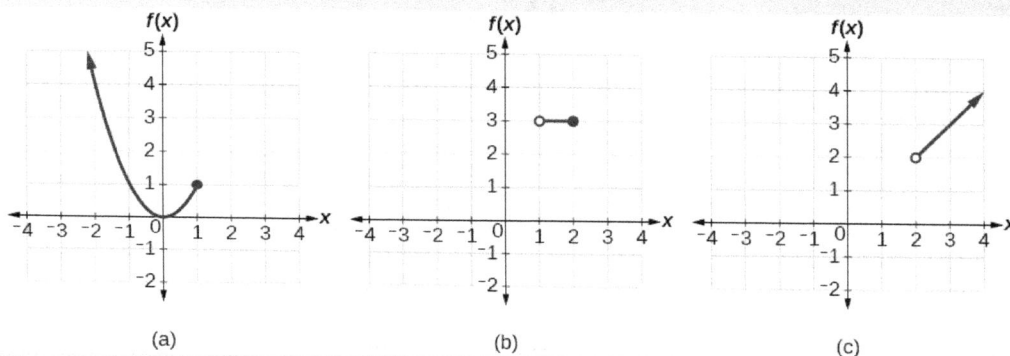

(a)   (b)   (c)

Figure 16 (a) $f(x) = x^2$ if $x \le 1$; (b) $f(x) = 3$ if $1 \text{¡} x \le 2$; (c) $f(x) = x$ if $x > 2$

Now that we have sketched each piece individually, we combine them in the same coordinate plane. See Figure 17.

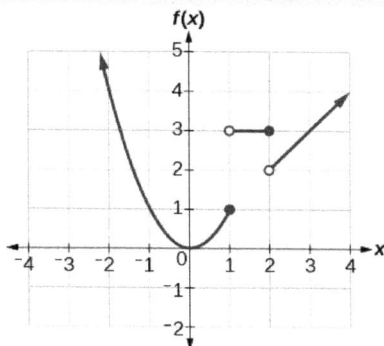

Figure 17

Analysis

Note that the graph does pass the vertical line test even at $x = 1$ and $x = 2$ because the points $(1, 3)$ and $(2, 2)$ are not part of the graph of the function, though $(1, 1)$ and $(2, \ 3)$ are.

TRY IT #8

Graph the following piecewise function.

$$f(x) = \begin{cases} x^3 & \text{if} & x < -1 \\ -2 & \text{if} & -1 < x < 4 \\ \sqrt{x} & \text{if} & x > 4 \end{cases}$$

Answer

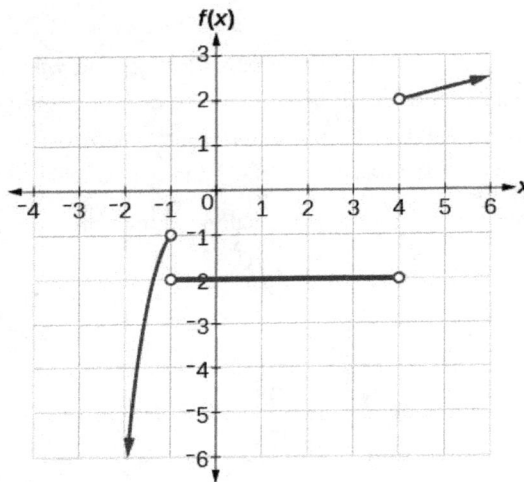

Q&A

Can more than one formula from a piecewise function be applied to a value in the domain?

No. Each value corresponds to one equation in a piecewise formula.

Access these online resources for additional instruction and practice with domain and range.

- Domain and Range of Square Root Functions
  - https://youtu.be/lj_JB8sfyIM

- Determining Domain and Range
  - https://youtu.be/FtJRstFMdhA

- Find Domain and Range Given the Graph
  - https://youtu.be/8jrkzZy04BQ

- Find Domain and Range Given a Table
  - https://youtu.be/GPBq18fCEv4

- Find Domain and Range Given Points on a Coordinate Plane

  - https://youtu.be/xOsYVyjTM0Q

KEY CONCEPTS

- The domain of a function includes all real input values that would not cause us to attempt an undefined mathematical operation, such as dividing by zero or taking the square root of a negative number.
- The domain of a function can be determined by listing the input values of a set of ordered pairs.
- The domain of a function can also be determined by identifying the input values of a function written as an equation.
- Interval values for functions of real numbers represented on a number line can be described using inequality notation, set-builder notation, and interval notation.
- For many functions, the domain and range can be determined from a graph.
- An understanding of toolkit functions can be used to find the domain and range of related functions.

Optional:

- A piecewise function is described by more than one formula.
- A piecewise function can be graphed using each algebraic formula on its assigned subdomain.

GLOSSARY

interval notation
  a method of describing a set that includes all numbers between a lower limit and an upper limit; the lower and upper values are listed between brackets or parentheses, a square bracket indicating inclusion in the set, and a parenthesis indicating exclusion

piecewise function
  a function in which more than one formula is used to define the output

set-builder notation
  a method of describing a set by a rule that all of its members obey; it takes the form $\{x|$ statement about $x\}$

1.3 Rates of Change and Behavior of Graphs

LEARNING OBJECTIVES

In this section, you will:

- Find the average rate of change of a function.
- Use a graph to determine where a function is increasing, decreasing, or constant.
- Use a graph to locate local maxima and local minima.
- Use a graph to locate the absolute maximum and absolute minimum.

Gasoline costs have experienced some wild fluctuations over the last several decades. Table 1[1] lists the average cost, in dollars, of a gallon of gasoline for the years 2005–2012. The cost of gasoline can be considered as a function of year.

Table 1

y	2005	2006	2007	2008	2009	2010	2011	2012
$C(y)$	2.31	2.62	2.84	3.30	2.41	2.84	3.58	3.68

If we were interested only in how the gasoline prices changed between 2005 and 2012, we could compute that the cost per gallon had increased from $2.31 to $3.68, an increase of $1.37. While this is interesting, it might be more useful to look at how much the price changed *per year*. In this section, we will investigate changes such as these.

Finding the Average Rate of Change of a Function

The price change per year is a **rate of change** because it describes how an output quantity changes relative to the change in the input quantity. We can see that the price of gasoline in Table 1 did not change by the same amount each year, so the rate of change was not constant. If we use only the beginning and ending data, we would be finding the average rate of change over the specified period of time. To find the average rate of change, we divide the change in the output value by the change in the input value.

If we consider the two points (x_1, y_1) and (x_2, y_2) on the graph of a function f, we can talk about the average rate of change on the interval of input values $[x_1, x_2]$.

---

1. http://www.eia.gov/totalenergy/data/annual/showtext.cfm?t=ptb0524. Accessed 3/5/2014.

$$\text{Average rate of change} = \frac{\text{Change in output}}{\text{Change in input}}$$

$$= \frac{y_2 - y_1}{x_2 - x_1} \qquad \text{using coordinates of the point,}$$

$$= \frac{f(x_2) - f(x_1)}{x_2 - x_1} \qquad \text{using function notation,}$$

$$= \frac{\Delta y}{\Delta x} \qquad \text{using delta notation.}$$

The Greek letter Δ (delta) signifies the change in a quantity; we read the ratio as "delta-y over delta-x" or "the change in y divided by the change in x." Occasionally we write Δf instead of Δy, which still represents the change in the function's output value resulting from a change to its input value. It does not mean we are changing the function into some other function.

You may also see the interval given as $[a, b]$. This means that our points would be $(a, f(a))$ and $(b, f(b))$. This would lead to the alternate form for the average rate of change shown below.

$$\text{Average rate of change} = \frac{f(b) - f(a)}{b - a}$$

You should be comfortable working with any of these presentations of the material.

In our example, the gasoline price increased by $1.37 from 2005 to 2012. Over 7 years, the average rate of change was

$$\frac{\Delta y}{\Delta x} = \frac{\$1.37}{7 \text{ years}} \approx 0.196 \text{ dollars per year}$$

On average, the price of gas increased by about 19.6¢ each year.

Other examples of rates of change include:

- A population of rats increasing by 40 rats per week
- A car traveling 68 miles per hour (distance traveled changes by 68 miles each hour as time passes)
- A car driving 27 miles per gallon (distance traveled changes by 27 miles for each gallon)
- The current through an electrical circuit increasing by 0.125 amperes for every volt of increased voltage
- The amount of money in a college account decreasing by $4,000 per quarter

---

DEFINITION

A **rate of change** describes how an output quantity changes relative to the change in the input quantity. The units on a rate of change are "output units per input units."

The average rate of change between two input values is the total change of the function output values $f(x_1)$ and $f(x_2)$ divided by the change in the input values x_1 and x_2.

$$\frac{\Delta y}{\Delta x} = \frac{f(x_2) - f(x_1)}{x_2 - x_1}$$

Given the value of a function at different points, calculate the average rate of change of a function for the interval between two input values x_1 and x_2.

1. Calculate the difference $y_2 - y_1 = \Delta y$.
2. Calculate the difference $x_2 - x_1 = \Delta x$.
3. Find the ratio $\frac{\Delta y}{\Delta x}$.

EXAMPLE 1: COMPUTING AN AVERAGE RATE OF CHANGE

Using the data in Table 1, find the average rate of change of the price of gasoline between 2007 and 2009.

Answer

In 2007, the price of gasoline was $2.84. In 2009, the cost was $2.41. The average rate of change is

$$\frac{\Delta y}{\Delta x} = \frac{y_2 - y_1}{x_2 - x_1}$$
$$= \frac{2.41 - 2.84}{2009 - 2007}$$
$$= \frac{-0.43}{2}$$
$$= -0.22$$

On average, the price of gasoline goes down $0.22 or 22 cents per year between 2007 and 2009.

Analysis

Note that a decrease is expressed by a negative change or "negative increase." A rate of change is negative when the output decreases as the input increases or when the output increases as the input decreases.

TRY IT #1

Using the data in Table 1, find the average rate of change between 2005 and 2010.

Answer

$\frac{\$2.84 - \$2.31}{5 \text{ years}} = \frac{\$0.53}{5 \text{ years}} = \$0.106$ per year.

EXAMPLE 2: COMPUTING AVERAGE RATE OF CHANGE FROM A GRAPH

Given the function $g\left(t\right)$ shown in Figure 1, find the average rate of change on the interval $\left[-1, 2\right]$.

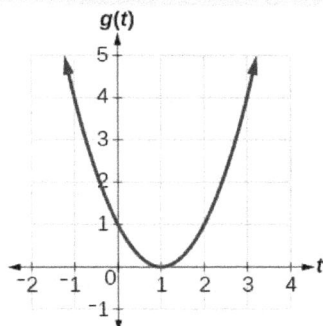

Figure 1

Answer

At $t = -1$, Figure 2 shows $g\left(-1\right) = 4$. At $t = 2$, the graph shows $g\left(2\right) = 1$.

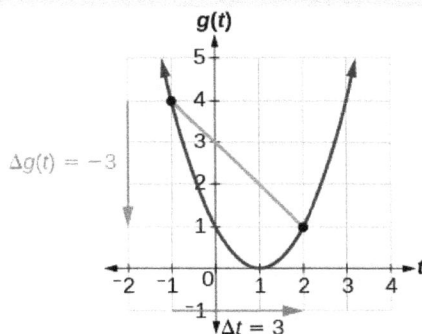

Figure 2

The horizontal change $\Delta t = 3$ is shown by the red arrow, and the vertical change $\Delta g\left(t\right) = -3$ is shown by the turquoise arrow. The output changes by –3 while the input changes by 3, giving an average rate of change of

$$\frac{1-4}{2-(-1)} = \frac{-3}{3} = -1$$

Analysis

Note that the order we choose is very important. If, for example, we use $\frac{y_2-y_1}{x_1-x_2}$, we will not get the correct answer. Decide which point will be 1 and which point will be 2, and keep the coordinates fixed as $\left(x_1, y_1\right)$ and $\left(x_2, y_2\right)$.

EXAMPLE 3: COMPUTING AVERAGE RATE OF CHANGE FROM A TABLE

After picking up a friend who lives 10 miles away, Anna records her distance from home over time. The values are shown in Table 2. Find her average speed over the first 6 hours.

Table 2

t (hours)	0	1	2	3	4	5	6	7
D(t) (miles)	10	55	90	153	214	240	292	300

Answer

Here, the average speed is the average rate of change. She traveled 282 miles in 6 hours, for an average speed of

$$\frac{292 - 10}{6 - 0} = \frac{282}{6}$$
$$= 47$$

The average speed is 47 miles per hour.

Analysis

Because the speed is not constant, the *average speed* depends on the interval chosen. For the interval [2,3], the average speed is 63 miles per hour

EXAMPLE 4: COMPUTING AVERAGE RATE OF CHANGE FOR A FUNCTION EXPRESSED AS A FORMULA

Compute the average rate of change of $f(x) = x^2 - \frac{1}{x}$ on the interval $[2, 4]$.

Answer

We can start by computing the function values at each endpoint of the interval.

$$f(2) = 2^2 - \frac{1}{2} \qquad\qquad f(4) = 4^2 - \frac{1}{4}$$
$$= 4 - \frac{1}{2} \qquad\qquad\qquad = 16 - \frac{1}{4}$$
$$= \frac{7}{2} \qquad\qquad\qquad\qquad = \frac{63}{4}$$

Now we compute the average rate of change from $x = 2$ to $x = 4$ using the formula $\frac{f(x_2)-f(x_1)}{x_2-x_1}$.

$$\frac{f(4) - f(2)}{4 - 2}$$ Plug 2 and 4 into the formula.

$$= \frac{\frac{63}{4} - \frac{7}{2}}{4 - 2}$$ Substitute values for $f(4)$ and $f(2)$.

$$= \frac{\frac{49}{4}}{2}$$ Simplify.

$$= \frac{49}{8}$$

TRY IT #2

Find the average rate of change of $f(x) = x - 2\sqrt{x}$ on the interval $[1, \ 9]$.

Answer

$\frac{1}{2}$

EXAMPLE 5: FINDING THE AVERAGE RATE OF CHANGE OF A FORCE

The electrostatic force F, measured in newtons, between two charged particles can be related to the distance between the particles d, in centimeters, by the formula $F(d) = \frac{2}{d^2}$. Find the average rate of change of force if the distance between the particles is increased from 2 cm to 6 cm.

Answer

We are computing the average rate of change of $F(d) = \frac{2}{d^2}$ on the interval $[2, 6]$.

$$\text{Average rate of change} = \frac{F(6) - F(2)}{6 - 2}$$

$$= \frac{\frac{2}{6^2} - \frac{2}{2^2}}{6 - 2}$$ Simplify.

$$= \frac{\frac{2}{36} - \frac{2}{4}}{4}$$

$$= \frac{-\frac{16}{36}}{4}$$ Combine numerator terms.

$$= -\frac{1}{9}$$ Simplify.

The average rate of change is $-\frac{1}{9}$ newton per centimeter.

EXAMPLE 6: FINDING AN AVERAGE RATE OF CHANGE AS AN EXPRESSION

Find the average rate of change of $g\left(t\right)=t^{2}+3t+1$ on the interval $\left[0,\ a\right]$. The answer will be an expression involving a.

Answer

We use the average rate of change formula with input values a and 0.

$$\frac{g\left(a\right)-g\left(0\right)}{a-0} \qquad \text{Evaluate.}$$

$$=\frac{\left(a^{2}+3a+1\right)-\left(0^{2}+3\left(0\right)+1\right)}{a-0} \qquad \text{Simplify.}$$

$$=\frac{a^{2}+3a+1-1}{a} \qquad \text{Simplify and factor.}$$

$$=\frac{a\left(a+3\right)}{a} \qquad \text{Cancel } a.$$

$$=a+3$$

This result tells us the average rate of change in terms of a between $t=0$ and any other point $t=a$. For example, on the interval $\left[0,5\right]$, the average rate of change would be $5+3=8$.

TRY IT #3

Find the average rate of change of $f\left(x\right)=x^{2}+2x-8$ on the interval $\left[5,\ a\right]$.

Answer

$a+7$

Using a Graph to Determine Where a Function is Increasing, Decreasing, or Constant

As part of exploring how functions change, we can identify intervals over which the function is changing in specific ways. We say that a function is increasing on an interval if the function values increase as the input values increase within that interval. Similarly, a function is decreasing on an interval if the function values decrease as the input values increase over that interval.

The average rate of change of an increasing function is positive, and the average rate of change of a decreasing function is negative. Figure 3 shows examples of increasing and decreasing intervals on a function.

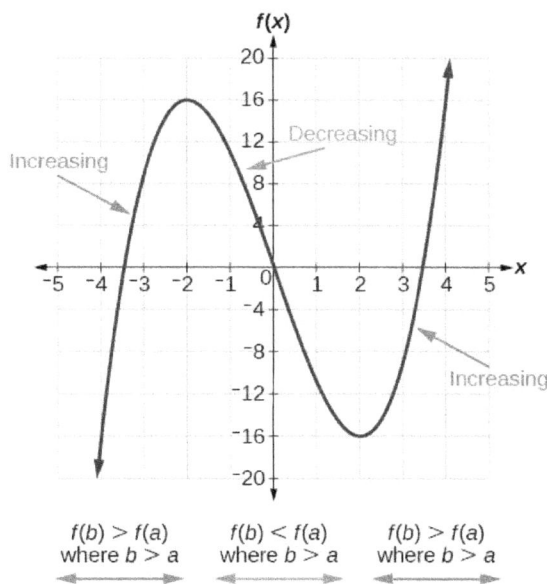

Figure 3 The function $f\left(x\right) = x^3 - 12x$ is increasing on $\left(-\infty,\ -2\right) \cup \left(2,\ \infty\right)$ and is decreasing on $\left(-2,\ 2\right)$.

While some functions are increasing (or decreasing) over their entire domain, many others are not. A value of the input where a function changes from increasing to decreasing (as we go from left to right, that is, as the input variable increases) is called a **local maximum**. If a function has more than one, we say it has local maxima. Similarly, a value of the input where a function changes from decreasing to increasing as the input variable increases is called a **local minimum**. The plural form is "local minima." Together, local maxima and minima are called **local extrema**, or local extreme values, of the function. (The singular form is "extremum.") Often, the term *local* is replaced by the term *relative*. In this text, we will use the term *local*.

Clearly, a function is neither increasing nor decreasing on an interval where it is constant. A function is also neither increasing nor decreasing at extrema. Note that we have to speak of *local* extrema, because any given local extremum as defined here is not necessarily the highest maximum or lowest minimum in the function's entire domain.

For the function whose graph is shown in Figure 4, the local maximum occurs when $x = -2$ The maximum value is the output value of 16. The local minimum occurs when $x = 2$. The minimum value is the output value of -16

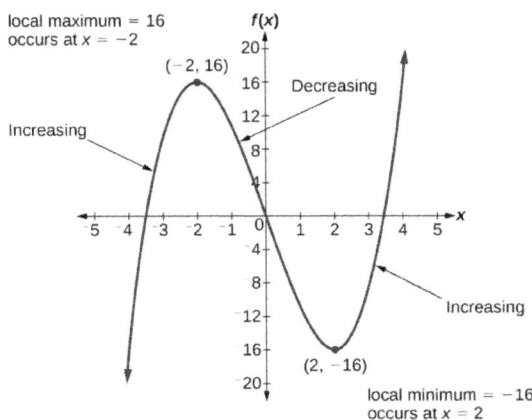

Figure 4

To locate the local maxima and minima from a graph, we need to observe the graph to determine where the graph attains its highest and lowest points, respectively, within an open interval. Like the summit of a roller coaster, the graph of a function is higher at a local maximum than at nearby points on both sides. The graph will also be lower at a local minimum than at neighboring points. Figure 5 illustrates these ideas for a local maximum.

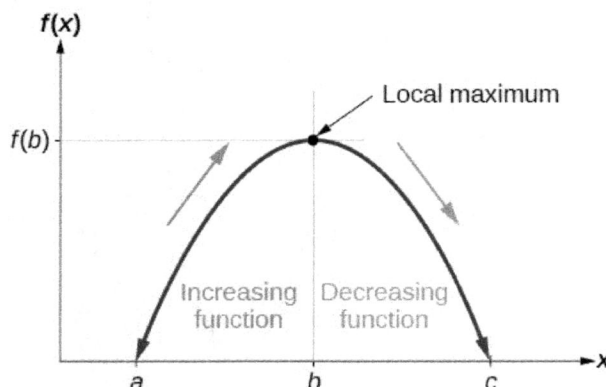

Figure 5 Definition of a local maximum

These observations lead us to a formal definition of local extrema.

DEFINITION

A function f is an **increasing function on an open interval** if $f(b) > f(a)$ for every two input values a and b in the interval where $b > a$.

A function f is a **decreasing function on an open interval** if $f(b) < f(a)$ for every two input values a and b in the interval where $b > a$.

A function f has a **local maximum** at a point b in an open interval (a, c) if $f(b)$ is greater than or equal to $f(x)$ for every point x (x does not equal b) in the interval. Likewise, f has a **local minimum** at a point b in (a, c) if $f(b)$ is less than or equal to $f(x)$ for every x (x does not equal b) in the interval.

EXAMPLE 7: FINDING INCREASING AND DECREASING INTERVALS ON A GRAPH

Given the function $p(t)$ in Figure 6, identify the intervals on which the function appears to be increasing and decreasing.

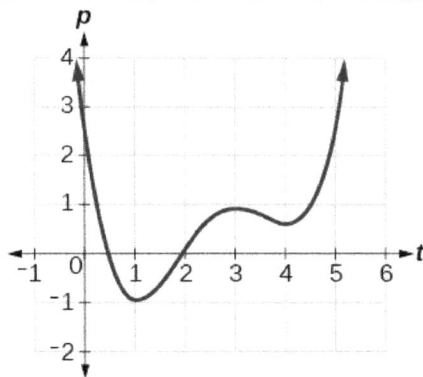

Figure 6

Answer

We see that the function is not constant on any interval. The function is increasing where it slants upward as we move to the right and decreasing where it slants downward as we move to the right. The function appears to be increasing from $t = 1$ to $t = 3$ and from $t = 4$ on. The function appears to be decreasing until $t = 1$ and then again from $t = 3$ to $t = 4$.

In interval notation, we would say the function appears to be increasing $(1,\ 3) \cup (4, \infty)$ and decreasing on $(-\infty,\ 1) \cup (3,\ 4)$.

Analysis

Notice in this example that we used open intervals (intervals that do not include the endpoints), because the function is neither increasing nor decreasing at $t = 1, t = 3$, and $t = 4$. These points are the local extrema (two minima and a maximum).

EXAMPLE 8: FINDING LOCAL EXTREMA FROM A GRAPH

Use technology to graph the function $f(x) = \frac{2}{x} + \frac{x}{3}$. Then use features of your graphing utility to estimate the local extrema of the function and to determine the intervals on which the function is increasing.

Answer

Using technology, we find that the graph of the function looks like that in Figure 7. It appears there is a low point, or local minimum, between $x = 2$ and $x = 3$, and a mirror-image high point, or local maximum, somewhere between $x = -3$ and $x = -2$.

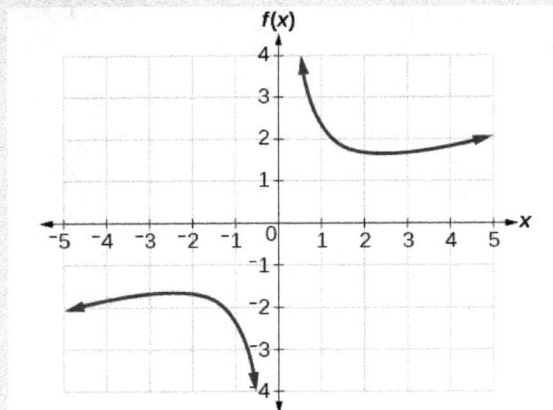

f(x)

Figure 7

Analysis

Most graphing calculators and graphing utilities can estimate the location of maxima and minima. Figure 8 provides screen images from two different technologies, showing the estimate for the local maximum and minimum.

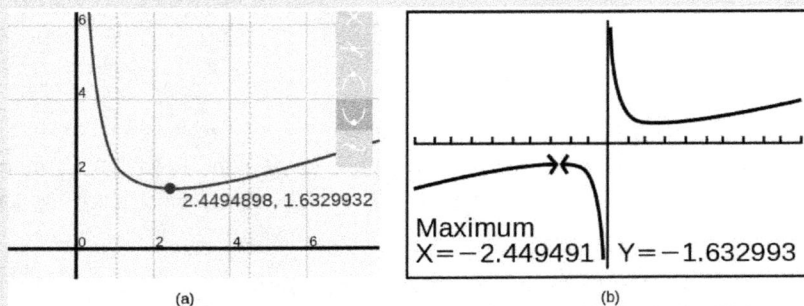

2.4494898, 1.6329932

(a)

Maximum
X=−2.449491   Y=−1.632993

(b)

Figure 8

Based on these estimates, the function is increasing on the interval $(-\infty, -2.449)$ and $(2.449, \infty)$. Notice that, while we expect the extrema to be symmetric, the two different technologies agree only up to four decimals due to the differing approximation algorithms used by each. (The exact location of the extrema is at $\sqrt{6}$, but determining this requires calculus.)

TRY IT #4

Use technology to graph the function $f(x) = x^3 - 6x^2 - 15x + 20$ and to estimate the local extrema of the function. Use these to determine the intervals on which the function is increasing and decreasing.

Answer

Using technology, we find the local maximum of 28 appears to occur at $(-1, 28)$, and the local minimum of -80 appears to occur at $(5, -80)$. The function is increasing on $(-\infty, -1) \cup (5, \infty)$ and decreasing on $(-1, 5)$.

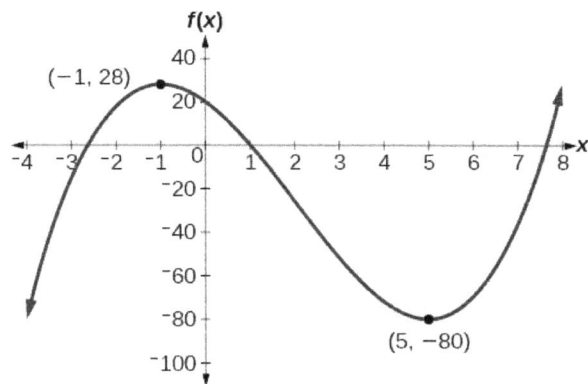

$f(x)$

$(-1, 28)$

$(5, -80)$

EXAMPLE 9: FINDING LOCAL MAXIMA AND MINIMA FROM A GRAPH

For the function f whose graph is shown in Figure 9, find all local maxima and minima.

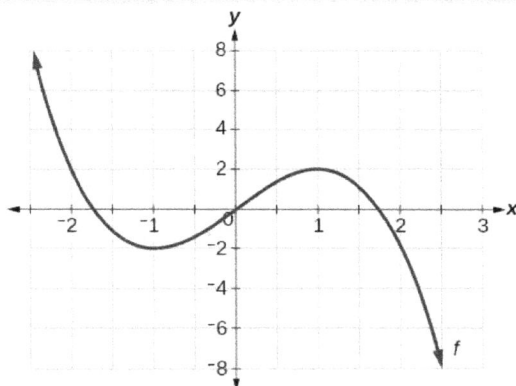

Figure 9

Answer

Observe the graph of f. The graph attains a local maximum at $x = 1$ because the highest point in an open interval occurs when $x = 1$. The local maximum is the y-coordinate at $x = 1$, which is 2.

The graph attains a local minimum at $x = -1$ because the lowest point in an open interval occurs when $x = -1$. The local minimum is the y-coordinate at $x = -1$, which is -2.

Analyzing the Toolkit Functions for Increasing or Decreasing Intervals

We will now return to our toolkit functions and discuss their graphical behavior in Figure 10, Figure 11, and Figure 12.

Function	Increasing/Decreasing	Example
Constant Function $f(x) = c$	Neither increasing nor decreasing	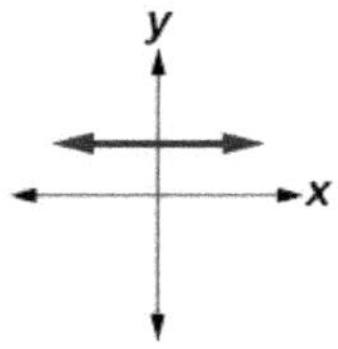
Identity Function $f(x) = x$	Increasing	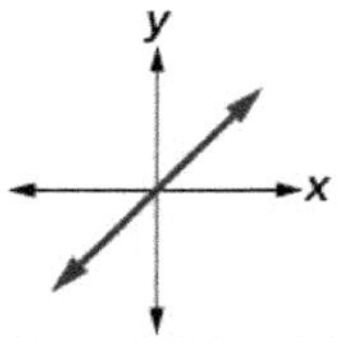
Quadratic Function $f(x) = x^2$	Increasing on $(0, \infty)$ Decreasing on $(-\infty, 0)$ Minimum at $x = 0$	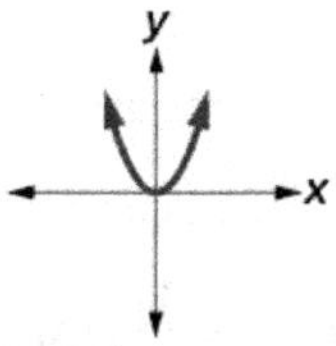

Figure 10

Function	Increasing/Decreasing	Example
Cubic Function $f(x) = x^3$	Increasing	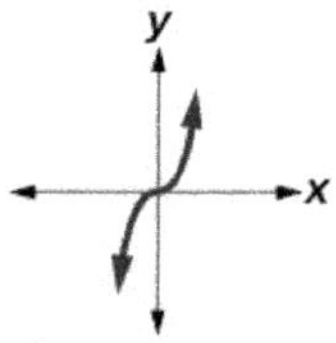
Reciprocal $f(x) = \frac{1}{x}$	Decreasing $(-\infty, 0) \cup (0, \infty)$	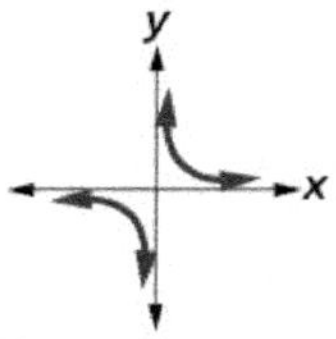
Reciprocal Squared $f(x) = \frac{1}{x^2}$	Increasing on $(-\infty, 0)$ Decreasing on $(0, \infty)$	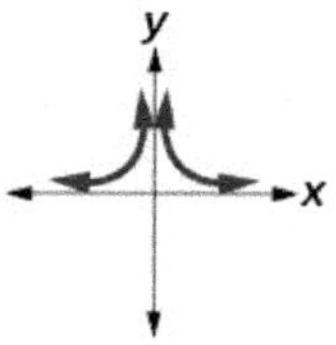

Figure 11

Function	Increasing/Decreasing	Example
Cube Root $f(x) = \sqrt[3]{x}$	Increasing	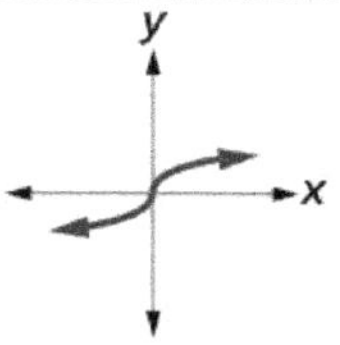
Square Root $f(x) = \sqrt{x}$	Increasing on $(0, \infty)$	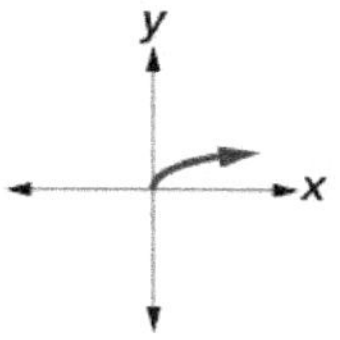
Absolute Value $f(x) = |x|$	Increasing on $(0, \infty)$ Decreasing on $(-\infty, 0)$	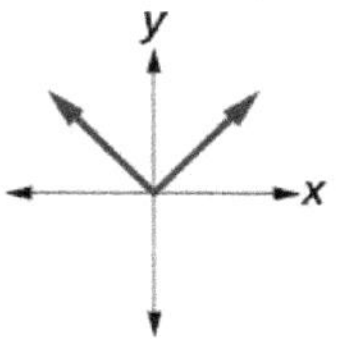

Figure 12

Use A Graph to Locate the Absolute Maximum and Absolute Minimum (Optional)

There is a difference between locating the highest and lowest points on a graph in a region around an open interval (locally) and locating the highest and lowest points on the graph for the entire domain. The y-coordinates (output) at the highest and lowest points are called the **absolute maximum** and **absolute minimum**, respectively.

To locate absolute maxima and minima from a graph, we need to observe the graph to determine where the graph attains it highest and lowest points on the domain of the function. See Figure 13.

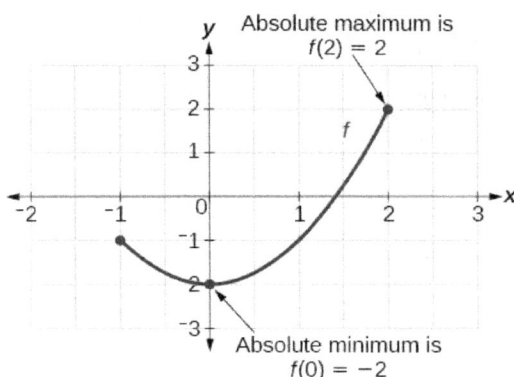

Figure 13

Not every function has an absolute maximum or minimum value. The toolkit function $f(x) = x^3$ is one such function.

DEFINITION

The **absolute maximum** of f at $x = c$ is $f(c)$ where $f(c) \geq f(x)$ for all x in the domain of f.

The **absolute minimum** of f at $x = d$ is $f(d)$ where $f(d) \leq f(x)$ for all x in the domain of f.

EXAMPLE 10: FINDING ABSOLUTE MAXIMA AND MINIMA FROM A GRAPH

For the function f shown in Figure 14, find all absolute maxima and minima.

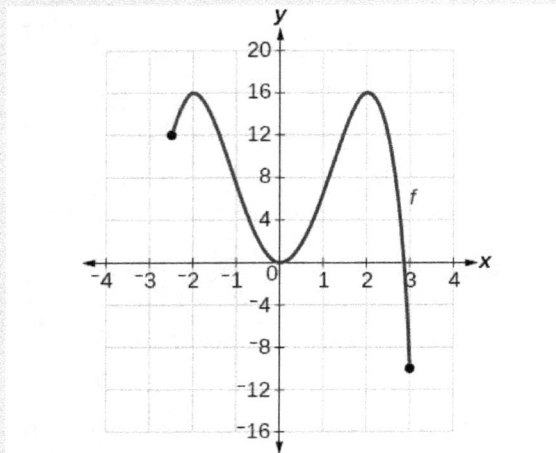

Figure 14

Answer

Observe the graph of f. The graph attains an absolute maximum in two locations, $x = -2$ and $x = 2$, because at these locations, the graph attains its highest point on the domain of the function. The absolute maximum is the y-coordinate at $x = -2$ and $x = 2$, which is 16.

The graph attains an absolute minimum at $x = 3$, because it is the lowest point on the domain of the function's graph. The absolute minimum is the y-coordinate at $x = 3$, which is -10.

Access this online resource for additional instruction and practice with rates of change.

- Average Rate of Change
  - https://youtu.be/F-7Poa3i1ZU

Key Equations

Average rate of change	$\dfrac{\Delta y}{\Delta x} = \dfrac{f(x_2) - f(x_1)}{x_2 - x_1}$

KEY CONCEPTS

- A rate of change relates a change in an output quantity to a change in an input quantity. The average rate of change is determined using only the beginning and ending data over an interval.
- Identifying points that mark the interval on a graph can be used to find the average rate of change.
- Comparing pairs of input and output values in a table can also be used to find the average rate of change.
- An average rate of change can also be computed by determining the function values at the endpoints of an interval described by a formula.
- The average rate of change can sometimes be determined as an expression.
- A function is increasing where its rate of change is positive and decreasing where its rate of change is negative.
- A local maximum is where a function changes from increasing to decreasing and has an output value larger (more positive or less negative) than output values at neighboring input values.
- A local minimum is where the function changes from decreasing to increasing (as the input increases) and has an output value smaller (more negative or less positive) than output values at neighboring input values.
- Minima and maxima are also called extrema.
- We can find local extrema from a graph.
- The highest and lowest points on a graph indicate the absolute maxima and minima.

GLOSSARY

absolute maximum
the greatest value of a function over an interval

absolute minimum
the lowest value of a function over an interval

average rate of change
the difference in the output values of a function found for two values of the input divided by the difference between the inputs

decreasing function
a function is decreasing in some open interval if $f(b) < f(a)$ for any two input values a and b in the given interval where $b > a$

increasing function
a function is increasing in some open interval if $f(b) > f(a)$ for any two input values a and b in the given interval where $b > a$

local extrema
collectively, all of a function's local maxima and minima

local maximum
a value of the input where a function changes from increasing to decreasing as the input value increases.

local minimum
a value of the input where a function changes from decreasing to increasing as the input value increases.

rate of change
the change of an output quantity relative to the change of the input quantity

1.4 Concavity

As part of exploring how functions change, it is interesting to explore the graphical behavior of functions. Concavity describes the shape of the function and how it is changing.

Consider the graphs below that show the total sales, in thousands of dollars, for two companies over 4 weeks.

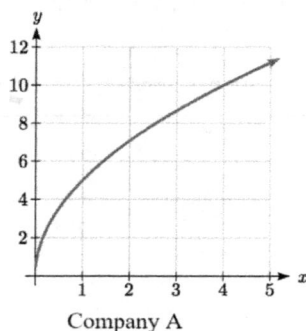

Company A

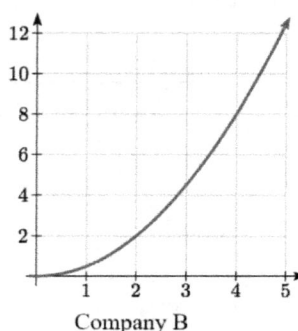

Company B

As you can see, the sales for each company are increasing, but they are increasing in very different ways. Company A has a lot of sales immediately which could represent the release of a much anticipated product and then the increase in sales levels off. Company B starts with slower sales perhaps representing an unknown product which then starts to sell more rapidly perhaps because of word of mouth. To describe the difference in behavior, we can investigate how the average rate of change varies over different intervals. Using tables of values, we can find the average rate of change between consecutive points. For example, in Company A, we can use the first pair of points to get the average rate of change $\frac{5-0}{1-0} = 5$ and the second pair of points to get the average rate of change $\frac{7.1-5}{2-1} = 2.1$.

Company A

Week	Sales	Rate of Change
0	0	
		5
1	5	
		2.1
2	7.1	
		1.6
3	8.7	
		1.3
4	10	

Company B

Week	Sales	Rate of Change
0	0	
		0.5
1	0.5	
		1.5
2	2	
		2.5
3	4.5	
		3.5
4	8	

From the tables, we can see that the rate of change for company A is *decreasing*, while the rate of change for company B is *increasing*.

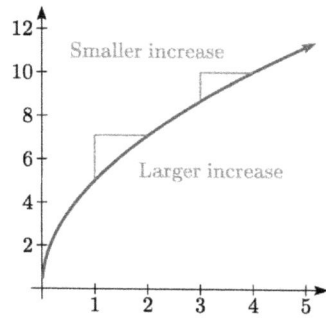

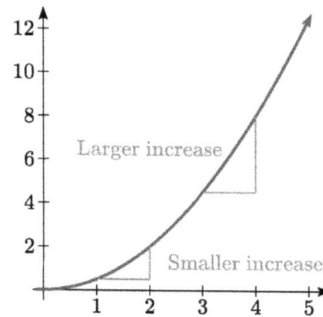

For an increasing function, when the rate of change is decreasing, as with Company A, we say the function is **concave down**. For an increasing function, when the rate of change is increasing, as with Company B, we say the function is **concave up**.

DEFINITION

A function is **concave up** if the rate of change is increasing.

A function is **concave down** if the rate of change is decreasing.

A point where a function changes from concave up to concave down or vice versa is called an **inflection point**.

EXAMPLE 1: DESCRIBE THE CONCAVITY

An object is thrown from the top of a building. The object's height in feet above ground after t seconds is given by the function $h(t) = 144 - 16t^2$ for $0 \leq t \leq 3$. Describe the concavity of the graph.

Answer

Sketching a graph of the function, we can see that the function is decreasing.

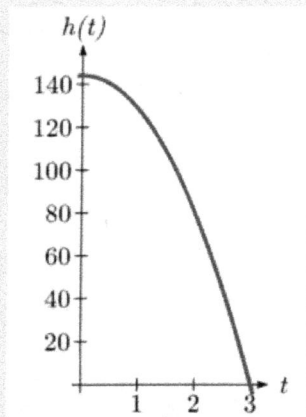

We can calculate some rates of change to explore the behavior. For example, the interval $t = 0$ to $t = 1$ has an average rate of change of $\frac{128-144}{1-0} = -16.$ The remaining intervals are shown in the table below.

t	$h(t)$	Rate of Change
0	144	
		-16
1	128	
		-48
2	80	
		-80
3	0	

Notice that the rates of change are becoming more negative, so the rates of change are *decreasing*. This means the function is concave down.

EXAMPLE 2: CONCAVITY FROM A TABLE OF VALUES

The value, V, of a car after t years is given in the table below. Is the value increasing or decreasing? Is the function concave up or concave down?

t	0	2	4	6	8
$V(t)$	28000	24342	21162	18397	15994

Answer

Since the values $V(t)$, are getting smaller as we let t increase, we can determine that the value of the car is decreasing. We can compute rates of change to determine concavity.

t		0	2	4	6	8
$V(t)$		28000	24342	21162	18397	15994
Rate of change			-1829	-1590	-1382.5	-1201.5

These rate of change values are becoming less negative since they are moving to the right on a number line, so the rates of change are *increasing* meaning this function is concave up.

TRY IT #1

Is the function described in the table below concave up or concave down?

x	0	5	10	15	20
g(x)	10000	9000	7000	4000	0

Answer

The average rates of change are decreasing so the function is concave down.

x	0	5	10	15	20
g(x)	10000	9000	7000	4000	0
Rate of change		-1000	-2000	-3000	-4000	

Graphically, concave down functions bend downwards like a frown, and concave up function bend upwards like a smile.

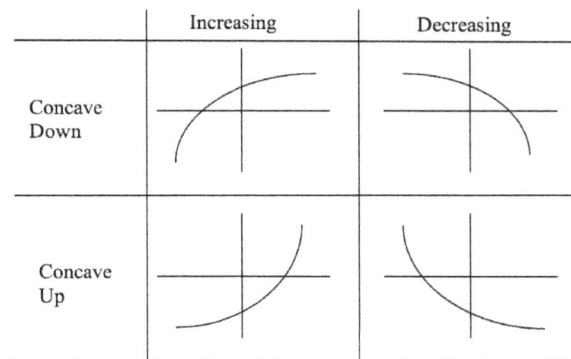

EXAMPLE 3: DETERMINE INTERVALS OF CONCAVITY FROM A GRAPH

From the graph shown, estimate the intervals on which the function is concave down and concave up.

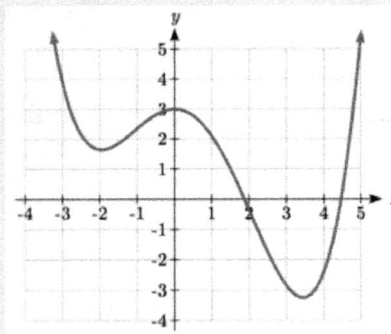

Answer

On the far left, the graph is decreasing but concave up, since it is bending upwards. It begins increasing at *x* = -2, but it continues to bend upwards until about *x* = -1.

From *x* = -1 the graph starts to bend downward, and continues to do so until about *x* = 2. The graph then begins curving upwards for the remainder of the graph shown.

From this, we can estimate that the graph is concave up on the intervals $(-\infty, -1)$ and $(2, \infty)$, and is concave down on the interval $(-1, 2)$. The graph has inflection points at *x* = -1 and *x* = 2.

TRY IT #2

Create a graph of $f(x) = x^3 - 6x^2 - 15x + 20$ and use it to estimate the intervals on which the function is concave up and concave down.

Answer

Looking at the graph of $f(x) = x^3 - 6x^2 - 15x + 20$, it appears the function is concave down on $(-\infty,\ 2)$ and concave up on $(2,\ \infty)$.

Behaviors of the Toolkit Functions

We will now return to our toolkit functions and discuss their graphical behavior.

Function	Increasing/Decreasing	Concavity
Constant Function $f(x) = c$	Neither increasing nor decreasing	Neither concave up nor down
Identity Function $f(x) = x$	Increasing	Neither concave up nor down
Quadratic Function	Increasing on $(0, \infty)$	Concave up

$f(x) = x^2$	Decreasing on $-\infty, 0)$ Minimum at $x = 0$	$(-\infty, \infty)$
Cubic Function $f(x) = x^3$	Increasing	Concave down on $(-\infty, 0)$ Concave up on $(0, \infty)$ Inflection point at (0,0)
Reciprocal $f(x) = \frac{1}{x}$	Decreasing $(-\infty, 0) \cup (0, \infty)$	Concave down on $(-\infty, 0)$ Concave up on $(0, \infty)$
Reciprocal squared $f(x) = \frac{1}{x^2}$	Increasing on $(-\infty, 0)$ Decreasing on $(0, \infty)$	Concave up on $(-\infty, 0) \cup (0, \infty)$
Cube Root $f(x) = \sqrt[3]{x}$	Increasing	Concave down on $(0, \infty)$ Concave up on $(-\infty, 0)$ Inflection point at (0,0)
Square Root $f(x) = \sqrt{x}$	Increasing on $(0, \infty)$	Concave down on $(0, \infty)$
Absolute Value $f(x) = |x|$	Increasing on $(0, \infty)$ Decreasing on $(-\infty, 0)$	Neither concave up or down

KEY CONCEPTS

- Concavity describes the shape of the curve. If the average rates are increasing on an interval then the function is concave up and if the average rates are decreasing on an interval then the function is concave down on the interval.
- A function has an inflection point when it switches from concave down to concave up or visa versa.
- Given a graph, intervals of concavity can be estimated by determining where the graph bends up versus where it bends down.
- Input values are used when describing intervals of concavity. Endpoint in the interval are not included so the notation uses parenthesis not square brackets.

1.5 Composition of Functions

Suppose we want to calculate how much it costs to heat a house on a particular day of the year. The cost to heat a house will depend on the average daily temperature, and in turn, the average daily temperature depends on the particular day of the year. Notice how we have just defined two relationships: The cost depends on the temperature, and the temperature depends on the day.

Using descriptive variables, we can notate these two functions. The function $C(T)$ gives the cost C of heating a house for a given average daily temperature in T degrees Celsius. The function $T(d)$ gives the average daily temperature on day d of the year. For any given day, $\text{Cost} = C(T(d))$ means that the cost depends on the temperature, which in turns depends on the day of the year. Thus, we can evaluate the cost function at the temperature $T(d)$. For example, we could evaluate $T(5)$ to determine the average daily temperature on the 5th day of the year. Then, we could evaluate the cost function at that temperature. We would write $C(T(5))$.

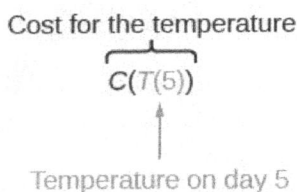

Cost for the temperature

$$C(T(5))$$

Temperature on day 5

By combining these two relationships into one function, we have performed function composition, which is the focus of this section.

Create a Function by Composition of Functions

We can create functions by composing functions. When we wanted to compute a heating cost from a day of the year, we created a new function that takes a day as input and yields a cost as output. The process of combining functions so that the output of one function becomes the input of another is known as a composition of functions. The resulting function is known as a **composite function**. We represent this combination by the following notation:

$$(f \circ g)(x) = f(g(x))$$

We read the left-hand side as "f composed with g at x," and the right-hand side as "f of g of x." The two sides of the equation have the same mathematical meaning and are equal. The open circle symbol $\circ$ is called the composition operator. We use this operator mainly when we wish to emphasize the relationship between the functions themselves without referring to any particular input value. Composition is a binary operation that takes two functions and forms a new function, much as addition or multiplication takes two numbers and gives a new number. It is important to realize that the product of functions fg is not the same as the function composition $f\left(g\left(x\right)\right)$, because, in general, $f\left(x\right)g\left(x\right) \neq f\left(g\left(x\right)\right)$.

It is also important to understand the order of operations in evaluating a composite function. We follow the usual convention with parentheses by starting with the innermost parentheses first, and then working to the outside. In the equation above, the function g takes the input x first and yields an output $g\left(x\right)$. Then the function f takes $g\left(x\right)$ as an input and yields an output $f\left(g\left(x\right)\right)$.

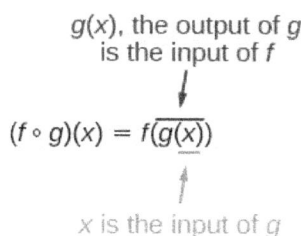

g(x), the output of g
is the input of f

↓

$(f \circ g)(x) = f(\overline{g(x)})$

↑

x is the input of g

In general, $f \circ g$ and $g \circ f$ are different functions. In other words, in many cases $f\left(g\left(x\right)\right) \neq g\left(f\left(x\right)\right)$ for all x. We will also see that sometimes two functions can be composed only in one specific order.

For example, if $f\left(x\right) = x^2$ and $g\left(x\right) = x + 2$, then

$$f\left(g\left(x\right)\right) = f\left(x + 2\right)$$
$$= \left(x + 2\right)^2$$
$$= x^2 + 4x + 4$$

but

$$g\left(f\left(x\right)\right) = g\left(x^2\right)$$
$$= x^2 + 2.$$

These expressions are not equal for all values of x, so the two functions are not equal. It is irrelevant that the expressions happen to be equal for the single input value $x = -\frac{1}{2}$.

Note that the range of the inside function (the first function to be evaluated) needs to be within the domain of the outside function. Less formally, the composition has to make sense in terms of inputs and outputs.

DEFINITION

When the output of one function is used as the input of another, we call the entire operation a **composition of functions**. For any input x and functions f and g, this action defines a composite function, which we write as $f \circ g$ such that

$$\left(f \circ g\right)\left(x\right) = f\left(g\left(x\right)\right)$$

The domain of the composite function $f \circ g$ is all x such that x is in the domain of g and $g\left(x\right)$ is in the domain of f.

EXAMPLE 1: DETERMINING WHETHER COMPOSITION OF FUNCTIONS IS COMMUTATIVE

Using the functions provided, find $f\left(g\left(4\right)\right)$ and $g\left(f\left(4\right)\right)$. Determine whether the composition of the functions is commutative.

$$f\left(x\right) = 2x + 1 \qquad\qquad g\left(x\right) = 3 - x$$

Answer

Let's begin by evaluating $g\left(4\right) = 3 - 4 = -1.$ The output of g is the input to f so

$$f\left(g\left(4\right)\right) = f\left(-1\right)$$
$$= 2\left(-1\right) + 1$$
$$= -1$$

For the second composition $g\left(f\left(4\right)\right)$, we begin by evaluating $f\left(4\right) = 2\left(4\right) + 1 = 9.$ The output of f will be the input to g so

$$g\left(f\left(4\right)\right) = g\left(9\right)$$
$$= 3 - 9$$
$$= -6$$

We find that $g\left(f\left(4\right)\right) \neq f\left(g\left(4\right)\right)$, so the operation of function composition is not commutative and order matters.

EXAMPLE 2: INTERPRETING COMPOSITE FUNCTIONS

The function $c\left(s\right)$ gives the number of calories burned completing s sit-ups, and $s\left(t\right)$ gives the number of sit-ups a person can complete in t minutes. Interpret $c\left(s\left(3\right)\right)$.

Answer

The inside expression in the composition is $s\left(3\right)$. Because the input to the s-function is time, $t = 3$ represents 3 minutes, and $s\left(3\right)$ is the number of sit-ups completed in 3 minutes.

Using $s\left(3\right)$ as the input to the function $c\left(s\right)$ gives us the number of calories burned during the number of sit-ups that can be completed in 3 minutes, or simply the number of calories burned in 3 minutes (by doing sit-ups).

EXAMPLE 3: INVESTIGATING THE ORDER OF FUNCTION COMPOSITION

Suppose $f(x)$ gives the number of miles that can be driven in x hours and $g(y)$ gives the number of gallons of gas used in driving y miles. Which of these expressions is meaningful: $f(g(y))$ or $g(f(x))$?

Answer

The function $y = f(x)$ is a function whose output is the number of miles driven corresponding to the number of hours driven.

$$\text{number of miles } = f \text{ (number of hours)}$$

The function $g(y)$ is a function whose output is the number of gallons used corresponding to the number of miles driven. This means:

$$\text{number of gallons } = g \text{ (number of miles)}$$

The expression $g(y)$ takes miles as the input and a number of gallons as the output. The function $f(x)$ requires a number of hours as the input. Trying to input a number of gallons does not make sense. The expression $f(g(y))$ is meaningless.

The expression $f(x)$ takes hours as input and a number of miles driven as the output. The function $g(y)$ requires a number of miles as the input. Using $f(x)$ (miles driven) as an input value for $g(y)$, where gallons of gas depends on miles driven, does make sense. The expression $g(f(x))$ makes sense, and will yield the number of gallons of gas used, g, driving a certain number of miles, $f(x)$, in x hours.

Q&A

Are there any situations where $f(g(y))$ and $g(f(x))$ would both be meaningful or useful expressions?

Yes. For many pure mathematical functions, both compositions make sense, even though they usually produce different new functions. In real-world problems, functions whose inputs and outputs have the same units also may give compositions that are meaningful in either order.

TRY IT #1

The gravitational force on a planet a distance r from the sun is given by the function $G(r)$. The acceleration of a planet subjected to any force F is given by the function $a(F)$. Form a meaningful composition of these two functions, and explain what it means.

Answer

A gravitational force is still a force, so $a\left(G\left(r\right)\right)$ makes sense as the acceleration of a planet at a distance r from the sun (due to gravity), but $G\left(a\left(F\right)\right)$ does not make sense.

Evaluating Composite Functions

Once we compose a new function from two existing functions, we need to be able to evaluate it for any input in its domain. We will do this with specific numerical inputs for functions expressed as tables, graphs, and formulas and with variables as inputs to functions expressed as formulas. In each case, we evaluate the inner function using the starting input and then use the inner function's output as the input for the outer function.

Evaluating Composite Functions Using Tables

When working with functions given as tables, we read input and output values from the table entries and always work from the inside to the outside. We evaluate the inside function first and then use the output of the inside function as the input to the outside function.

EXAMPLE 4: USING A TABLE TO EVALUATE A COMPOSITE FUNCTION

Using Table 1, evaluate $f\left(g\left(3\right)\right)$ and $g\left(f\left(3\right)\right)$.

Table 1

x	$f\left(x\right)$	$g\left(x\right)$
1	6	3
2	8	5
3	3	2
4	1	7

Answer

To evaluate $f\left(g\left(3\right)\right)$, we start from the inside with the input value 3. We then evaluate the inside expression $g\left(3\right)$ using the table that defines the function $g : g\left(3\right) = 2.$ We can then use that result as the input to the function f, so $g\left(3\right)$ is replaced by 2 and we get $f\left(2\right)$. Then, using the table that defines the function f, we find that $f\left(2\right) = 8.$

$$g\left(3\right) = 2$$
$$f\left(g\left(3\right)\right) = f\left(2\right) = 8$$

To evaluate $g\left(f\left(3\right)\right)$, we first evaluate the inside expression $f\left(3\right)$ using the first table: $f\left(3\right) = 3.$ Then, using the table for g, we can evaluate

$$g\left(f\left(3\right)\right) = g\left(3\right) = 2$$

Table 2 shows the composite functions $f \circ g$ and $g \circ f$ as tables.

Table 2

x	$g(x)$	$f(g(x))$	$f(x)$	$g(f(x))$
3	2	8	3	2

TRY IT #2

Using Table 1, evaluate $f(g(1))$ and $g(f(4))$.

Answer

$$f(g(1)) = f(3) = 3 \text{ and } g(f(4)) = g(1) = 3$$

Evaluating Composite Functions Using Graphs

When we are given individual functions as graphs, the procedure for evaluating composite functions is similar to the process we use for evaluating tables. We read the input and output values, but this time, from the x- and y-axes of the graphs.

HOW TO

Given a composite function and graphs of its individual functions, evaluate it using the information provided by the graphs.

1. Locate the given input to the inner function on the x-axis of its graph.
2. Read off the output of the inner function from the y-axis of its graph.
3. Locate the inner function output on the x-axis of the graph of the outer function.
4. Read the output of the outer function from the y-axis of its graph. This is the output of the composite function.

EXAMPLE 5: USING A GRAPH TO EVALUATE A COMPOSITE FUNCTION

Using Figure 1, evaluate $f(g(1))$.

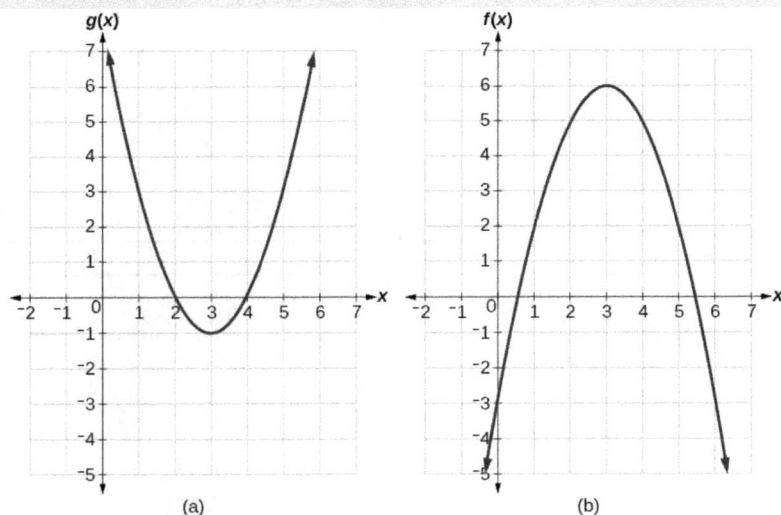

(a)   (b)

Figure 1

Answer

To evaluate $f\left(g\left(1\right)\right)$, we start with the inside evaluation. See Figure 2.

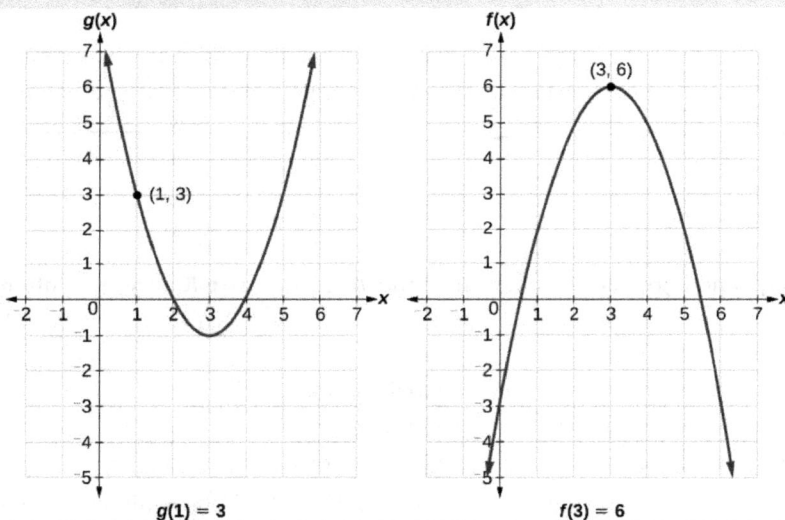

g(1) = 3   f(3) = 6

Figure 2

We evaluate $g\left(1\right)$ using the graph of $g\left(x\right)$, finding the input of 1 on the x-axis and finding the output value of the graph at that input. Here, $g\left(1\right)=3.$ We use this value as the input to the function $f.$

$$f\left(g\left(1\right)\right)=f\left(3\right)$$

We can then evaluate the composite function by looking to the graph of $f\left(x\right)$, finding the input of 3 on the x-axis and reading the output value of the graph at this input. Here, $f\left(3\right)=6,$ so $f\left(g\left(1\right)\right)=6.$

Analysis

Figure 3 shows how we can mark the graphs with arrows to trace the path from the input value to the output value.

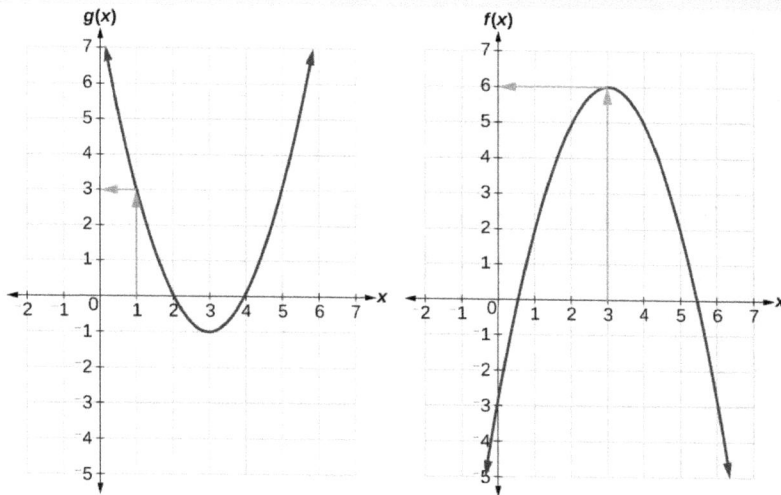

Figure 3

TRY IT #3

Using Figure 1, evaluate $g\left(f\left(2\right)\right).$

Answer

$$g\left(f\left(2\right)\right) = g\left(5\right) = 3$$

Evaluating Composite Functions Using Formulas

When evaluating a composite function where we have either created or been given formulas, the rule of working from the inside out remains the same. The input value to the outer function will be the output of the inner function, which may be a numerical value, a variable name, or a more complicated expression.

While we can compose the functions for each individual input value, it is sometimes helpful to find a single formula that will calculate the result of a composition $f\left(g\left(x\right)\right)$. To do this, we will extend our idea of function evaluation. Recall that, when we evaluate a function like $f\left(t\right) = t^2 - t,$ we substitute the value inside the parentheses into the formula wherever we see the input variable.

HOW TO

Given a formula for a composite function, evaluate the function.

1. Evaluate the inside function using the input value or variable provided.

2. Use the resulting output as the input to the outside function.

EXAMPLE 6: EVALUATING A COMPOSITION OF FUNCTIONS EXPRESSED AS FORMULAS

Given $f(t) = t^2 - t$ and $h(x) = 3x + 2$, evaluate

1. $f(h(1))$
2. $f(h(x))$
3. $h(f(t))$

Answer

1. Because the inside expression is $h(1)$, we start by evaluating $h(x)$ at 1.

$$h(1) = 3(1) + 2$$
$$= 5$$

Then $f(h(1)) = f(5)$, so we evaluate $f(t)$ at an input of 5.

$$f(h(1)) = f(5)$$
$$= 5^2 - 5$$
$$= 20$$

2. The inside expression for $f(h(x))$ is $h(x)$ so we use its output, $3x + 2$, as the input to the function f. We then must evaluate $f(3x + 2)$. Begin by replacing each variable t, with $3x + 2$ in the function f to get

$$f(3x + 2) = (3x + 2)^2 - (3x + 2).$$

Then simplify to get

$$f(h(x)) = 9x^2 + 6x + 6x + 4 - 3x - 2 = 9x^2 + 9x + 2.$$

3. The inside expression for $h(f(t))$ is $f(t)$, so we use its output, $t^2 - t$, as the input to the function h. We then must evaluate $h(t^2 - t)$. Begin by replacing each variable x, with $t^2 - t$ in the function h to get

$$h(t^2 - t) = 3(t^2 - t) + 2.$$

Then simplify to get

$$h(f(t)) = 3t^2 - 3t + 2.$$

EXAMPLE 7: DETERMINING WHETHER COMPOSITION OF FUNCTIONS IS COMMUTATIVE

Using the functions provided, find $f\left(g\left(x\right)\right)$ and $g\left(f\left(x\right)\right)$. Determine whether the composition of the functions is commutative.

$$f\left(x\right) = 2x + 1 \qquad g\left(x\right) = 3 - x$$

Answer

Let's begin by substituting $g\left(x\right) = 3 - x$ into $f\left(x\right)$.

$$f\left(g\left(x\right)\right) = f\left(3 - x\right)$$
$$= 2\left(3 - x\right) + 1$$
$$= 6 - 2x + 1$$
$$= 7 - 2x$$

Now we can substitute $f\left(x\right) = 2x + 1$ into $g\left(x\right)$.

$$g\left(f\left(x\right)\right) = g\left(2x + 1\right)$$
$$= 3 - \left(2x + 1\right)$$
$$= 3 - 2x - 1$$
$$= -2x + 2$$

We find that $g\left(f\left(x\right)\right) \neq f\left(g\left(x\right)\right)$, so the operation of function composition is not commutative.

TRY IT #4

Given $f\left(t\right) = t^2 - t$ and $h\left(x\right) = 3x + 2$, evaluate

a. $h\left(f\left(2\right)\right)$
b. $h\left(f\left(-2\right)\right)$

Answer

a. 8; b. 20

Decomposing a Composite Function into its Component Functions

In some cases, it is necessary to decompose a complicated function. In other words, we can write it as a composition of two simpler functions. There may be more than one way to decompose a composite function, so we may choose the decomposition that appears to be most expedient. However, in calculus, you will be studying the chain rule in order to find the derivative of a composite function. While we can't begin to try to define the concept of a derivative at this point, we can help you begin to think about an "inner" and "outer" function in a composition. When you study the chain rule, you will think of the outer function as $f(x)$ and the inner function as $g(x)$.

Let's consider the function $h(x) = (x+10)^5$. This reminds us of the function $f(x) = x^5$ where the input x has been replaced by $x+10$. We can see this as though we have a place holder where we would normally see the x. This would look like (where we are thinking about the "outer" function as being the function that raises the "inside" to the 5th power. We would let the outside function be $f(x) = x^5$ and the "inside" function be $g(x) = x+10$. Can you see that $f(g(x)) = (x+10)^5$?

Now let's try a function that isn't as obvious with outer and inner functions. Consider $k(x) = 2^{3x+4}$. This reminds us of the function $f(x) = 2^x$ where the x has been replaced by $3x+4$. Let's let our outer function be $f(x) = 2^x$, and our inner function be $g(x) = 3x+4$. Now form the composite function $f(g(x))$. You get $k(x) = 2^{3x+4}$.

Sometimes the easiest way to see the "inside" and "outside" is to consider what a simpler function would look like if the input were simply x and not some more complex expression.

Let's do one more. Consider $r(x) = \frac{3}{\sqrt{2x+5}}$. We can see that this fits the form of a simpler function $f(x) = \frac{3}{\sqrt{x}}$ where the x has been replaced by $2x+5$. Our outer function would be $f(x) = \frac{3}{\sqrt{x}}$ and our inner function would be $g(x) = 2x+5$. If you form the composite function $f(g(x))$, you will get $r(x) = \frac{3}{\sqrt{2x+5}}$.

EXAMPLE 8: DECOMPOSING A FUNCTION

Write $f(x) = \sqrt{5-x^2}$ as the composition of two functions.

Answer

We are looking for two functions, g and h, so $f(x) = g(h(x))$. To do this, we look for a function inside a function in the formula for $f(x)$. As one possibility, we might notice that the expression $5-x^2$ is the inside of the square root. We could then decompose the function as

$$h(x) = 5-x^2 \text{ and } g(x) = \sqrt{x}$$

We can check our answer by recomposing the functions.

$$g(h(x)) = g(5-x^2) = \sqrt{5-x^2}$$

TRY IT #5

Write $f(x) = \frac{4}{3-\sqrt{4+x^2}}$ as the composition of two functions.

Answer

Possible answer:

$$g(x) = \sqrt{4 + x^2}$$

$$h(x) = \frac{4}{3-x}$$

$$f = h \circ g$$

Finding the Domain of a Composite Function (Optional)

As we discussed previously, the domain of a composite function such as $f \circ g$ is dependent on the domain of g and the domain of f. It is important to know when we can apply a composite function and when we cannot, that is, to know the domain of a function such as $f \circ g$. Let us assume we know the domains of the functions f and g separately. If we write the composite function for an input x as $f(g(x))$, we can see right away that x must be a member of the domain of g in order for the expression to be meaningful, because otherwise we cannot complete the inner function evaluation. However, we also see that $g(x)$ must be a member of the domain of f, otherwise the second function evaluation in $f(g(x))$ cannot be completed, and the expression is still undefined. Thus the domain of $f \circ g$ consists of only those inputs in the domain of g that produce outputs from g belonging to the domain of f. Note that the domain of f composed with g is the set of all x such that x is in the domain of g and $g(x)$ is in the domain of f.

DEFINITION

The **domain of a composite function** $f(g(x))$ is the set of those inputs x in the domain of g for which $g(x)$ is in the domain of f.

HOW TO

Given a function composition $f(g(x))$, determine its domain.

1. Find the domain of g.
2. Find the domain of f.
3. Find those inputs x in the domain of g for which $g(x)$ is in the domain of f. That is, exclude those inputs x from the domain of g for which $g(x)$ is not in the domain of f. The resulting set is the domain of $f \circ g$.

EXAMPLE 9: FINDING THE DOMAIN OF A COMPOSITE FUNCTION

Find the domain of

$(f \circ g)(x)$ where $f(x) = \frac{5}{x-1}$ and $g(x) = \frac{4}{3x-2}$

Answer

The domain of $g(x)$ consists of all real numbers except $x = \frac{2}{3}$, since that input value would cause us to divide by 0. Likewise, the domain of f consists of all real numbers except 1. So we need to exclude from the domain of $g(x)$ that value of x for which $g(x) = 1$.

$$\frac{4}{3x-2} = 1$$
$$4 = 3x - 2$$
$$6 = 3x$$
$$x = 2$$

So the domain of $f \circ g$ is the set of all real numbers except $\frac{2}{3}$ and 2. This means that

$$x \neq \frac{2}{3} \quad \text{or} \quad x \neq 2$$

We can write this in interval notation as $\left(-\infty, \frac{2}{3}\right) \cup \left(\frac{2}{3}, 2\right) \cup (2, \infty)$

EXAMPLE 10: FINDING THE DOMAIN OF A COMPOSITE FUNCTION INVOLVING RADICALS

Find the domain of $(f \circ g)(x)$ where $f(x) = \sqrt{x+2}$ and $g(x) = \sqrt{3-x}$.

Answer

Because we cannot take the square root of a negative number, the domain of g is $(-\infty, 3]$. Now we check the domain of the composite function

$$(f \circ g)(x) = \sqrt{\sqrt{3-x} + 2}$$

For $(f \circ g)(x) = \sqrt{\sqrt{3-x} + 2}$, we know that $\sqrt{3-x} + 2 \geq 0$, since the radicand of a square root must be positive. Since square roots are positive, $\sqrt{3-x} \geq 0$. Squaring both sides gives us $3 - x \geq 0$. Therefore, $x \leq 3$ which gives a domain of $(-\infty, 3]$.

Analysis

This example shows that knowledge of the range of functions (specifically the inner function) can also be helpful in finding the domain of a composite function. It also shows that the domain of $f \circ g$ can contain values that are not in the domain of f, though they must be in the domain of g. Note that the domain of f is $(-2, \infty)$.

Access these online resources for additional instruction and practice with composite functions.

- Composite Functions
    - https://youtu.be/qxBmISCJSME

- Composite Function Notation Application
    - https://youtu.be/VI2kJp69jNg

- Composite Functions Using Graphs
    - https://youtu.be/b-i7N0hE-Ys

- Decompose Functions
    - https://youtu.be/gFSSk8jaAwA

- Composite Function Values
    - https://youtu.be/y2kJI9XnyLY

Key Equation

Composite function	$(f \circ g)(x) = f(g(x))$

KEY CONCEPTS

- When functions are combined, the output of the first (inner) function becomes the input of the second (outer) function.
- The function produced by combining two functions is a composite function.
- The order of function composition must be considered when interpreting the meaning of composite functions.
- A composite function can be evaluated by evaluating the inner function using the given input value and then evaluating the outer function taking as its input the output of the inner function.
- A composite function can be evaluated from a table.
- A composite function can be evaluated from a graph.
- A composite function can be evaluated from a formula.

- Just as functions can be combined to form a composite function, composite functions can be decomposed into simpler functions.
- Functions can often be decomposed in more than one way.
- (Optional)The domain of a composite function consists of those inputs in the domain of the inner function that correspond to outputs of the inner function that are in the domain of the outer function.

GLOSSARY

commutative property
the order of the operations being preformed does not matter if the commutative property holds

composite function
the new function formed by function composition, when the output of one function is used as the input of another

1.6 Transformation of Functions

LEARNING OBJECTIVES

In this section, you will:

- Describe and apply vertical and horizontal shifts and reflections of graphs, tables and function formulas.
- Use function notation to express horizontal and vertical shifts and reflections of functions.
- Determine whether a function is even, odd or neither from its algebraic formula or graph.

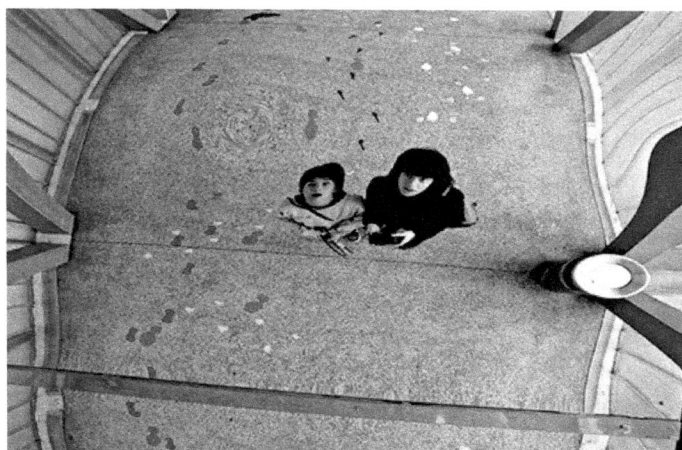

Figure 1 (credit: "Misko"/Flickr)

We all know that a flat mirror enables us to see an accurate image of ourselves and whatever is behind us. When we tilt the mirror, the images we see may shift horizontally or vertically. But what happens when we bend a flexible mirror? Like a carnival funhouse mirror, it presents us with a distorted image of ourselves, stretched or compressed horizontally or vertically. In a similar way, we can distort or transform mathematical functions to better adapt them to describing objects or processes in the real world. In this section, we will take a look at several kinds of transformations.

Graphing Functions Using Vertical and Horizontal Shifts

Often when given a problem, we try to model the scenario using mathematics in the form of words, tables, graphs, and equations. One method we can employ is to adapt the basic graphs of the toolkit functions to build new models for a given scenario. There are systematic ways to alter functions to construct appropriate models for the problems we are trying to solve.

Identifying Vertical Shifts

One simple kind of transformation involves shifting the entire graph of a function up, down, right, or left. The simplest shift is a **vertical shift**, moving the graph up or down, because this transformation involves adding a positive or negative constant to the function. In other words, we add the same constant to the output value of the function regardless of the input. For a function $g(x) = f(x) + k$, the function $f(x)$ is shifted vertically k units. See Figure 2 for an example.

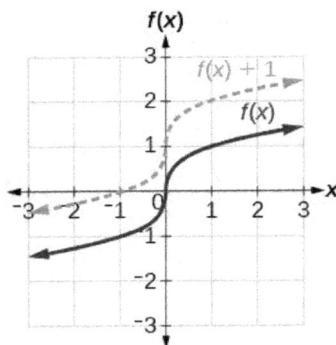

Figure 2 Vertical shift by $k = 1$ of the cube root function $f(x) = \sqrt[3]{x}.$

To help you visualize the concept of a vertical shift, consider that $y = f(x)$. Therefore, $f(x) + k$ is equivalent to $y + k$. Every unit of y is replaced by $y + k$, so the y-value increases or decreases depending on the value of k. The result is a shift upward or downward.

DEFINITION

Given a function $f(x)$, a new function $g(x) = f(x) + k$, where k is a constant, is a **vertical shift** of the function $f(x)$. All the output values change by k units. If k is positive, the graph will shift up. If k is negative, the graph will shift down.

EXAMPLE 1: ADDING A CONSTANT TO A FUNCTION

To regulate temperature in a green building, airflow vents near the roof open and close throughout the day. Figure 3 shows the area of open vents V (in square feet) throughout the day in hours after midnight, t. During the summer, the facilities manager decides to try to better regulate temperature by increasing the amount of open vents by 20 square feet throughout the day and night. Sketch a graph of this new function.

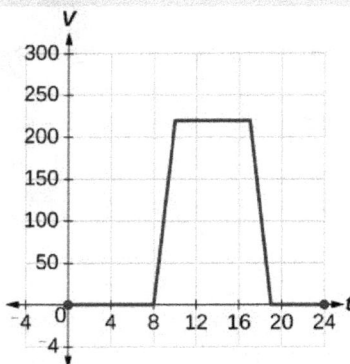

Figure 3

Answer

We can sketch a graph of this new function by adding 20 to each of the output values of the original function. This will have the effect of shifting the graph vertically up, as shown in Figure 4.

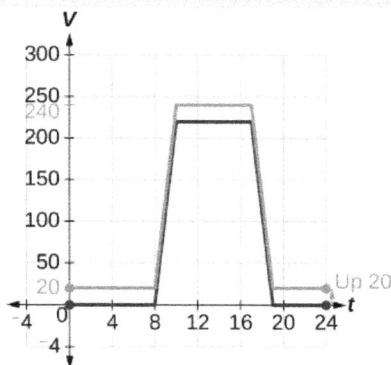

Figure 4

Notice that in Figure 4, for each input value, the output value has increased by 20, so if we call the new function $S(t)$, we could write

$$S(t) = V(t) + 20$$

This notation tells us that, any value of $S(t)$ can be found by evaluating the function V at the same input and then adding 20 to the result. This defines S as a transformation of the function V, in this case a vertical shift up 20 units. Notice that, with a vertical shift, the input values stay the same and only the output values change. See Table 1.

Table 1

t	0	8	10	17	19	24
$V(t)$	0	0	220	220	0	0
$S(t)$	20	20	240	240	20	20

HOW TO

Given a tabular function, create a new row to represent a vertical shift.

1. Identify the output row or column.
2. Determine the magnitude of the shift.
3. Add the shift to the value in each output cell. Add a positive value for up or a negative value for down.

EXAMPLE 2: SHIFTING A TABULAR FUNCTION VERTICALLY

A function $f(x)$ is given in Table 2. Create a table for the function $g(x) = f(x) - 3$.

Table 2

x	2	4	6	8
$f(x)$	1	3	7	11

Answer

The formula $g(x) = f(x) - 3$ tells us that we can find the output values of g by subtracting 3 from the output values of f. For example:

$$f(2) = 1 \qquad\qquad \text{Given}$$
$$g(x) = f(x) - 3 \qquad\qquad \text{Given transformation}$$
$$g(2) = f(2) - 3$$
$$= 1 - 3$$
$$= -2$$

Subtracting 3 from each $f(x)$ value, we can complete a table of values for $g(x)$ as shown in Table 3.

Table 3

x	2	4	6	8
$f(x)$	1	3	7	11
$g(x)$	-2	0	4	8

Analysis

As with the earlier vertical shift, notice the input values stay the same and only the output values change.

TRY IT #1

The function $h(t) = -4.9t^2 + 30t$ gives the height h of a ball (in meters) thrown upward from the ground after t seconds. Suppose the ball was instead thrown from the top of a 10 meter building. Relate this new height function $b(t)$ to $h(t)$, and then find a formula for $b(t)$.

Answer

$$b(t) = h(t) + 10 = -4.9t^2 + 30t + 10$$

Identifying Horizontal Shifts

We just saw that the vertical shift is a change to the output or outside of the function. We will now look at how changes to input or the inside of the function change its graph and meaning.

A change to the **input** results in a movement of the graph of an original function left or right in what is known as a **horizontal shift**. We will be creating a new function $g(x)$ which is based on an original function $f(x)$ using the following function notation: $g(x) = f(x - h)$ where h is a constant.

For example, if $f(x) = x^2$, then we can create a function in terms of f by writing $g(x) = f(x - 2)$, which is equivalent to $g(x) = (x - 2)^2$. Think about what happens carefully.

We can read the statement $g(x) = f(x - 2)$ as saying that the output for g at x will be the same as the output we get for the original function f evaluated two units earlier. Perhaps an easier way to see this is to recognize that if x is 5, then $g(5) = f(5 - 2) = f(3)$. We get the same output for g at the input of 5 as we did for the function f for an input two untis earlier. Therefore, in order to produce the graph of g, we will shift our original function $f(x)$ to the right by two units.

What if h is negative? Let's consider the graph of $f(x) = \sqrt[3]{x}$. If we let $h = -1$, then we can consider a new function $m(x) = f(x - (-1)) = f(x + 1)$. This is equivalent to $m(x) = \sqrt[3]{x + 1}$. Notice again that it is our input which has changed. We can read this statement as saying that the output for m evaluated at x will be the same as the output we get for the original function f evaluated one unit later. Therefore, $m(5) = f(5 + 1) = f(6)$. In order to produce the graph of m, we will shift the original function $f(x)$ to the left by one unit. Consider the picture shown in Figure 5.

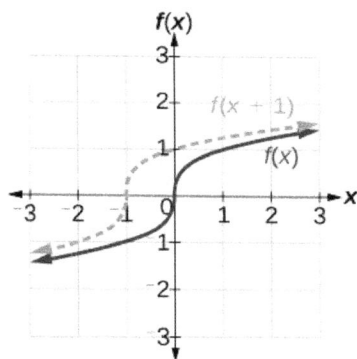

Figure 5 Horizontal shift of the function $f(x) = \sqrt[3]{x}$. Note that $f(x + 1)$ shifts the graph to the left by one unit.

DEFINITION

Given a function f, a new function $g(x) = f(x - h)$, where h is a constant, is a **horizontal shift** of the function f. If h is positive, the graph will shift right. If h is negative, the graph will shift left.

EXAMPLE 3: ADDING A CONSTANT TO AN INPUT

Returning to our building airflow example from Figure 3, suppose that in autumn the facilities manager decides that the original venting plan starts too late, and wants to begin the entire venting program 2 hours earlier. Sketch a graph of the new function.

Answer

We can set $V(t)$ to be the original program and $F(t)$ to be the revised program.

$$V(t) = \text{ the original venting plan}$$

$$F(t) = \text{ starting 2 hrs sooner}$$

In the new graph, at each time, the airflow is the same as the original function V was 2 hours later. For example, in the original function V, the airflow starts to change at 8 a.m., whereas for the function F, the airflow starts to change at 6 a.m. The comparable function values are $F(6) = V(8)$. See Figure 6. Notice also that the vents first opened to 220 ft^2 at 10 a.m. under the original plan, while under the new plan the vents reach 220 ft^2 at 8 a.m., so $F(8) = V(10)$.

In both cases, we see that, because $F(t)$ starts 2 hours sooner, $h = -2$. That means that the same output values are reached when $F(t) = V(t - (-2)) = V(t + 2)$.

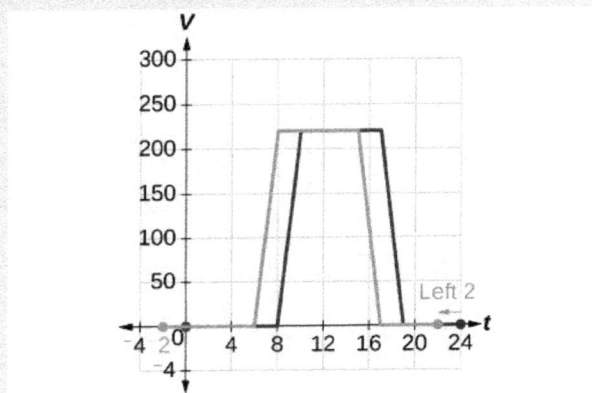

Figure 6

Analysis

Note that $V(t + 2)$ has the effect of shifting the graph to the *left*.

Horizontal changes or "inside changes" affect the domain of a function (the input) instead of the range and often seem counterintuitive. The new function $F(t)$ uses the same outputs as $V(t)$, but matches those outputs to inputs 2 hours earlier than those of $V(t)$. Said another way, we must add 2 hours to the input of V to find the corresponding output for F : $F(t) = V(t + 2)$.

Given a tabular function, create a new row to represent a horizontal shift.

1. Identify the input row or column.
2. Determine the magnitude of the shift.
3. Add the shift to the value in each input cell.

EXAMPLE 4: SHIFTING A TABULAR FUNCTION HORIZONTALLY

A function $f(x)$ is given in Table 4. Create a table for the function $g(x) = f(x - 3)$.

Table 4

x	$f(x)$
2	1
4	3
6	7
8	11

Answer

The formula $g(x) = f(x - 3)$ tells us that the output values of g are the same as the output value of f when the input value is 3 less. For example, we know that $f(2) = 1$. To get the same output from the function g, we will need an input value that is 3 *larger*. We input a value that is 3 larger for $g(x)$ because the function takes 3 away before evaluating the function f.

$$g(5) = f(5 - 3)$$
$$= f(2)$$
$$= 1$$

We continue with the other values to create Table 5. In our table for $g(x)$, we need to increase each input value for f by 3.

Table 5

x	$g(x) = f(x - 3)$
5	$g(5) = f(5 - 3) = f(2) = 1$
7	$g(7) = f(7 - 3) = f(4) = 3$
9	$g(9) = f(9 - 3) = f(6) = 7$

| 11 | $g(11) = f(11 - 3) = f(8) = 11$ |

The result is that the function $g(x)$ has been shifted to the right by 3. Notice the output values for $g(x)$ remain the same as the output values for $f(x)$, but the corresponding input values, x, have shifted to the right by 3. Specifically, 2 shifted to 5, 4 shifted to 7, 6 shifted to 9, and 8 shifted to 11.

Analysis

Figure 7 represents both of the functions. We can see the horizontal shift in each point.

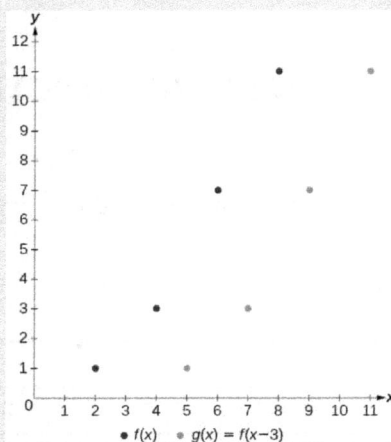

Figure 7

EXAMPLE 5: IDENTIFYING A HORIZONTAL SHIFT OF A TOOLKIT FUNCTION

Figure 8 represents a transformation of the toolkit function $f(x) = x^2$. Relate this new function $g(x)$ to $f(x)$, and then find a formula for $g(x)$.

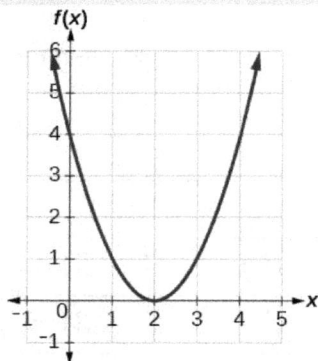

Figure 8

Answer

Notice that the graph is identical in shape to the $f(x) = x^2$ function, but the x-values are shifted to the right 2 units. The vertex used to be at (0,0), but now the vertex is at (2,0). The graph is the basic quadratic function shifted 2 units to the right, so

$$g(x) = f(x - 2)$$

Notice how we must input the value $x = 2$ to get the output value $y = 0$; the x-values must be 2 units larger because of the shift to the right by 2 units. We can then use the definition of the $f(x)$ function to write a formula for $g(x)$ by evaluating $f(x - 2)$.

$$f(x) = x^2$$
$$g(x) = f(x - 2)$$
$$g(x) = f(x - 2) = (x - 2)^2$$

Analysis

To determine whether the shift is $+2$ or -2, consider a single reference point on the graph. For a quadratic, looking at the vertex point is convenient. In the original function, $f(0) = 0$. In our shifted function, $g(2) = 0$. To obtain the output value of 0 from the function f, we need to decide whether a plus or a minus sign will work to satisfy $g(2) = f(x - 2) = f(0) = 0$. For this to work, we will need to *subtract* 2 units from our input values.

EXAMPLE 6: INTERPRETING HORIZONTAL VERSUS VERTICAL SHIFTS

The function $G(m)$ gives the number of gallons of gas required to drive m miles. Interpret $G(m) + 10$ and $G(m + 10)$.

Answer

$G(m) + 10$ can be interpreted as adding 10 to the output, gallons. This is the gas required to drive m miles, plus another 10 gallons of gas. The graph would indicate a vertical shift.

$G(m + 10)$ can be interpreted as adding 10 to the input, miles. So this is the number of gallons of gas required to drive 10 miles more than m miles. The graph would indicate a horizontal shift.

TRY IT #2

Given the function $f(x) = \sqrt{x}$, graph the original function $f(x)$ and the transformation $g(x) = f(x + 2)$ on the same axes. Is this a horizontal or a vertical shift? Which way is the graph shifted and by how many units?

Answer

The graphs of $f\left(x\right)$ and $g\left(x\right)$ are shown below. The transformation is a horizontal shift. The function is shifted to the left by 2 units.

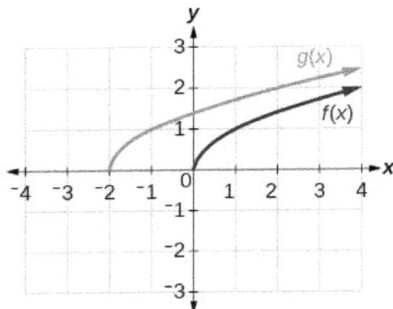

Combining Vertical and Horizontal Shifts

Now that we have two transformations, we can combine them together. Vertical shifts are outside changes that affect the output, y- axis values and shift the function up or down. Horizontal shifts are inside changes that affect the input, x- axis values and shift the function left or right. Combining the two types of shifts will cause the graph of a function to shift up or down *and* right or left.

HOW TO

Given a function and both a vertical and a horizontal shift, sketch the graph.

1. Identify the vertical and horizontal shifts from the formula.
2. The vertical shift results from a constant added to the output. Move the graph up for a positive constant and down for a negative constant.
3. The horizontal shift results from a constant subtracted from the input. Move the graph right for a positive constant and left for a negative constant.
4. Apply the shifts to the graph in either order.

EXAMPLE 7: GRAPHING COMBINED VERTICAL AND HORIZONTAL SHIFTS

Given $f\left(x\right)=\left|x\right|,$ sketch a graph of $h\left(x\right)=f\left(x+1\right)-3.$

Answer

The function f is our toolkit absolute value function. We know that this graph has a V shape, with the point at the origin. The graph of h has transformed f in two ways: $f(x+1)$ is a change on the inside of the function, giving a horizontal shift left by 1 since $h=-1$, and the subtraction by 3 in $f(x+1)-3$ is a change to the outside of the function, giving a vertical shift down by 3. The transformation of the graph is illustrated in Figure 9.

Let us follow one point of the graph of $f(x) = |x|$.

- The point $(0,0)$ is transformed first by shifting left 1 unit: $(0,0) \rightarrow (-1,0)$
- The point $(-1,0)$ is transformed next by shifting down 3 units: $(-1,0) \rightarrow (-1,-3)$

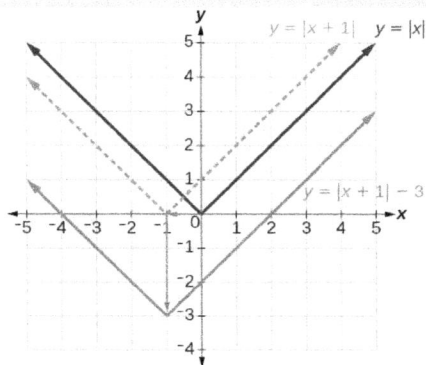

Figure 9

Figure 10 shows the graph of h.

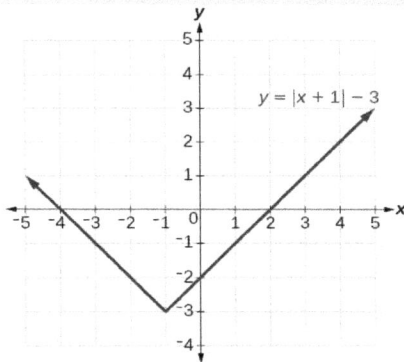

Figure 10

TRY IT #3

Given $f(x) = |x|$, sketch a graph of $h(x) = f(x-2) + 4$.

Answer

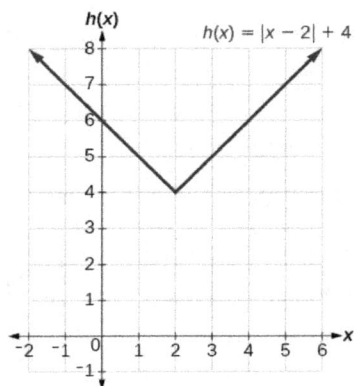

$h(x) = |x - 2| + 4$

EXAMPLE 8: IDENTIFYING COMBINED VERTICAL AND HORIZONTAL SHIFTS

Write a formula for the graph shown in Figure 11, which is a transformation of the toolkit square root function.

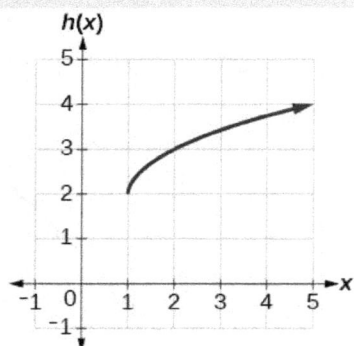

Figure 11

Answer

The graph of the toolkit function starts at the origin, so this graph has been shifted 1 to the right and up 2. In function notation, we could write that as

$$h(x) = f(x - 1) + 2$$

Using the formula for the square root function, we can write

$$h(x) = \sqrt{x - 1} + 2$$

Analysis

Note that this transformation has changed the domain and range of the function. This new graph has domain $[1, \infty)$ and range $[2, \infty)$.

Graphing Functions Using Reflections about the Axes

Another transformation that can be applied to a function is a reflection over the $x-$ or y-axis. A **vertical reflection** reflects a graph vertically across the x-axis, while a **horizontal reflection** reflects a graph horizontally across the y-axis. The reflections are shown in Figure 12.

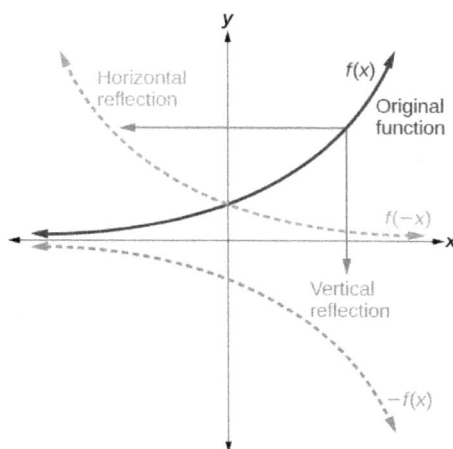

Figure 12 Vertical and horizontal reflections of a function.

Notice that the vertical reflection produces a new graph that is a mirror image of the base or original graph about the x-axis. The horizontal reflection produces a new graph that is a mirror image of the base or original graph about the y-axis.

DEFINITIONS

Given a function $f(x)$, a new function $g(x) = -f(x)$ is a **vertical reflection** of the function $f(x)$, sometimes called a reflection about (or over, or through) the x-axis.

Given a function $f(x)$, a new function $g(x) = f(-x)$ is a **horizontal reflection** of the function $f(x)$, sometimes called a reflection about the y-axis.

HOW TO

Given a function, reflect the graph both vertically and horizontally.

1. Multiply all outputs by –1 for a vertical reflection. The new graph is a reflection of the original graph about the x-axis.
2. Multiply all inputs by –1 for a horizontal reflection. The new graph is a reflection of the original graph about the y-axis.

EXAMPLE 9: REFLECTING A GRAPH HORIZONTALLY AND VERTICALLY

Reflect the graph of $s\left(t\right)=\sqrt{t}$ (a) vertically and (b) horizontally.

Answer

a. Reflecting the graph vertically means that each output value will be reflected over the horizontal t-axis as shown in Figure 13.

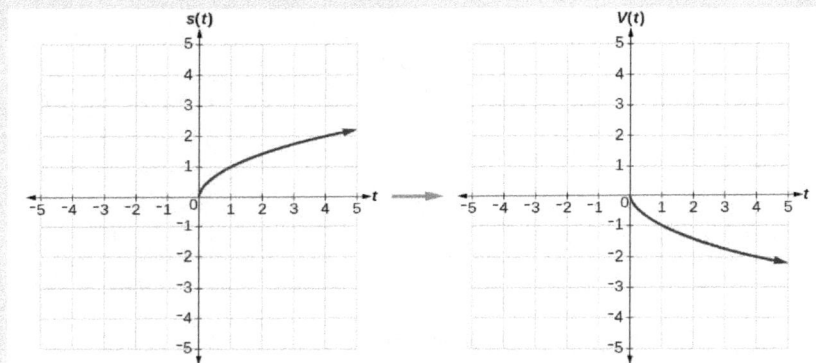

Figure 13 Vertical reflection of the square root function

Because each output value is the opposite of the original output value, we can write

$$V\left(t\right)=-s\left(t\right) \ \text{ or } \ V\left(t\right)=-\sqrt{t}$$

Notice that this is an outside change, or vertical reflection, that affects the output $s\left(t\right)$ values, so the negative sign belongs outside of the function.

b. Reflecting horizontally means that each input value will be reflected over the vertical axis as shown in Figure 14.

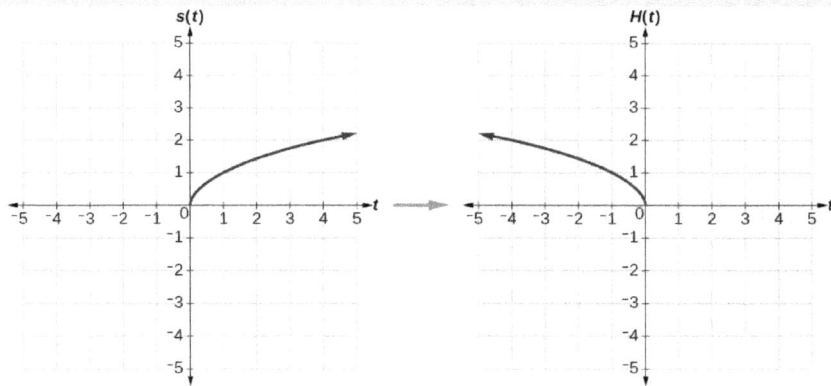

Figure 14 Horizontal reflection of the square root function

Because each input value is the opposite of the original input value, we can write

$$H\left(t\right) = s\left(-t\right) \ \text{or} \ H\left(t\right) = \sqrt{-t}.$$

Notice that this is an inside change or horizontal change that affects the input values, so the negative sign is on the inside of the function.

Note that these transformations can affect the domain and range of the functions. While the original square root function has domain $[0, \infty)$ and range $[0, \infty)$, the vertical reflection gives the $V\left(t\right)$ function the range $(-\infty, \ 0]$ and the horizontal reflection gives the $H\left(t\right)$ function the domain $(-\infty, \ 0]$.

TRY IT #5

Reflect the graph of $f\left(x\right) = |x - 1|$ (a) vertically and (b) horizontally.

Answer

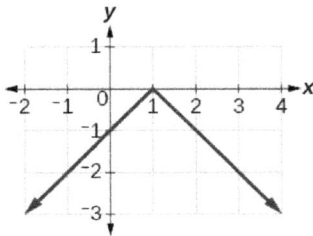

a.

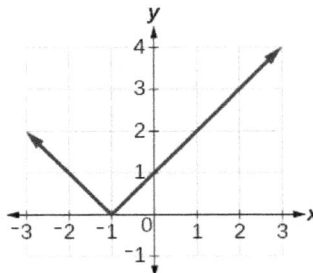

b.

EXAMPLE 10: REFLECTING A TABULAR FUNCTION HORIZONTALLY AND VERTICALLY

A function $f(x)$ is given as Table 6. Create a table for the functions below.

a. $g(x) = -f(x)$
b. $h(x) = f(-x)$

Table 6

x	2	4	6	8
$f(x)$	1	3	7	11

Answer

a. For $g(x)$, the negative sign outside the function indicates a vertical reflection, so the x-values stay the same and each output value will be the opposite of the original output value. See Table 7.

Table 7

x	2	4	6	8
$g(x)$	−1	−3	−7	−11

b. For $h(x)$, the negative sign inside the function indicates a horizontal reflection, so each input value will be the opposite of the original input value and the $h(x)$ values stay the same as the $f(x)$ values. See Table 8.

Table 8

x	−2	−4	−6	−8
$h(x)$	1	3	7	11

TRY IT #6

A function $f(x)$ is given as Table 9. Create a table for the functions below.

a. $g(x) = -f(x)$
b. $h(x) = f(-x)$

Table 9

x	−2	0	2	4
$f(x)$	5	10	15	20

Answer

a. $g(x) = -f(x)$

x	-2	0	2	4
$g(x)$	-5	-10	-15	-20

b. $h(x) = f(-x)$

x	-2	0	2	4
$h(x)$	15	10	5	unknown

EXAMPLE 11: APPLYING A LEARNING MODEL EQUATION

A common model for learning has an equation similar to $k(t) = -2^{-t} + 1$, where k is the percentage of mastery that can be achieved after t practice sessions. This is a transformation of the function $f(t) = 2^t$ shown in Figure 15. Sketch a graph of $k(t)$.

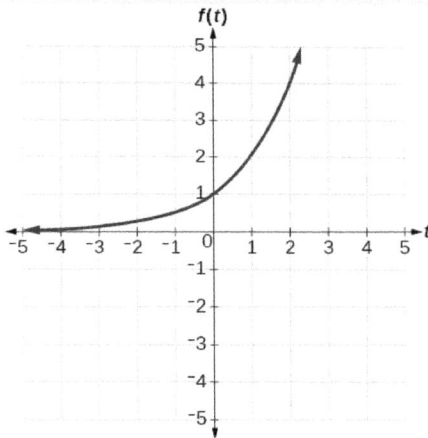

Figure 15

Answer

This equation combines three transformations into one equation.

- A horizontal reflection: $f(-t) = 2^{-t}$
- followed by a vertical reflection: $-f(-t) = -2^{-t}$
- and finally vertical shift: $-f(-t) + 1 = -2^{-t} + 1$

We can sketch a graph by applying these transformations one at a time to the original function. Let us follow two points through each of the three transformations. We will choose the points (0, 1) and (1, 2).

1. First, we apply a horizontal reflection to (0,1) and (1,2) by negating the input value to get (0, 1) and (-1, 2) respectively.
2. Then, we apply a vertical reflection by negating the second coordinate to get (0, -1) and (-1, -2) respectively.
3. Finally, we apply a vertical shift by adding 1 giving the points (0, 0) and (-1, -1) on the function $k(t)$.

This means that the original points, (0,1) and (1,2) become (0,0) and (-1,-1) after we apply the transformations.

In Figure 16, the first graph results from a horizontal reflection. The second results from a vertical reflection. The third results from a vertical shift up 1 unit.

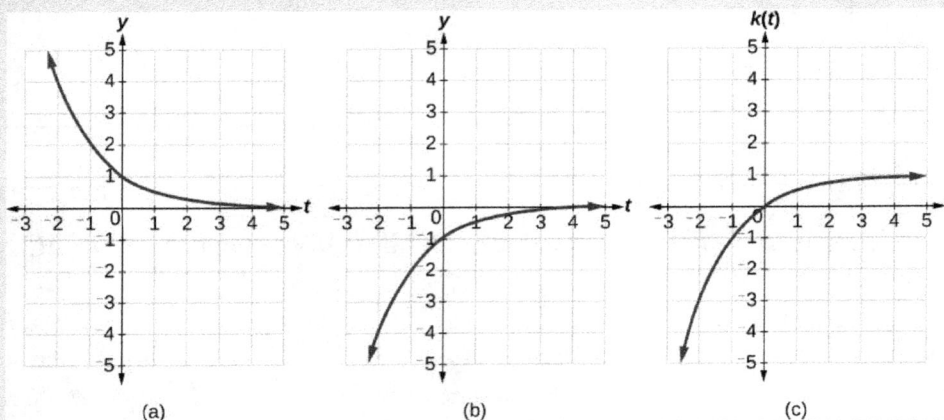

Figure 16

Analysis

As a model for learning, this function would be limited to a domain of $t \geq 0$, with corresponding range $[0, 1)$.

TRY IT #7

Given the toolkit function $f(x) = x^2$, graph $g(x) = -f(x)$ and $h(x) = f(-x)$. Take note of any surprising behavior for these functions.
Answer

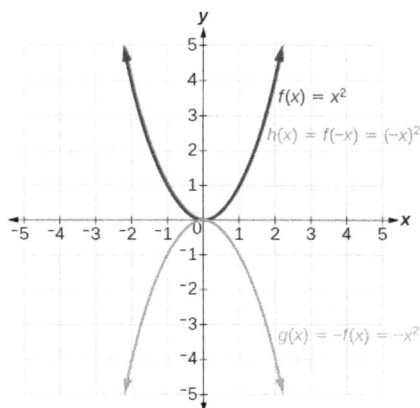

Notice: $g\left(x\right)=f\left(-x\right)$ looks the same as $f\left(x\right)$.

Determining Even and Odd Functions

Some functions exhibit symmetry so that reflections result in the original graph. For example, horizontally reflecting the toolkit functions $f\left(x\right)=x^2$ or $f\left(x\right)=\left|x\right|$ will result in the original graph. We say that these types of graphs are symmetric about the y-axis. Functions whose graphs are symmetric about the y-axis are called **even functions.**

If the graphs of $f\left(x\right)=x^3$ or $f\left(x\right)=\frac{1}{x}$ were reflected over *both* axes, the result would be the original graph, as shown in Figure 17.

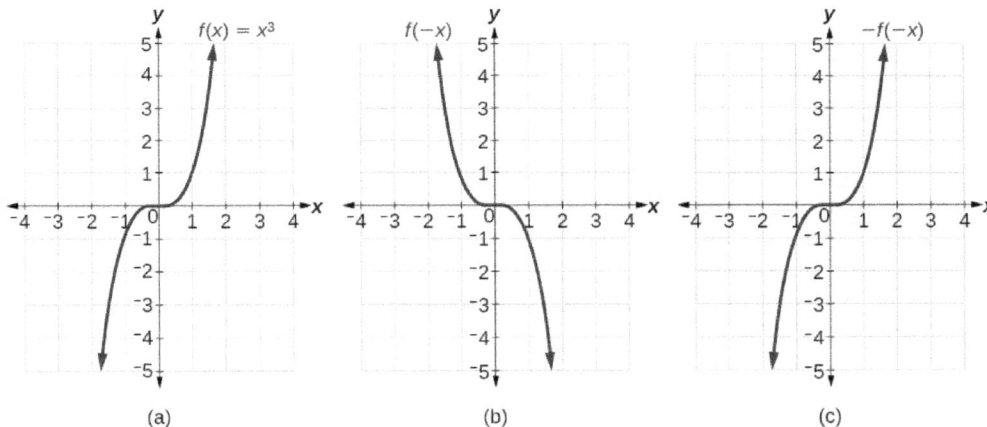

Figure 17 (a) The cubic toolkit function (b) Horizontal reflection of the cubic toolkit function (c) Horizontal and vertical reflections reproduce the original cubic function.

We say that these graphs are symmetric about the origin. A function with a graph that is symmetric about the origin is called an **odd function**.

Note: A function can be neither even nor odd if it does not exhibit either symmetry. For example, $f\left(x\right)=2^x$ is neither even nor odd. Also, the only function that is both even and odd is the constant function $f\left(x\right)=0.$

DEFINITION

A function is called an **even function** if for every input x

$$f(x) = f(-x).$$

The graph of an even function is symmetric about the y-axis.

A function is called an **odd function** if for every input x

$$f(x) = -f(-x) \text{ or equivalently } f(-x) = -f(x)$$

The graph of an odd function is symmetric about the origin.

HOW TO

Given the formula for a function, determine if the function is even, odd, or neither.

1. Determine whether the function satisfies $f(x) = f(-x)$. If it does, it is even.
2. Determine whether the function satisfies $f(x) = -f(-x)$. If it does, it is odd. Note that you can also show the equivalent statement $f(-x) = -f(x)$.
3. If the function does not satisfy either rule, it is neither even nor odd.

EXAMPLE 12: DETERMINING WHETHER A FUNCTION IS EVEN, ODD, OR NEITHER

Is the function $f(x) = x^3 + 2x$ even, odd, or neither?

Answer

Without looking at a graph, we can determine whether the function is even or odd by finding formulas for the reflections and determining if they return us to the original function. Let's begin with the rule for even functions.

$$f(-x) = (-x)^3 + 2(-x) = -x^3 - 2x$$

This does not return us to the original function, so this function is not even. We can now test the rule for odd functions.

$$-f(-x) = -\left(-x^3 - 2x\right) = x^3 + 2x$$

Because $-f(-x) = f(x)$, this is an odd function.

Analysis

Consider the graph of f in Figure 18. Notice that the graph is symmetric about the origin. For every point (x, y) on the graph, the corresponding point $(-x, -y)$ is also on the graph. For example, (1, 3) is on the graph of f, and the corresponding point $(-1, -3)$ is also on the graph.

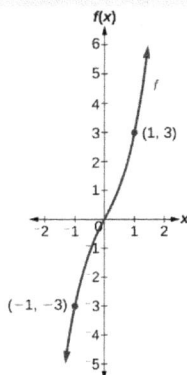

Figure 18

TRY IT #8

Is the function $f(s) = s^4 + 3s^2 + 7$ even, odd, or neither?

Answer

even

Access this online resource for additional instruction and practice with transformation of functions.

- Function Transformations

Key Equations

Vertical shift	$g(x) = f(x) + k$ (up for $k > 0$)
Horizontal shift	$g(x) = f(x - h)$ (right for $h > 0$)
Vertical reflection	$g(x) = -f(x)$
Horizontal reflection	$g(x) = f(-x)$
Vertical stretch	$g(x) = af(x)$ ($a > 1$)
Vertical compression	$g(x) = af(x)$ ($0 < a < 1$)

Horizontal stretch	$g(x) = f(bx) \, (0 < b < 1)$
Horizontal compression	$g(x) = f(bx) \, (b > 1)$

KEY CONCEPTS

- A function can be shifted vertically by adding a constant to the output.
- A function can be shifted horizontally by adding a constant to the input.
- Relating the shift to the context of a problem makes it possible to compare and interpret vertical and horizontal shifts.
- Vertical and horizontal shifts are often combined.
- A vertical reflection reflects a graph about the x-axis. A graph can be reflected vertically by multiplying the output by –1.
- A horizontal reflection reflects a graph about the y-axis. A graph can be reflected horizontally by multiplying the input by –1.
- A graph can be reflected both vertically and horizontally. The order in which the reflections are applied does not affect the final graph.
- A function presented in tabular form can also be reflected by multiplying the values in the input and output rows or columns accordingly.
- A function presented as an equation can be reflected by applying transformations one at a time.
- Even functions are symmetric about the y-axis, whereas odd functions are symmetric about the origin.
- Even functions satisfy the condition $f(x) = f(-x)$.
- Odd functions satisfy the condition $f(x) = -f(-x)$.
- A function can be odd, even, or neither.

GLOSSARY

even function
a function whose graph is unchanged by horizontal reflection, $f(x) = f(-x)$, and is symmetric about the y-axis

horizontal reflection
a transformation that reflects a function's graph across the y-axis by multiplying the input by -1

horizontal shift
a transformation that shifts a function's graph left or right by adding a positive or negative constant to the input

odd function
a function whose graph is unchanged by combined horizontal and vertical reflection, $f(x) = -f(-x)$, and is symmetric about the origin

vertical reflection
a transformation that reflects a function's graph across the x-axis by multiplying the output by -1

vertical shift
a transformation that shifts a function's graph up or down by adding a positive or negative constant to the output

1.7 Transformations: Stretches and Compressions

Graphing Functions Using Stretches and Compressions

Adding a constant to the inputs or outputs of a function changed the position of a graph with respect to the axes, but it did not affect the shape of a graph. We now explore the effects of multiplying the inputs or outputs by some quantity.

We can transform the inside (input values) of a function or we can transform the outside (output values) of a function. Each change has a specific effect that can be seen graphically.

Vertical Stretches and Compressions

When we multiply a function by a positive constant, we get a function whose graph is stretched vertically away from or compressed vertically toward the x-axis in relation to the graph of the original function. If the constant is greater than 1, we get a **vertical stretch**; if the constant is between 0 and 1, we get a **vertical compression**. Figure 1 shows a function multiplied by constant factors 2 and 0.5 and the resulting vertical stretch and compression.

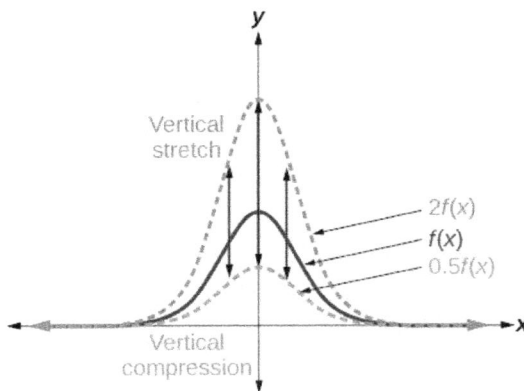

Figure 1 Vertical stretch and compression

DEFINITION

Given a function $f(x)$, a new function $g(x) = af(x)$, where a is a constant, is a **vertical stretch** or **vertical compression** of the function $f(x)$.

- If $|a| > 1$, then the graph will be stretched away from the x-axis.
- If $0 < |a| < 1$, then the graph will be compressed toward the x-axis.
- If $a < 0$, then there will be combination of a vertical stretch or compression with a vertical reflection.

HOW TO

Given a function, graph its vertical stretch or compression.

1. Identify the value of a.
2. Multiply all range values by a.

3. If $|a| > 1$, the graph is stretched by a factor of a.

   If $0 < |a| < 1$, the graph is compressed by a factor of a.

   If $a < 0$, the graph is either stretched or compressed and also reflected about the x-axis.

EXAMPLE 1: GRAPHING A VERTICAL STRETCH

A function $P(t)$ models the population of fruit flies. The graph is shown in Figure 2.

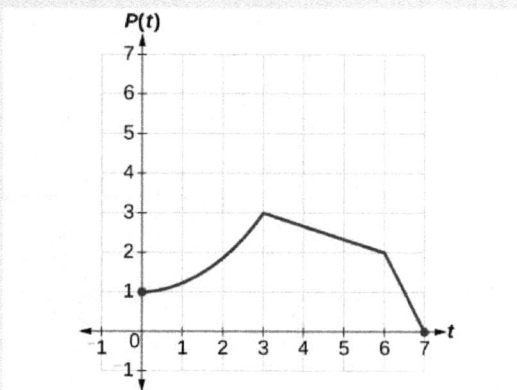

Figure 2

A scientist is comparing this population to another population, Q, whose growth follows the same pattern, but is twice as large. Sketch a graph of this population.

Answer

Because the population is always twice as large, the new population's output values are always twice the original function's output values. Graphically, this is shown in Figure 3.

If we choose four reference points, (0, 1), (3, 3), (6, 2) and (7, 0) we will multiply all of the outputs by 2.

The following shows where the new points for the new graph will be located.

$$(0,\ 1) \rightarrow (0,\ 2)$$
$$(3,\ 3) \rightarrow (3,\ 6)$$
$$(6,\ 2) \rightarrow (6,\ 4)$$
$$(7,\ 0) \rightarrow (7,\ 0)$$

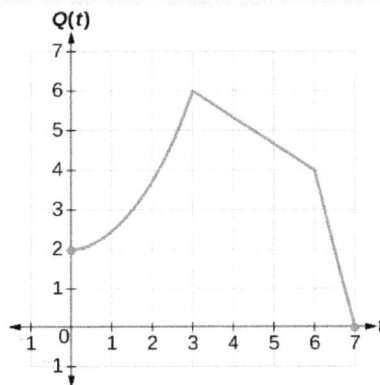

Figure 3

Symbolically, the relationship is written as

$$Q(t) = 2P(t).$$

This means that for any input t, the value of the function Q is twice the value of the function P. Notice that the effect on the graph is a vertical stretching of the graph, where every point doubles its distance from the horizontal axis. The input values, t, stay the same while the output values are twice as large as before.

HOW TO

Given a tabular function and assuming that the transformation is a vertical stretch or compression, create a table for the transformation.

1. Determine the value of a.
2. Multiply all of the output values by a.

EXAMPLE 2: FINDING A VERTICAL COMPRESSION OF A TABULAR FUNCTION

A function f is given as Table 1. Create a table for the function $g(x) = \frac{1}{2}f(x)$.

Table 1

x	2	4	6	8
$f(x)$	1	3	7	11

Answer

The formula $g(x) = \frac{1}{2}f(x)$ tells us that the output values of g are half of the output values of f with the same inputs. For example, we know that $f(4) = 3$. Then

$$g(4) = \tfrac{1}{2}f(4) = \tfrac{1}{2}(3) = \tfrac{3}{2}.$$

We do the same for the other values to produce Table 2.

Table 2

x	2	4	6	8
$g(x)$	$\frac{1}{2}$	$\frac{3}{2}$	$\frac{7}{2}$	$\frac{11}{2}$

Analysis

The result is that the function $g(x)$ has been compressed vertically by a factor of $\frac{1}{2}$. Each output value is divided in half, so the graph is half the original height.

TRY IT #1

A function f is given as Table 3. Create a table for the function $g(x) = \frac{3}{4}f(x)$.

Table 3

x	2	4	6	8
$f(x)$	12	16	20	0

Answer

x	2	4	6	8
$g(x)$	9	12	15	0

EXAMPLE 3: RECOGNIZING A VERTICAL STRETCH

The graph in Figure 4 is a transformation of the toolkit function $f(x) = x^3$. Relate this new function $g(x)$ to $f(x)$, and then find a formula for $g(x)$.

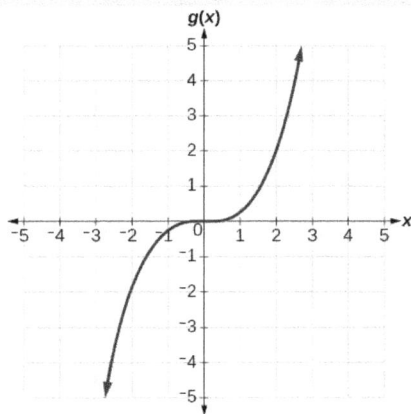

Figure 4

Answer

When trying to determine a vertical stretch or compression, it is helpful to look for a point on the graph that is relatively clear. In this graph, it appears that $g(2) = 2$. With the basic cubic function at the same input, $f(2) = 2^3 = 8$. Based on that, it appears that the outputs of g are $\frac{1}{4}$ the outputs of the function f because $g(2) = \frac{1}{4}f(2)$. From this we can fairly safely conclude that $g(x) = \frac{1}{4}f(x)$.

We can write a formula for g by using the definition of the function f.

$$g(x) = \tfrac{1}{4}f(x) = \tfrac{1}{4}x^3$$

TRY IT #2

Write the formula for the function that we get when we vertically stretch the identity toolkit function by a factor of 3, and then shift it down by 2 units.

Answer

$$g(x) = 3x - 2$$

Horizontal Stretches and Compressions

Now we consider the changes that occur to a function if we multiply the input of an original function $f(x)$ by some constant. Notice that we are changing the inside of a function. When we multiply a function's input by a positive constant, we get a function whose graph is stretched horizontally away from or compressed horizontally toward the vertical axis in relation to the graph of the original function. If the constant is between 0 and 1, we get a **horizontal stretch**; if the constant is greater than 1, we get a **horizontal compression** of the function. Let's consider an example.

Suppose a scientist is comparing a population of fruit flies to a population that progresses through its lifespan twice as fast as the original population. Let's let our original population be P and our new population be R. Our new population, R, will progress in 1 hour the same amount as the original population P does in 2 hours, and in 2 hours, the new population R will progress as much as the original population P does in 4 hours. Sketch a graph of this population.

Symbolically, we could write

$$R(1) = P(2),$$
$$R(2) = P(4), \text{ and in general,}$$
$$R(t) = P(2t).$$

See Figure 5 for a graphical comparison of the original population and the compressed population.

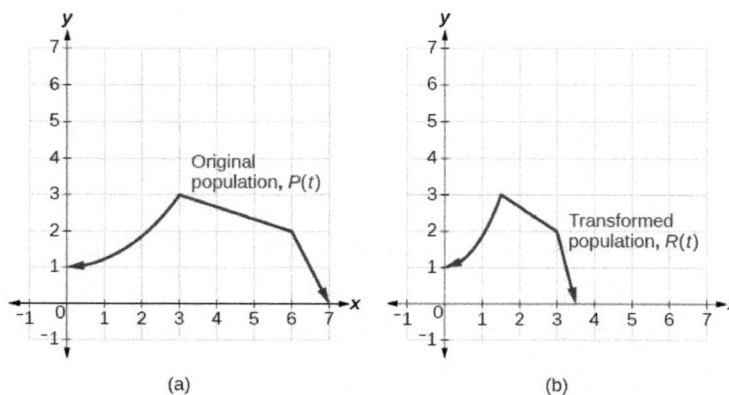

(a)   (b)

Figure 5 (a) Original population graph (b) Compressed population graph

Notice that the effect on the graph is a horizontal compression towards the vertical axis where all input values for our new function R are half of the original input value for P. You can clearly see that $R(3) = P(6)$. We have therefore compressed the original graph of $P(t)$ towards the vertical axis by a factor of ½ in order to create the graph of our new function $R(t)$.

Another way to think about this is that if the point (6, 2) is on the graph of P, then the point (3, 2) will be a point on the graph of R. We get the same outputs, but the inputs for R are half as large as the corresponding input for P. This results in a horizontal compression towards the vertical axis.

You should spend some time convincing yourself that if we multiplied our original input by a value between 0 and 1 that we would get a horizontal stretch away from the vertical axis. Think about what would happen if we considered another population of fruit flies who progress through their life span half as fast as those represented by $P(t)$. We can consider $S(t) = P\left(\frac{1}{2}t\right)$. This means that $S(4) = P(2)$ and that $S(6) = P(3)$. Can you see in order to get the same outputs that our inputs for the new function S are twice as large as the inputs for the original function P? This means there would be a horizontal stretch by a factor of 2 away from the vertical axis. Our outputs for S are the same as the outputs for P when the inputs for S are double the inputs for P. If P has the point (3, 3) on the graph, then S will have the point (6, 3) on the graph.

Given a function $f\left(x\right)$, a new function $g\left(x\right) = f\left(bx\right)$, where b is a constant, is a **horizontal stretch** or **horizontal compression** of the function $f\left(x\right)$.

- If $\left|b\right| > 1$, then the graph will be compressed by a factor of $\frac{1}{b}$ toward the y-axis.
- If $0 < \left|b\right| < 1$, then the graph will be stretched by a factor $\frac{1}{b}$ away from the y-axis.
- If $b < 0$, then there will be combination of a horizontal stretch or compression with a horizontal reflection.

HOW TO

Given a description of a function, sketch a horizontal compression or stretch.

1. Write a formula to represent the function.
2. Set $g\left(x\right) = f\left(bx\right)$ where $b > 1$ for a compression or $0 < b < 1$ for a stretch.

EXAMPLE 4: FINDING A HORIZONTAL STRETCH FOR A TABULAR FUNCTION

A function $f\left(x\right)$ is given as Table 4. Create a table for the function $g\left(x\right) = f\left(\frac{1}{2}x\right)$.

Table 4

x	2	4	6	8
$f\left(x\right)$	1	3	7	11

Answer

The formula $g\left(x\right) = f\left(\frac{1}{2}x\right)$ tells us that the output values for g are the same as the output values for the function f at an input half the size. Notice that we do not have enough information to determine $g\left(2\right)$ because $g\left(2\right) = f\left(\frac{1}{2} \cdot 2\right) = f\left(1\right)$, and we do not have a value for $f\left(1\right)$ in our table. Our input values to g will need to be twice as large to get inputs for f that we can evaluate. For example, we can determine $g\left(4\right)$.

$$g\left(4\right) = f\left(\tfrac{1}{2} \cdot 4\right) = f\left(2\right) = 1$$

We do the same for the other values to produce Table 5.

Table 5

x	4	8	12	16
$g(x)$	1	3	7	11

Figure 6 shows the graphs of both of these sets of points.

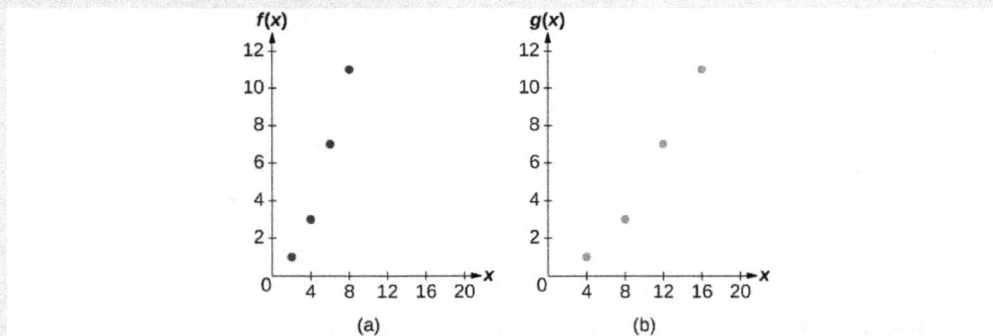

Figure 6

Analysis

Because each input value has been doubled, the result is that the function $g(x)$ has been stretched horizontally by a factor of 2.

EXAMPLE 5: RECOGNIZING A HORIZONTAL COMPRESSION ON A GRAPH

Relate the function $g(x)$ to $f(x)$ in Figure 7.

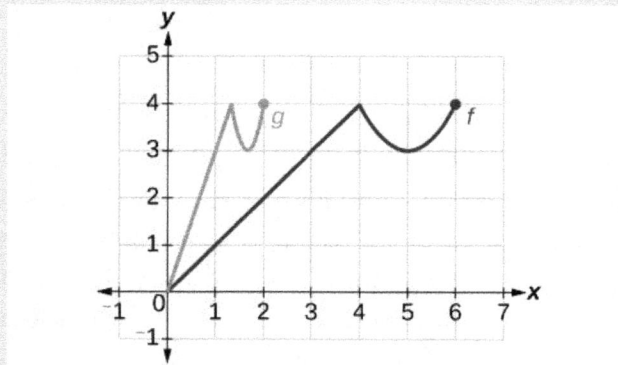

Figure 7

Answer

The graph of $g(x)$ looks like the graph of $f(x)$ horizontally compressed. Because $f(x)$ ends at $(6, 4)$ and $g(x)$ ends at $(2, 4)$, we can see that the x- values have been compressed by a factor of $\frac{1}{3}$, because $6\left(\frac{1}{3}\right) = 2$. We might also notice that $g(2) = f(6)$ and $g(1) = f(3)$. Either way, we can describe this relationship as $g(x) = f(3x)$. This is a horizontal compression by a factor of $\frac{1}{3}$.

Analysis

Notice that the coefficient needed for a horizontal stretch or compression is the reciprocal of the stretch or compression. So to stretch the graph horizontally by a scale factor of 4, we need a coefficient of $\frac{1}{4}$ in our function: $f\left(\frac{1}{4}x\right)$. This means that the input values must be four times larger to produce the same result, requiring the input to be larger, causing the horizontal stretching.

TRY IT #3

Write a formula for the toolkit square root function horizontally stretched by a factor of 3.

Answer

$g(x) = f\left(\frac{1}{3}x\right)$ so using the square root function we get $g(x) = \sqrt{\frac{1}{3}x}$.

Performing a Sequence of Transformations

When combining transformations, it is very important to consider the order of the transformations. For example, vertically shifting by 3 and then vertically stretching by a factor of 2 does not create the same graph as vertically stretching by a factor of 2 and then vertically shifting by 3, because when we shift first, both the original function and the shift get stretched, while only the original function gets stretched when we stretch first.

When we see an expression such as $2f(x) + 3,$ which transformation should we start with? The answer here follows nicely from the order of operations. Given the output value of $f(x)$, we first multiply by 2, causing the vertical stretch, and then add 3, causing the vertical shift. In other words, multiplication before addition.

Horizontal transformations are a little trickier to think about. When we write $g(x) = f(2x + 3)$, for example, we have to think about how the inputs to the function g relate to the inputs to the function f. Suppose we know $f(7) = 12$. What input to g would produce that output? In other words, what value of x will allow $g(x) = f(2x + 3) = 12$? We would need $2x + 3 = 7$. To solve for x, we would first subtract 3, resulting in a horizontal shift, and then divide by 2, causing a horizontal compression.

This format ends up being very difficult to work with, because it is usually much easier to horizontally stretch or compress a graph before shifting. We can work around this by factoring inside the function.

$$f(bx + p) = f\left(b\left(x + \tfrac{p}{b}\right)\right)$$

Let's work through an example.

$$f(x) = (2x + 4)^2$$

We can factor out a 2.

$$f(x) = (2(x+2))^2$$

Now we can more clearly observe a horizontal shift to the left 2 units and a horizontal compression. Factoring in this way allows us to horizontally stretch first and then shift horizontally.

HOW TO

Given a transformation, determine the order in which they should be preformed.

1. When combining vertical transformations written in the form $af(x) + k$, first vertically stretch or compress by a factor of a and then vertically shift by k.
2. When combining horizontal transformations written in the form $f(bx - h)$, first horizontally shift by h and then horizontally stretch or compress by a factor of $\frac{1}{b}$.
3. When combining horizontal transformations written in the form $f(b(x - h))$, first horizontally stretch or compress by a factor of $\frac{1}{b}$ and then horizontally shift by h.
4. Horizontal and vertical transformations are independent. It does not matter whether horizontal or vertical transformations are performed first.

EXAMPLE 6: FINDING A TRIPLE TRANSFORMATION OF A TABULAR FUNCTION

Given Table 6 for the function $f(x)$, create a table of values for the function $g(x) = 2f(3x) + 1$.

Table 6

x	6	12	18	24
$f(x)$	10	14	15	17

Answer

There are three steps to this transformation, and we will work from the inside out. Starting with the horizontal transformations, $f(3x)$ is a horizontal compression by a factor of $\frac{1}{3}$, which means we multiply each x-value by $\frac{1}{3}$. See Table 7.

Table 7

x	2	4	6	8
$f(3x)$	10	14	15	17

Looking now to the vertical transformations, we start with the vertical stretch, which will multiply the output values by 2. We apply this to the previous transformation. See Table 8.

Table 8

x	2	4	6	8
$2f\left(3x\right)$	20	28	30	34

Finally, we can apply the vertical shift, which will add 1 to all the output values. See Table 9.

Table 9

x	2	4	6	8
$g\left(x\right)=2f\left(3x\right)+1$	21	29	31	35

EXAMPLE 7: FINDING A TRIPLE TRANSFORMATION OF A GRAPH

Use the graph of $f\left(x\right)$ in Figure 8 to sketch a graph of $k\left(x\right)=f\left(\frac{1}{2}x+1\right)-3.$

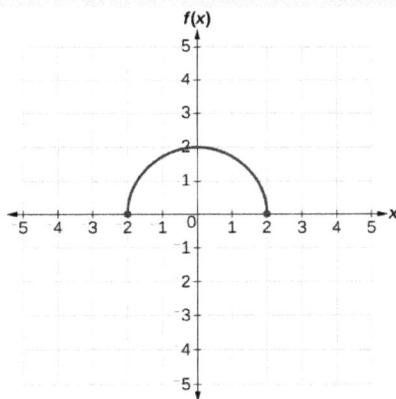

Figure 8

Answer

To simplify, let's start by factoring out the inside of the function.

$$f\left(\tfrac{1}{2}x+1\right)-3=f\left(\tfrac{1}{2}\left(x+2\right)\right)-3$$

By factoring the inside, we can first horizontally stretch by a factor of 2, as indicated by the $\frac{1}{2}$ on the inside of the function. Remember that twice the size of 0 is still 0, so the point (0,2) remains at (0,2) while the point (2,0) will stretch to (4,0). See Figure 9.

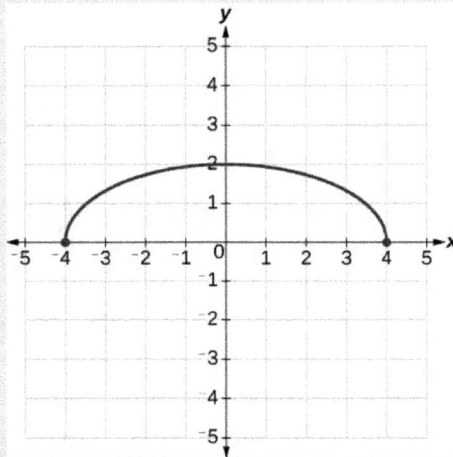

Figure 9

Next, we horizontally shift left by 2 units, as indicated by $x + 2$. See Figure 10.

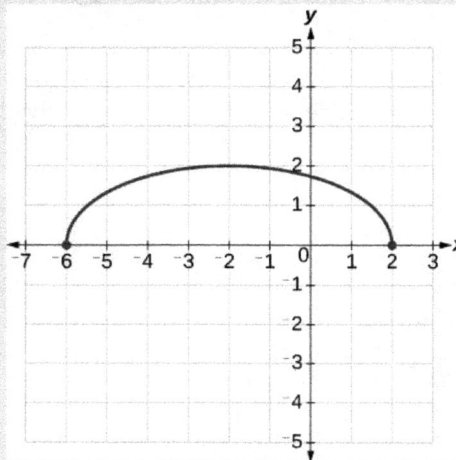

Figure 10

Last, we vertically shift down by 3 to complete our sketch, as indicated by the -3 on the outside of the function. See Figure 11.

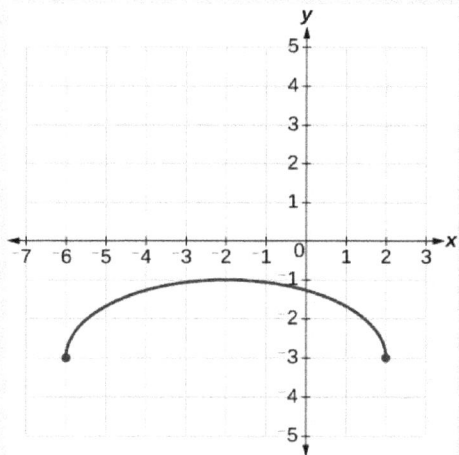

Figure 11

EXAMPLE 8: WRITING THE EQUATION OF A QUADRATIC FUNCTION FROM THE GRAPH

Write an equation for the quadratic function g in Figure 12 as a transformation of $f\left(x\right) = x^2$,. First, write your solutions in the **vertex form of a quadratic function** $g\left(x\right) = a(x - h)^2 + k$ where $\left(h,\ k\right)$ is the vertex. Then expand the formula, and simplify terms to write the equation in general form.

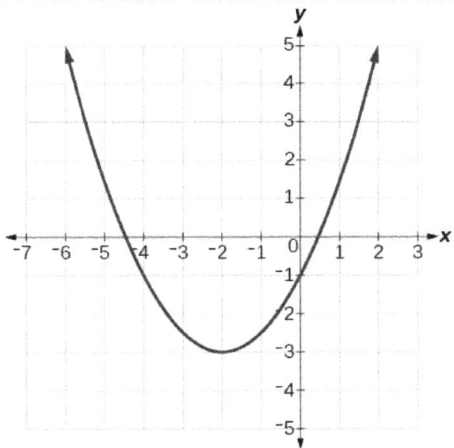

Figure 12

Answer

We can see the graph of g is the graph of $f\left(x\right) = x^2$ shifted to the left 2 and down 3, giving a formula in the form

$$g\left(x\right) = a(x + 2)^2 - 3.$$

Substituting the coordinates of a point on the curve, such as $(0, -1)$, we can solve for the vertical stretch or compression factor.

$$-1 = a(0+2)^2 - 3$$
$$2 = 4a$$
$$a = \frac{1}{2}$$

In vertex form, the algebraic model for this graph is $g(x) = \frac{1}{2}(x+2)^2 - 3$.

To write this in general polynomial form, we can expand the formula and simplify terms.

$$\begin{aligned} g(x) &= \frac{1}{2}(x+2)^2 - 3 \\ &= \frac{1}{2}(x+2)(x+2) - 3 \\ &= \frac{1}{2}(x^2 + 4x + 4) - 3 \\ &= \frac{1}{2}x^2 + 2x + 2 - 3 \\ &= \frac{1}{2}x^2 + 2x - 1 \end{aligned}$$

Notice that the horizontal and vertical shifts of the basic graph of the quadratic function determine the location of the vertex of the parabola; the vertex is unaffected by stretches and compressions.

Analysis

We can check our work using the table feature on a graphing utility. First enter $Y1 = \frac{1}{2}(x+2)^2 - 3$. Next, select TBLSET, then use $\mathrm{TblStart} = -6$ and $\Delta\mathrm{Tbl} = 2$, and select TABLE. See Table 10.

Table 10

x	−6	−4	−2	0	2
y	5	−1	−3	−1	5

The ordered pairs in the table correspond to points on the graph.

TRY IT #4

A coordinate grid has been superimposed over the quadratic path of a basketball in Figure 13. Find an equation for the path of the ball. Does the shooter make the basket?

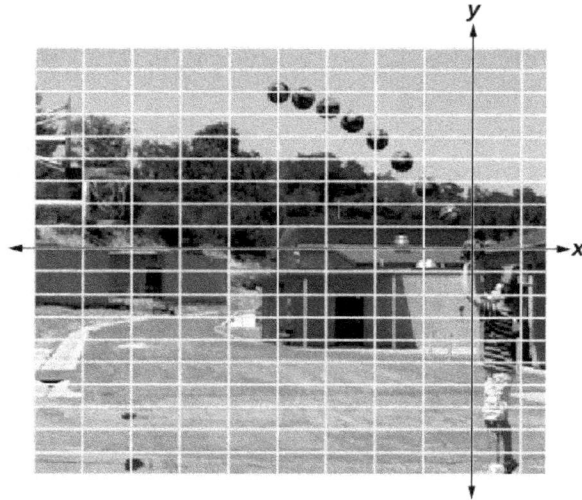

Figure 13. (credit: modification of work by Dan Meyer)

Answer

The path passes through the origin and has vertex at $(-4,\ 7)$, so $(h)\ x = -\frac{7}{16}(x+4)^2 + 7$. To make the shot, $h(-7.5)$ would need to be about 4 but $h(-7.5) \approx 1.64$; he doesn't make it.

Access this online resource for additional instruction and practice with transformation of functions.

- Function Transformations

Key Equations

Vertical shift	$g(x) = f(x) + k$ (up for $k > 0$)
Horizontal shift	$g(x) = f(x - h)$ (right for $h > 0$)
Vertical reflection	$g(x) = -f(x)$
Horizontal reflection	$g(x) = f(-x)$
Vertical stretch	$g(x) = af(x)$ ($a > 1$)
Vertical compression	$g(x) = af(x)$ $(0 < a < 1)$
Horizontal stretch	$g(x) = f(bx)$ $(0 < b < 1)$
Horizontal compression	$g(x) = f(bx)$ ($b > 1$)

KEY CONCEPTS

- A function can be compressed or stretched vertically by multiplying the output by a constant.
- A function can be compressed or stretched horizontally by multiplying the input by a constant.
- The order in which different transformations are applied does affect the final function. Shifts and stretches must be applied in the order given. However, vertical transformations may be done before or after horizontal transformations.

GLOSSARY

horizontal compression
a transformation that compresses a function's graph horizontally, by multiplying the input by a constant $b > 1$

horizontal shift
a transformation that shifts a function's graph left or right by adding a positive or negative constant to the input

odd function
a function whose graph is unchanged by combined horizontal and vertical reflection, $f(x) = -f(-x)$, and is symmetric about the origin

vertical reflection
a transformation that reflects a function's graph across the x-axis by multiplying the output by -1

vertical shift
a transformation that shifts a function's graph up or down by adding a positive or negative constant to the output

EXPONENTIAL AND LOGARITHMIC FUNCTIONS

2.1 Inverse Functions

A reversible heat pump is a climate-control system that is an air conditioner and a heater in a single device. Operated in one direction, it pumps heat out of a house to provide cooling. Operating in reverse, it pumps heat into the building from the outside, even in cool weather, to provide heating. As a heater, a heat pump is several times more efficient than conventional electrical resistance heating.

If some physical machines can run in two directions, we might ask whether some of the function "machines" we have been studying can also run backwards. Figure 1 provides a visual representation of this question. In this section, we will consider the reverse nature of functions.

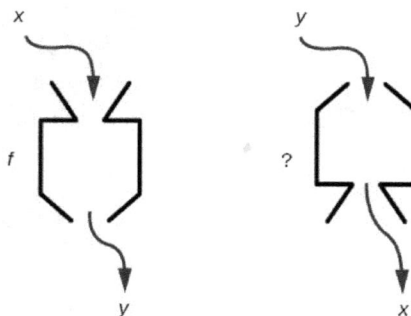

Figure 1. Can a function "machine" operate in reverse?

Verifying That Two Functions Are Inverse Functions

Suppose a fashion designer traveling to Milan for a fashion show wants to know what the temperature will be. He is not familiar with the Celsius scale. To get an idea of how temperature measurements are related, he asks his assistant, Betty, to convert 75 degrees Fahrenheit to degrees Celsius. She finds the formula

$$C = \tfrac{5}{9}\left(F - 32\right)$$

and substitutes 75 for F to calculate

$$\tfrac{5}{9}\left(75 - 32\right) \approx 24$$

Knowing that a comfortable 75 degrees Fahrenheit is about 24 degrees Celsius, he sends his assistant the week's weather forecast for Milan (Figure 2), and asks her to convert all of the temperatures to degrees Fahrenheit.

Mon	Tue	Wed	Thu
26°C | 19°C	29°C | 19°C	30°C | 20°C	26°C | 18°C

Figure 2.

At first, Betty considers using the formula she has already found to complete the conversions. After all, she knows her algebra, and can easily solve the equation for F after substituting a value for C. For example, to convert 26 degrees Celsius, she could write C. For example, to convert 26 degrees Celsius, she could write

$$26 = \frac{5}{9}(F - 32)$$

$$26 \cdot \frac{9}{5} = F - 32$$

$$F = 26 \cdot \frac{9}{5} + 32 \approx 79$$

After considering this option for a moment, however, she realizes that solving the equation for each of the temperatures will be awfully tedious. She realizes that since evaluation is easier than solving, it would be much more convenient to have a different formula, one that takes the Celsius temperature as input and outputs the Fahrenheit temperature.

The formula for which Betty is searching corresponds to the idea of an **inverse function**, which is a function for which the input of the original function becomes the output of the inverse function and the output of the original function becomes the input of the inverse function.

Given a function $f(x)$, we represent its inverse as $f^{-1}(x)$, read as "f inverse of x." The raised -1 is part of the notation. It is not an exponent; it does not imply a power of -1. In other words, $f^{-1}(x)$ does *not* mean $\frac{1}{f(x)}$ because $\frac{1}{f(x)}$ is the reciprocal of f and not the inverse.

The "exponent-like" notation comes from an analogy between function composition and multiplication: just as for multiplicative inverses, or reciprocals, $a^{-1}a = 1$ (1 is the identity element for multiplication) for any nonzero number a, so for composition of inverse functions, $f^{-1} \circ f$ equals the identity function, that is,

$$\left(f^{-1} \circ f\right)(x) = f^{-1}(f(x)) = f^{-1}(y) = x$$

This holds for all x in the domain of f. Informally, this means that inverse functions "undo" each other. However, just as the real number zero does not have a **reciprocal**, some functions do not have inverses.

Given a function $f(x)$, we can verify whether some other function $g(x)$ is the inverse of $f(x)$ by checking whether either $g(f(x)) = x$ or $f(g(x)) = x$ is true. We can test whichever equation is more convenient to work with because they are logically equivalent (that is, if one is true, then so is the other.)

For example, $y = 4x$ and $y = \frac{1}{4}x$ are inverse functions.

$$\left(f^{-1} \circ f\right)(x) = f^{-1}(4x) = \frac{1}{4}(4x) = x$$

and
$$\left(f \circ f^{-1}\right)(x) = f\left(\tfrac{1}{4}x\right) = 4\left(\tfrac{1}{4}x\right) = x$$

A few coordinate pairs from the graph of the function $y = 4x$ are (–2, –8), (0, 0), and (2, 8). A few coordinate pairs from the graph of the function $y = \tfrac{1}{4}x$ are (–8, –2), (0, 0), and (8, 2). If we interchange the input and output of each coordinate pair of a function, the interchanged coordinate pairs would appear on the graph of the inverse function.

DEFINITION

For any **one-to-one function** $f(x) = y$, a function $f^{-1}(x)$ is an **inverse function** of f if $f^{-1}(y) = x$. This can also be written as $f^{-1}(f(x)) = x$ for all x in the domain of f. It also follows that $f\left(f^{-1}(x)\right) = x$ for all x in the domain of f^{-1} if f^{-1} is the inverse of f.

The notation f^{-1} is read "f inverse." Like any other function, we can use any variable name as the input for f^{-1}, so we will often write $f^{-1}(x)$, which we read as "f inverse of x." Keep in mind that

$$f^{-1}(x) \neq \tfrac{1}{f(x)}$$

and not all functions have inverses.

EXAMPLE 1: IDENTIFYING AN INVERSE FUNCTION FOR A GIVEN INPUT-OUTPUT PAIR

If for a particular one-to-one function $f(2) = 4$ and $f(5) = 12$, what are the corresponding input and output values for the inverse function?

Answer

The inverse function reverses the input and output quantities, so if

$$f(2) = 4, \ \text{then} \ f^{-1}(4) = 2;$$
$$f(5) = 12, \ \text{then} \ f^{-1}(12) = 5.$$

Alternatively, if we want to name the inverse function g, then $g(4) = 2$ and $g(12) = 5$.

Analysis

Notice that if we show the coordinate pairs in a table form, the input and output are clearly reversed. See Table 1.

Table 1

$(x, f(x))$	$(x, g(x))$
$(2, 4)$	$(4, 2)$

$(x, f(x))$	$(x, g(x))$
$(5, 12)$	$(12, 5)$

TRY IT #1

Given that $h^{-1}(6) = 2,$ what are the corresponding input and output values of the original function $h?$

Answer

$h(2) = 6$

HOW TO

Given two functions $f(x)$ and $g(x)$, test whether the functions are inverses of each other.

1. Determine whether $f(g(x)) = x$ or $g(f(x)) = x.$
2. If both statements are true, then $g = f^{-1}$ and $f = g^{-1}.$ If either statement is false, then both are false, and $g \neq f^{-1}$ and $f \neq g^{-1}.$

EXAMPLE 2: TESTING INVERSE RELATIONSHIPS ALGEBRAICALLY

If $f(x) = \frac{1}{x+2}$ and $g(x) = \frac{1}{x} - 2,$ is $g = f^{-1}?$

Answer

$$g(f(x)) = \frac{1}{\left(\frac{1}{x+2}\right)} - 2$$
$$= x + 2 - 2$$
$$= x$$

so

$$g = f^{-1} \text{ and } f = g^{-1}$$

This is enough to answer yes to the question, but we can also verify the other formula.

$$f\left(g\left(x\right)\right) = \frac{1}{\frac{1}{x} - 2 + 2}$$

$$= \frac{1}{\frac{1}{x}}$$

$$= x$$

Analysis

Notice the inverse operations are in reverse order of the operations of the original function.

TRY IT #2

If $f\left(x\right) = x^3 - 4$ and $g\left(x\right) = \sqrt[3]{x + 4}$, is $g = f^{-1}$?

Answer

Yes

EXAMPLE 3: DETERMINING INVERSE RELATIONSHIPS FOR POWER FUNCTIONS

If $f\left(x\right) = x^3$ (the cube function) and $g\left(x\right) = \frac{1}{3}x$, is $g = f^{-1}$?

Answer

$f\left(g\left(x\right)\right) = \frac{x^3}{27} \neq x$. No, the functions are not inverses.

Analysis

The correct inverse to the cube is, of course, the cube root $\sqrt[3]{x} = x^{\frac{1}{3}}$, that is, the one-third is an exponent, not a multiplier.

Finding Domain and Range of Inverse Functions

The outputs of the function f are the inputs to f^{-1}, so the range of f is also the domain of f^{-1}. Likewise, because the inputs to f are the outputs of f^{-1}, the domain of f is the range of f^{-1}. We can visualize the situation as in Figure 3.

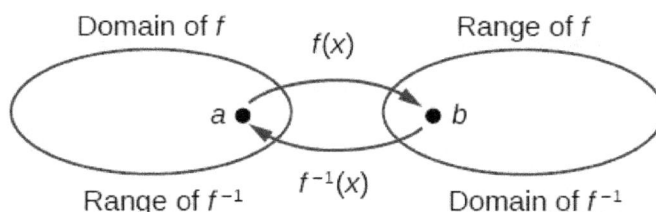

Figure 3. Domain and range of a function and its inverse.

Functions Whose Inverses Are Not Functions and Domain Restrictions

If a function is not one-to -one on its domain, then it does not have an inverse which is a function. When a function has no inverse function, it is possible to create a new function where that new function on a limited domain does have an inverse function.

For example, the inverse of $f(x) = \sqrt{x}$ is $f^{-1}(x) = x^2$, because a square "undoes" a square root; but the square is only the inverse of the square root on the domain $[0, \infty)$, since that is the range of $f(x) = \sqrt{x}$.

We can look at this problem from the other side, starting with the square (toolkit quadratic) function $f(x) = x^2$. If we want to construct an inverse to this function, we run into a problem, because for every given output of the quadratic function, there are two corresponding inputs (except when the input is 0). For example, the output 9 from the quadratic function corresponds to the inputs 3 and −3. But an output from a function is an input to its inverse; if this inverse input corresponds to more than one inverse output (input of the original function), then the "inverse" is not a function at all! To put it differently, the quadratic function is not a one-to-one function; it fails the horizontal line test, so it does not have an inverse function.

In many cases, if a function is not one-to-one, we can still restrict the function to a part of its domain on which it is one-to-one. For example, we can make a restricted version of the square function $f(x) = x^2$ with its domain limited to $[0, \infty)$, which is a one-to-one function (it passes the horizontal line test meaning that each output is paired with exactly one input) and which has an inverse (the square-root function).

If $f(x) = (x-1)^2$ on $[1, \infty)$, then the inverse function is $f^{-1}(x) = \sqrt{x} + 1$.

- The domain of f equals the range of f^{-1} : $[1, \infty)$.
- The domain of f^{-1} equals the range of f : $[0, \infty)$.

Is it possible for a function to have more than one inverse?

No. If two supposedly different functions, say, g and h, both meet the definition of being inverses of another function f, then you can prove that $g = h$. We have just seen that some functions only have inverses if we restrict the domain of the original function. In these cases, there may be more than one way to restrict the domain, leading to different inverses. However, on any one domain, the original function still has only one unique inverse.

DOMAIN AND RANGE OF INVERSE FUNCTIONS

The range of a function $f(x)$ is the domain of the inverse function $f^{-1}(x)$.

The domain of $f(x)$ is the range of $f^{-1}(x)$.

HOW TO

Given a function, find the domain and range of its inverse.

1. If the function is one-to-one, write the range of the original function as the domain of the inverse, and write the domain of the original function as the range of the inverse.
2. If the domain of the original function needs to be restricted to make it one-to-one, then this restricted domain becomes the range of the inverse function.

EXAMPLE 4: FINDING THE INVERSES OF TOOLKIT FUNCTIONS

Identify which of the toolkit functions besides the quadratic function are not one-to-one, and find a restricted domain on which each function is one-to-one, if any. The toolkit functions are reviewed in Table 2. We restrict the domain in such a fashion that the function assumes all y-values exactly once.

Table 2

Constant	Identity	Quadratic	Cubic	Reciprocal
$f(x) = c$	$f(x) = x$	$f(x) = x^2$	$f(x) = x^3$	$f(x) = \frac{1}{x}$
Reciprocal squared	Cube root	Square root	Absolute value	
$f(x) = \frac{1}{x^2}$	$f(x) = \sqrt[3]{x}$	$f(x) = \sqrt{x}$	$f(x) = |x|$	

Answer

The constant function is not one-to-one, and there is no domain (except a single point) on which it could be one-to-one, so the constant function has no meaningful inverse.

The absolute value function can be restricted to the domain $[0, \infty)$, where it is equal to the identity function.

The reciprocal-squared function can be restricted to the domain $(0, \infty)$.

Analysis

We can see that these functions (if unrestricted) are not one-to-one by looking at their graphs, shown in Figure 4. They both would fail the horizontal line test. However, if a function is restricted to a certain domain so that it passes the horizontal line test, then in that restricted domain, it can have an inverse.

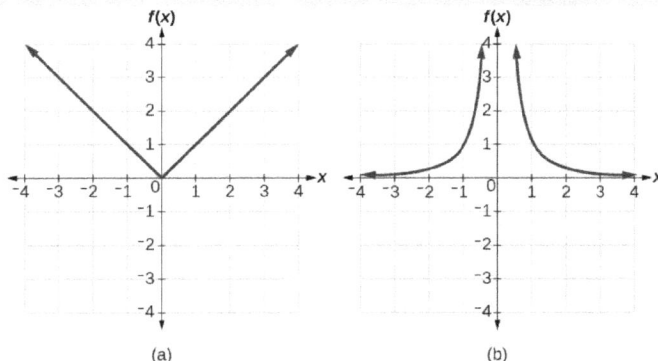

Figure 4. (a) Absolute value (b) Reciprocal squared

TRY IT #4

The domain of function f is $(1, \infty)$ and the range of function f is $(-\infty, -2)$. Find the domain and range of the inverse function.

Finding and Evaluating Inverse Functions

Once we have a one-to-one function, we can evaluate its inverse at specific inverse function inputs or construct a complete representation of the inverse function in many cases.

Inverting Tabular Functions

Suppose we want to find the inverse of a function represented in table form. Remember that the domain of a function is the range of the inverse and the range of the function is the domain of the inverse. So we need to interchange the domain and range.

Each row (or column) of inputs becomes the row (or column) of outputs for the inverse function. Similarly, each row (or column) of outputs becomes the row (or column) of inputs for the inverse function.

EXAMPLE 5: INTERPRETING THE INVERSE OF A TABULAR FUNCTION

A function $f(t)$ is given in Table 3, showing distance in miles that a car has traveled in t minutes. Find and interpret $f^{-1}(70)$.

Table 3

t (minutes)	30	50	70	90
$f(t)$ (miles)	20	40	60	70

Answer

The inverse function takes an output of f and returns an input for f. So in the expression $f^{-1}(70)$, 70 is an output value of the original function, representing 70 miles. The inverse will return the corresponding input of the original function f, 90 minutes, so $f^{-1}(70) = 90$. The interpretation of this is that, to drive 70 miles, it took 90 minutes.

Alternatively, recall that the definition of the inverse was that if $f(a) = b$, then $f^{-1}(b) = a$. By this definition, if we are given $f^{-1}(70) = a$, then we are looking for a value a so that $f(a) = 70$. In this case, we are looking for a t so that $f(t) = 70$, which is when $t = 90$.

TRY IT #5

Using Table 4, find and interpret (a) $f(60)$, and (b) $f^{-1}(60)$.

Table 4

t (minutes)	30	50	60	70	90
$f(t)$ (miles)	20	40	50	60	70

Answer

a. $f(60) = 50.$ In 60 minutes, 50 miles are traveled.

b. $f^{-1}(60) = 70.$ To travel 60 miles, it will take 70 minutes.

Evaluating the Inverse of a Function, Given a Graph of the Original Function

We saw in Section 1.1, Functions and Function Notation, that the domain of a function can be read by observing the horizontal extent of its graph. We find the domain of the inverse function by observing the *vertical* extent of the graph of the original function, because this corresponds to the horizontal extent of the inverse function. Similarly, we find the range of the inverse function by observing the *horizontal* extent of the graph of the original function, as this is the vertical extent of the inverse function. If we want to evaluate an inverse function, we find its input within its domain, which is all or part of the vertical axis of the original function's graph.

HOW TO

Given the graph of a function, evaluate its inverse at specific points.

1. Find the desired input on the *y*-axis of the given graph.
2. Read the inverse function's output from the *x*-axis of the given graph.

EXAMPLE 6: EVALUATING A FUNCTION AND ITS INVERSE FROM A GRAPH AT SPECIFIC POINTS

A function $g(x)$ is given in Figure 5. Find $g(3)$ and $g^{-1}(3)$.

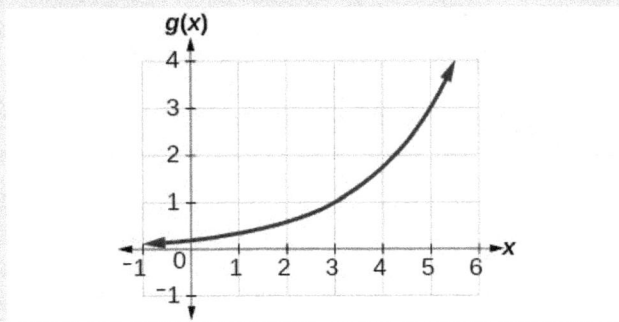

Figure 5.

Answer

To evaluate $g(3)$, we find 3 on the x-axis and find the corresponding output value on the y-axis. The point $(3, 1)$ tells us that $g(3) = 1$.

To evaluate $g^{-1}(3)$, recall that by definition $g^{-1}(3)$ means the value of x for which $g(x) = 3$. By looking for the output value 3 on the vertical axis, we find the point $(5, 3)$ on the graph, which means $g(5) = 3$, so by definition, $g^{-1}(3) = 5$. See Figure 6.

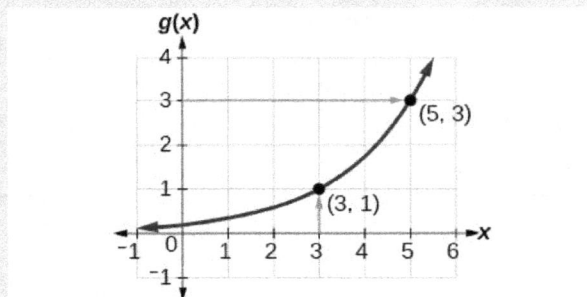

Figure 6.

TRY IT #6

Using the graph in Figure 6, (a) find $g^{-1}(1)$, and (b) estimate $g^{-1}(4)$.

Answer

a. 3; b. 5.6

Finding Inverses of Functions Represented by Formulas

Sometimes we will need to know an inverse function for all elements of its domain, not just a few. If the original function is given as a formula— for example, y as a function of x we can often find the inverse function by solving to obtain x as a function of y.

HOW TO

Given a function represented by a formula, find the inverse.

1. Make sure f is a one-to-one function.
2. Solve for the input variable.
3. Write using the appropriate function notation.

EXAMPLE 7: INVERTING THE FAHRENHEIT-TO-CELSIUS FUNCTION

Find a formula for the inverse function that gives Fahrenheit temperature as a function of Celsius temperature.

$$C = \tfrac{5}{9}(F - 32)$$

Answer

$$C = \frac{5}{9}(F - 32)$$

$$C \cdot \frac{9}{5} = F - 32$$

$$F = \frac{9}{5}C + 32$$

By solving in general, we have uncovered the inverse function. If

$$C = h(F) = \tfrac{5}{9}(F - 32),$$

then

$$F = h^{-1}(C) = \tfrac{9}{5}C + 32.$$

In this case, we introduced a function h to represent the conversion because the input and output variables are descriptive, and writing C^{-1} could get confusing.

TRY IT #7

Solve for x in terms of y given $y = \tfrac{1}{3}(x - 5)$.

Answer

$$x = 3y + 5$$

EXAMPLE 8: SOLVING TO FIND AN INVERSE FUNCTION

Find the inverse of the function $f(x) = \frac{2}{x-3} + 4$.

Answer

$$y = \frac{2}{x-3} + 4 \quad \text{Set up an equation.}$$

$$y - 4 = \frac{2}{x-3} \quad \text{Subtract 4 from both sides.}$$

$$x - 3 = \frac{2}{y-4} \quad \text{Multiply both sides by } x - 3 \text{ and divide by } y - 4.$$

$$x = \frac{2}{y-4} + 3 \quad \text{Add 3 to both sides.}$$

So $f^{-1}(y) = \frac{2}{y-4} + 3$.

Analysis

The domain and range of f exclude the values 3 and 4, respectively. f and f^{-1} are equal at two points but are not the same function, as we can see by creating Table 5.

Table 5

x	1	2	5	$f^{-1}(y)$
$f(x)$	3	2	5	y

EXAMPLE 9: SOLVING TO FIND AN INVERSE WITH RADICALS

Find the inverse of the function $f(x) = 2 + \sqrt{x-4}$.

Answer

$$y = 2 + \sqrt{x - 4} \qquad \text{Replace } f(x) \text{ with } y.$$
$$y - 2 = \sqrt{x - 4} \qquad \text{Subtract 2 from both sides.}$$
$$(y - 2)^2 = x - 4 \qquad \text{Square both sides.}$$
$$x = (y - 2)^2 + 4 \qquad \text{Add 4 to both sides.}$$

So $f^{-1}(y) = (y - 2)^2 + 4$.

The domain of f is $[4, \infty)$ so the range of f^{-1} is $[4, \infty)$. Notice that the range of f is $[2, \infty)$, so this means that the domain of the inverse function f^{-1} is also $[2, \infty)$.

Analysis

The formula we found for $f^{-1}(y)$ looks like it would be valid for all real y. However, f^{-1} itself must have an inverse (namely, f) so we have to restrict the domain of f^{-1} to $[2, \infty)$ in order to make f^{-1} a one-to-one function. This domain of f^{-1} is exactly the range of f.

TRY IT #8

What is the inverse of the function $f(x) = 2 - \sqrt{x}$? State the domains of both the function and the inverse function.

Answer

$$f^{-1}(y) = (2 - y)^2; \quad \text{domain of } f: \quad [0, \infty); \quad \text{domain of } f^{-1}: \quad (-\infty, 2]$$

Finding Inverse Functions and Their Graphs

Now that we can find the inverse of a function, we will explore the graphs of functions and their inverses. Let us return to the quadratic function $f(x) = x^2$ restricted to the domain $[0, \infty)$, on which this function is one-to-one, and graph it as in Figure 7.

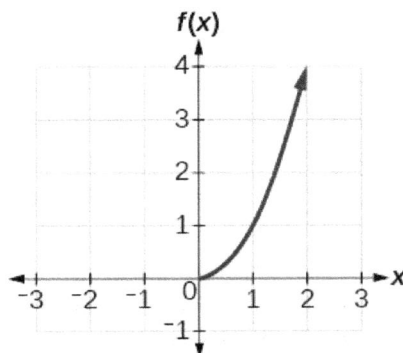

Figure 7. Quadratic function with domain restricted to [0, ∞).

Restricting the domain to $[0, \infty)$ makes the function one-to-one (it will obviously pass the horizontal line test), so it has an inverse on this restricted domain.

We already know that the inverse of the toolkit quadratic function is the square root function, that is, $f^{-1}(x) = \sqrt{x}$. What happens if we graph both f and f^{-1} on the same set of axes, using the x-axis for the input to both f and f^{-1}?

We notice a distinct relationship: The graph of $f^{-1}(x)$ is the graph of $f(x)$ reflected about the diagonal line $y = x$, which we will call the identity line, shown in Figure 8.

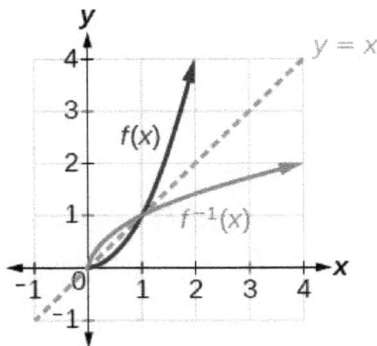

Figure 8. Square and square-root functions on the non-negative domain.

This relationship will be observed for all one-to-one functions, because it is a result of the function and its inverse swapping inputs and outputs. This is equivalent to interchanging the roles of the vertical and horizontal axes.

EXAMPLE 10: FINDING THE INVERSE OF A FUNCTION USING REFLECTION ABOUT THE IDENTITY LINE

Given the graph of $f(x)$ in Figure 9, sketch a graph of $f^{-1}(x)$.

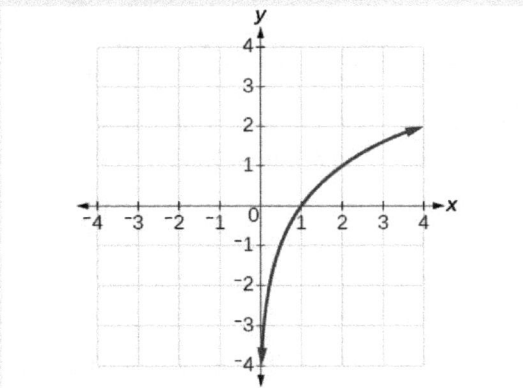

Figure 9.

Answer

This is a one-to-one function, so we will be able to sketch an inverse. Note that the graph shown has an apparent domain of $(0, \infty)$ and range of $(-\infty, \infty)$, so the inverse will have a domain of $(-\infty, \infty)$ and range of $(0, \infty)$.

If we reflect this graph over the line $y = x$, the point $(1, 0)$ reflects to $(0, 1)$ and the point $(4, 2)$ reflects to $(2, 4)$. Sketching the inverse on the same axes as the original graph gives Figure 10.

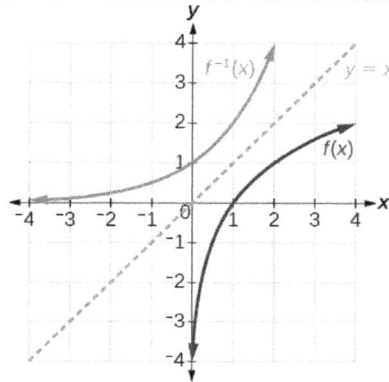

Figure 10. The function and its inverse, showing reflection about the identity line

TRY IT #9

Draw graphs of the functions f and f^{-1} from Example 8 where $f(x) = \frac{2}{x-3} + 4.$

Answer

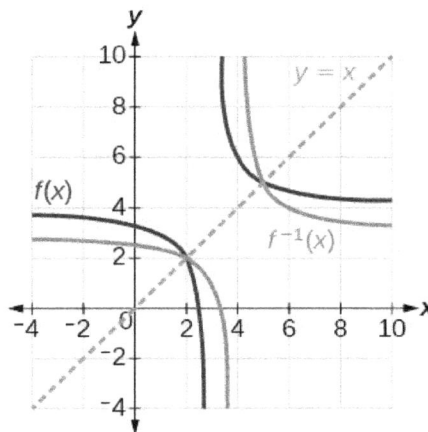

Figure 11

Q&A

Is there any function that is equal to its own inverse?

Yes. If $f = f^{-1}$, then $f(f(x)) = x$, and we can think of several functions that have this property. The identity function does, and so does the reciprocal function, because

$$\frac{1}{\frac{1}{x}} = x$$

Any function $f(x) = c - x$, where C is a constant, is also equal to its own inverse.

Media

Access these online resources for additional instruction and practice with inverse functions.

- Inverse Functions
- Inverse Function Values Using Graph
- Restricting the Domain and Finding the Inverse

Visit this website for additional practice questions from Learningpod.

KEY CONCEPTS

- If $g(x)$ is the inverse of $f(x)$, then $g(f(x)) = f(g(x)) = x$.
- Some of the toolkit functions have an inverse function.
- For a function to have an inverse, it must be one-to-one (pass the horizontal line test).
- A function that is not one-to-one over its entire domain may be one-to-one on part of its domain.
- For a tabular function, exchange the input and output rows to obtain the inverse.
- The inverse of a function can be determined at specific points on its graph.
- To find the inverse of a formula, solve the equation $y = f(x)$ for x as a function of y.
- The graph of an inverse function is the reflection of the graph of the original function across the line $y = x$.

GLOSSARY

inverse function

for any one-to-one function $f(x)$, the inverse is a function $f^{-1}(x)$ such that $f^{-1}(f(x)) = x$ for all x in the domain of f; this also implies that $f(f^{-1}(x)) = x$ for all x in the domain of f^{-1}

2.2 Exponential Functions

India is the second most populous country in the world with a population of about 1.37 billion people in 2019. The population is growing at a rate of about 1.08% each year[1]. If this rate continues, the population of India will exceed China's population by the year 2031. When populations grow rapidly, we often say that the growth is "exponential," meaning that something is growing very rapidly. To a mathematician, however, the term *exponential growth* has a very specific meaning. In this section, we will take a look at *exponential functions*, which model this kind of rapid growth.

Identifying Exponential Functions

When exploring linear growth, we observed a constant rate of change—a constant number by which the output increased for each unit increase in the input. For example, in the equation $f\left(x\right) = 3x + 4,$ the slope tells us the output increases by 3 each time the input increases by 1. The scenario in the India population example is different because we have a *percent* change per unit time (rather than a constant change) in the number of people.

Defining an Exponential Function

A study found that the percent of the population who are vegans in the United States doubled from 2009 to 2011. In 2011, 2.5% of the population was vegan, adhering to a diet that does not include any animal products—no meat, poultry, fish, dairy, or eggs. If this rate continues, vegans will make up 10% of the U.S. population in 2015, 40% in 2019, and 80% in 2021.

What exactly does it mean to *grow exponentially*? What does the word *double* have in common with *percent increase*? People toss these words around errantly. Are these words used correctly? The words certainly appear frequently in the media.

- **Percent change** refers to a *change* based on a *percent* of the original amount.
- **Exponential growth** refers to an *increase* based on a constant multiplicative rate of change over equal increments of time, that is, a *percent* increase of the original amount over time.
- **Exponential decay** refers to a *decrease* based on a constant multiplicative rate of change over equal increments of time, that is, a *percent* decrease of the original amount over time.

---

1. http://www.worldometers.info/world-population/. Accessed April 22, 2019.

For us to gain a clear understanding of **exponential growth**, let us contrast exponential growth with linear growth. We will construct two functions. The first function is exponential. We will start with an input of 0, and increase each input by 1. We will double the corresponding consecutive outputs. The second function is linear. We will start with an input of 0, and increase each input by 1. We will add 2 to the corresponding consecutive outputs. See Table 1.

Table 1

x	$f(x) = 2^x$	$g(x) = 2x$
0	1	0
1	2	2
2	4	4
3	8	6
4	16	8
5	32	10
6	64	12

From Table 1 we can infer that for these two functions, exponential growth dominates linear growth.

- **Exponential growth** refers to the original value from the range increases by the *same percentage* over equal increments found in the domain. Another way to say this that there is a constant multiplier over equal increments in the domain.
- **Linear growth** refers to the original value from the range increases by the *same amount* over equal increments found in the domain.

Apparently, the difference between "the same percentage" and "the same amount" is quite significant. For exponential growth, over equal increments, the constant multiplicative rate of change resulted in doubling the output whenever the input increased by one. For linear growth, the constant additive rate of change over equal increments resulted in adding 2 to the output whenever the input was increased by one.

The general form of the **exponential function** is $f(x) = ab^x$, where a is any nonzero number, and b is a positive real number not equal to 1.

- If $b > 1$, the function grows at a rate proportional to its size.
- If $0 < b < 1$, the function decays at a rate proportional to its size.

Let's look at the function $f(x) = 2^x$ from our example. We will create a table to determine the corresponding outputs over an interval in the domain from -3 to 3. See Table 2.

Table 2

x	-3	-2	-1	0	1	2	3
$f(x) = 2^x$	$2^{-3} = \frac{1}{8}$	$2^{-2} = \frac{1}{4}$	$2^{-1} = \frac{1}{2}$	$2^0 = 1$	$2^1 = 2$	$2^2 = 4$	$2^3 = 8$

Let us examine the graph of f by plotting the ordered pairs we observe on the table, and then make a few observations. See Figure 1.

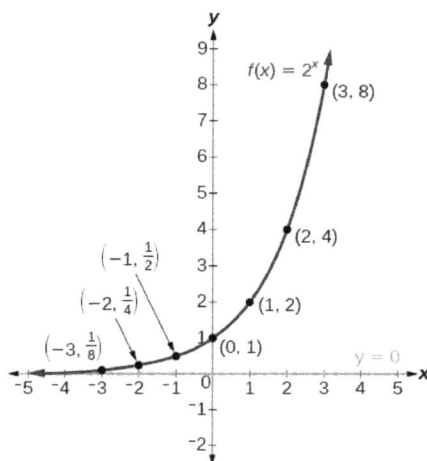

Figure 1.

Let's define the behavior of the graph of the exponential function $f(x) = 2^x$ and highlight some its key characteristics.

- The domain is $(-\infty, \infty)$.
- The range is $(0, \infty)$.
- As x gets larger and larger so does $f(x)$.
- The graph of $f(x)$ will never touch the x-axis because base two raised to any exponent never has the result of zero.
- $f(x)$ is always increasing.
- The vertical intercept is $(0, 1)$.

EXPONENTIAL FUNCTION

For any real number x, an exponential function is a function with the form

$$f(x) = ab^x$$

where

- a is a non-zero real number called the initial value or initial condition and
- b is any positive real number such that $b \neq 1$.
- The domain of f is all real numbers.
- The range of f is all positive real numbers if $a > 0$.
- The range of f is all negative real numbers if $a < 0$.
- The vertical intercept is $(0, a)$.

EXAMPLE 1: IDENTIFYING EXPONENTIAL FUNCTIONS

Which of the following equations are *not* exponential functions?

- $f(x) = 4^{3(x-2)}$
- $g(x) = x^3$
- $h(x) = \left(\frac{1}{3}\right)^x$
- $j(x) = (-2)^x$

Answer

By definition, an exponential function has a constant base and an independent variable as an exponent. Thus, $g(x) = x^3$ does not represent an exponential function because the base is an independent variable. In fact, $g(x) = x^3$ is a power function which will be studied later.

Recall that the base b of an exponential function is always a positive constant, and $b \neq 1$. Thus, $j(x) = (-2)^x$ does not represent an exponential function because the base, -2, is less than 0.

The function $f(x) = 4^{3(x-2)}$ is a transformation of 4^x. 4^x is compressed by a factor of 1/3 followed by a shift two units right. Therefore, it is an exponential function. The function $h(x) = \left(\frac{1}{3}\right)^x$ is also an exponential function with base 1/3 and an initial condition of 1.

TRY IT #1

Which of the following equations represent exponential functions?

- $f(x) = 2x^2 - 3x + 1$
- $g(x) = 0.875^x$
- $h(x) = 1.75x + 2$
- $j(x) = 1095.6^{-2x}$

Answer

$g(x) = 0.875^x$ and $j(x) = 1095.6^{-2x}$ represent exponential functions.

$f(x)$ is a quadratic function and $h(x)$ is a linear function and therefore they are not exponential functions.

Writing Formulas for Exponential Functions and Evaluating Them

Recall that the base of an exponential function must be a positive real number other than 1. Why do we limit the base b to positive values? To ensure that the outputs will be real numbers. Observe what happens if the base is not positive:

- Let $b = -9$ and $x = \frac{1}{2}$. Then $f\left(\frac{1}{2}\right) = (-9)^{\frac{1}{2}} = \sqrt{-9}$, which is not a real number.

Why do we limit the base to positive values other than 1? Because base 1 results in the constant function. Observe what happens if the base is 1 :

- Let $b = 1$. Then $f(x) = 1^x = 1$ for any value of x.

To evaluate an exponential function with the form $f(x) = b^x$, we simply substitute x with the given value, and calculate the resulting power. For example:

Let $f(x) = 2^x$. What is $f(3)$?

$$f(x) = 2^x$$
$$f(3) = 2^3 \qquad \text{Substitute } x = 3.$$
$$= 8 \qquad \text{Evaluate the power.}$$

To evaluate an exponential function with a form other than the basic form, it is important to follow the order of operations. For example:

Let $f(x) = 30(2)^x$. What is $f(3)$?

$$f(x) = 30(2)^x$$
$$f(3) = 30(2)^3 \qquad \text{Substitute } x = 3.$$
$$= 30(8) \qquad \text{Simplify the power first.}$$
$$= 240 \qquad \text{Multiply.}$$

Note that if the order of operations were not followed, the result would be incorrect:

$$f(3) = 30(2)^3 \neq 60^3 = 216{,}000$$

EXAMPLE 2: EVALUATING EXPONENTIAL FUNCTIONS

Let $f(x) = 5(3)^{x+1}$. Evaluate $f(2)$ without using a calculator.

Answer

Follow the order of operations. Be sure to pay attention to the parentheses.

$$f(x) = 5(3)^{x+1}$$
$$f(2) = 5(3)^{2+1} \qquad \text{Substitute } x = 2.$$
$$= 5(3)^3 \qquad \text{Add the exponents.}$$
$$= 5(27) \qquad \text{Simplify the power.}$$
$$= 135 \qquad \text{Multiply.}$$

TRY IT #2

Let $f(x) = 8(1.2)^{x-5}$. Evaluate $f(3)$ using a calculator. Round to four decimal places.
Answer

5.5556

Defining Exponential Growth and Decay

Because the output of exponential functions increases very rapidly, the term "exponential growth" is often used in everyday language to describe anything that grows or increases rapidly. However, exponential growth can be defined more precisely in a mathematical sense. If the growth rate is proportional to the amount present, the function models exponential growth or decay.

EXPONENTIAL GROWTH OR DECAY

A function that models **exponential growth or decay** grows by a rate proportional to the amount present. For any real number x and any positive real numbers a and b such that $b \neq 1$, an exponential growth function has the form

$$f(x) = ab^x = a(1+r)^x$$

where

- a is the initial or starting value of the function.
- $b = 1 + r$ is the growth factor or growth multiplier per unit x.
- $r = b - 1$ is the percent increase or decrease expressed as a decimal.

In more general terms, we have an *exponential function*, in which a constant base is raised to a variable exponent. To differentiate between linear and exponential functions, let's consider two companies, A and B. Company A has 100 stores and expands by opening 50 new stores a year, so its growth can be represented by the function $A(x) = 100 + 50x$. Company B has 100 stores and expands by increasing the number of stores by 50% each year, so its growth can be represented by the function $B(x) = 100(1 + 0.5)^x$.

A few years of growth for these companies are illustrated in Table 3.

Table 3

Year, x	Stores, Company A		Stores, Company B
0	$100 + 50(0) = 100$	Starting with 100 each	$100(1 + 0.5)^0 = 100$
1	$100 + 50(1) = 150$	Both grow by 50 stores in the first year.	$100(1 + 0.5)^1 = 150$
2	$100 + 50(2) = 200$	Company A grows by 50 stores and company B by 75 stores.	$100(1 + 0.5)^2 = 225$
3	$100 + 50(3) = 250$	Company A grows by 50 stores and company B by 112.5 stores.	$100(1 + 0.5)^3 = 337.5$
x	$A(x) = 100 + 50x$		$B(x) = 100(1 + 0.5)^x$

The graphs comparing the number of stores for each company over a five-year period are shown in Figure 2. We can see that, with exponential growth, the number of stores increases much more rapidly than with linear growth.

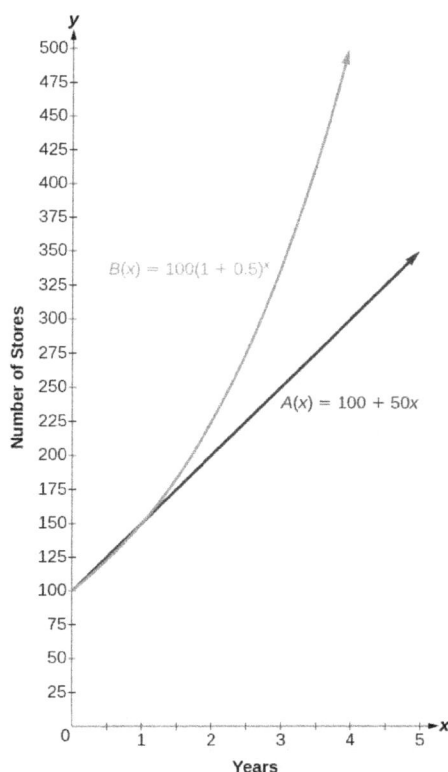

Figure 2.

Notice that the contextual domain for both functions is $[0, \infty)$, and the range for both functions is $[100, \infty)$. After year 1, Company B always has more stores than Company A.

Now we will turn our attention to the function representing the number of stores for Company B,

$$B\left(x\right) = 100(1 + 0.5)^{x}.$$

In this exponential function, 100 represents the initial number of stores, 0.50 represents the **growth rate**, r, and $b = 1 + 0.5 = 1.5$ represents the **growth factor**. Generalizing further, we can write this function as

$$B\left(x\right) = 100(1.5)^{x}.$$

EXAMPLE 3: EVALUATING A REAL-WORLD EXPONENTIAL MODEL

At the beginning of this section, we learned that the population of India was about 1.37 billion in the year 2019, with an annual growth rate of about 1.08%. Let t be the number of years since 2019. Find an exponential function to model this growth. Then, to the nearest thousandth, determine what will the population of India be in 2031.

Answer

The growth rate is given as 1.08% so $r = 0.0108$. Since the growth factor is $b = 1 + r$, we have that $b = 1 + 0.0108 = 1.0108$. There are approximately 1.37 billion people at our initial time $t = 0$ which represents 2019 so $a = 1.37$. Note that we do not add a bunch of zeros to the equation but rather keep the billions unit in mind when we interpret the function's output. Finally our function is $P\left(t\right) = 1.37(1.0108)^{t}$.

To estimate the population in 2031, we evaluate the model for $t = 12$ because 2031 is 12 years after 2019. Rounding to the nearest thousandth,

$$P(12) = 1.37(1.0108)^{12} \approx 1.558$$

There will be about 1.558 billion people in India in the year 2031.

EXAMPLE 4: REAL WORLD EXPONENTIAL MODELS

Bismuth-210 is an isotope that radioactively decays by about 13% each day, meaning 13% of the remaining Bismuth-210 transforms into another atom (polonium-210 in this case) each day. If you begin with 100 mg of Bismuth-210, how much remains after one week?

Answer

With radioactive decay, instead of the quantity increasing at a percent rate, the quantity is decreasing at a percent rate. Our initial quantity is $a = 100$ mg, and our growth rate will be negative 13%, since we are decreasing: $r = -0.13$. This gives the equation:

$$Q(d) = 100(10.13)^d = 100(0.87)^d$$

This can also be explained by recognizing that if 13% decays, then 87% remains.

Next, we are asked to find how much remains after one week so we evaluate the function when $d = 7$. Therefore,

$Q(7) = 100(0.87)^7 = 37.73.$ After one week, 37.73 mg of Bismuth-210 remains.

TRY IT #3

The population of China was about 1.39 billion in the year 2013, with an annual growth rate of about 0.6%. Find an exponential function that models this situation where t represents the number of years since 2013. To the nearest thousandth, what will the population of China be for the year 2031?

Answer

This situation is represented by the growth function $P(t) = 1.39(1.006)^t$, where t is the number of years since 2013.

China's population will be about 1.548 billion people in 2031.

TRY IT #4

A population of 1000 animals is decreasing 3% each year. Find the population in 30 years.

Answer

The function modeling this situation is $P\left(t\right) = 1000(0.97)^{t}$ and in 30 years the population will be 401 animals.

Finding Equations of Exponential Functions From Data

In the previous examples, we were given an exponential growth or decay rate, which we used to find a formula to model the situation and then evaluated for a given input. Sometimes we are given data points for an exponential function and we must use them to first find the growth factor and then the formula for the function before we can answer questions about the situation.

HOW TO

Given two data points with the initial value known, write an exponential model.

1. Identify the initial value from the data point of the form $\left(0, a\right)$. Then a is the initial value.
2. Using a, substitute the second point into the equation $f\left(x\right) = a(b)^{x}$, and solve for $b.$ To do this, divide both sides by a and then raise both sides to the appropriate power to solve the equation.
3. Write the equation using the values you found for a and b in the form $f\left(x\right) = a(b)^{x}.$

EXAMPLE 5: WRITING AN EXPONENTIAL MODEL WHEN THE INITIAL VALUE IS KNOWN

In 2006, 80 deer were introduced into a wildlife refuge. By 2012, the population had grown to 180 deer. The population was growing exponentially. Write an algebraic function $N\left(t\right)$ representing the population $\left(N\right)$ of deer over time $t.$

Answer

We let our independent variable t be the number of years after 2006. Thus, the information given in the problem can be written as input-output pairs: (0, 80) and (6, 180). Notice that by choosing our input variable to be measured as years after 2006, we have given ourselves the initial value for the function, $a = 80.$ We can now substitute the second point into the equation $N\left(t\right) = 80b^{t}$ to find b :

$$N(t) = 80b^t$$

$$180 = 80b^6 \qquad \text{Substitute using point } (6, 180).$$

$$\frac{9}{4} = b^6 \qquad \text{Divide and write in lowest terms.}$$

$$b = \left(\frac{9}{4}\right)^{\frac{1}{6}} \qquad \text{Isolate } b \text{ using properties of exponents.}$$

$$b \approx 1.1447 \qquad \text{Round to 4 decimal places.}$$

NOTE: *Unless otherwise stated, do not round any intermediate calculations. Then round the final answer to four places for the remainder of this section.*

The exponential model for the population of deer is $N(t) = 80(1.1447)^t$. (Note that this exponential function models short-term growth. As the inputs gets large, the output will get increasingly larger, so much so that the model may not be useful in the long term.)

We can graph our model to observe the population growth of deer in the refuge over time. Notice that the graph in Figure 3 passes through the initial points given in the problem, $(0,\ 80)$ and $(6,\ 180)$. We can also see that the domain for the function is $[0, \infty)$, and the range for the function is $[80, \infty)$.

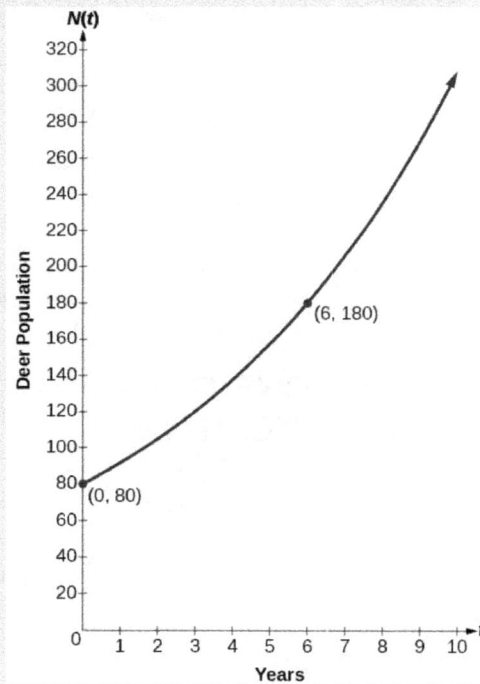

Figure 3.

TRY IT #5

A wolf population is growing exponentially. In 2011, 129 wolves were counted. By 2013, the population had reached 236 wolves. Let $t = 0$ represent the year 2011. What two points can be used to derive an exponential equation modeling this situation? Write the equation representing the population, N, of wolves over time t.

Answer

$(0, 129)$ and $(2, 236)$; $\quad N\left(t\right) = 129(1.3526)^t$

HOW TO

Given two data points, write an exponential model with the initial value unknown.

Method 1: Substitution

1. Substitute both points into two equations with the form $f\left(x\right) = a(b)^x$.
2. Solve the first equation for a in terms of b.
3. Substitute this expression for a into the second equation and then solve for b.
4. Once you have found b, substitute a data point and b into the general equation and solve for a.
5. Using the a and b found in the steps above, write the exponential function in the form $f\left(x\right) = a(b)^x$.

Method 2: Ratios

1. Choose two points to work with: $\left(c, f\left(c\right)\right)$ and $\left(d, f\left(d\right)\right)$. Order your points so $c < d$.
2. Set up your ratio: $\frac{f(d)}{f(c)} = \frac{ab^d}{ab^c}$. Note that the values on the left are gotten from your output values of the data points and the values on the right come from the general equation with c and d filled in.
3. Simplify. Note that $\frac{a}{a} = 1$ so essentially they always cancel and you will always find b first. Remember that $\frac{b^d}{b^c} = b^{d-c}$. You will then need to find the (d-c)th root of both sides.
4. Once you have found b, substitute a data point and b into the general equation and solve for a.
5. Write your equation using the values of a and b in the form $f\left(x\right) = a(b)^x$.

EXAMPLE 6: WRITING AN EXPONENTIAL MODEL WHEN THE INITIAL VALUE IS NOT KNOWN

Find an exponential function that passes through the points $\left(-2, 6\right)$ and $\left(2, 1\right)$.

Answer

Method 1: Because we don't have the initial value, we substitute both points into the equation of the form $f(x) = ab^x$, and then solve the system for a and b.

- Substituting $(-2, 6)$ gives $6 = ab^{-2}$ (Equation 1)
- Substituting $(2, 1)$ gives $1 = ab^2$ (Equation 2)

Use the first equation to solve for a in terms of b :

$$6 = ab^{-2}$$

$$\frac{6}{b^{-2}} = a \qquad \text{Divide by } b^{-2}.$$

$$a = 6b^2 \qquad \text{Use properties of exponents to rewrite the denominator.}$$

Substitute a in the second equation, and solve for b :

$$1 = ab^2$$

$$1 = \left(6b^2\right) b^2 = 6b^4 \qquad \text{Substitute } a.$$

$$b = \left(\frac{1}{6}\right)^{\frac{1}{4}} \qquad \text{Use properties of exponents to isolate } b.$$

$$b \approx 0.6389 \qquad \text{Round 4 decimal places.}$$

Use the value of b in the first equation to solve for the value of a :

$$a = 6b^2 \approx 6\left(0.6389\right)^2 \approx 2.4492$$

Thus, the equation is $f(x) = 2.4492(0.6389)^x$.

Method 2: The two points $(-2, 6)$ and $(2, 1)$ were given so we start with step 2; setting up the ratio.

$$\frac{1}{6} = \frac{ab^2}{ab^{-2}}.$$

Next we simplify using the properties of exponents.

$$\frac{1}{6} = b^{2-(-2)}$$

$$\frac{1}{6} = b^4$$

$$\left(\frac{1}{6}\right)^{\frac{1}{4}} = \left(b^4\right)^{\frac{1}{4}}$$

$$b \approx 0.6389$$

Then we find a by substituting in a point $(-2, 6)$ and the known value of $b \approx 0.6389$ into $f(x) = ab^x$. We get

$$6 = a\left(0.6389\right)^{-2}$$

$$6 = 2.44982027a$$

$$a \approx 2.4492.$$

Finally, the equation is $f(x) = 2.4492(0.6389)^x$.

Check: We can graph our model to check our work. Notice that the graph in Figure 4 passes through the initial points given in the problem, $(-2, \ 6)$ and $(2, \ 1)$. The graph is an example of an **exponential decay** function.

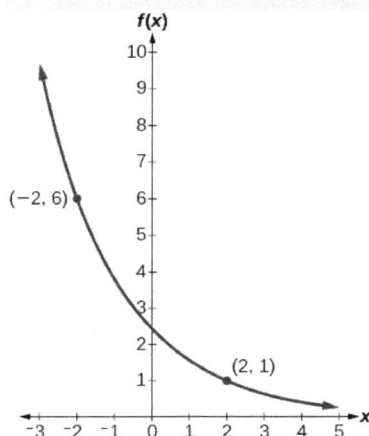

Figure 4.

TRY IT #6

Given the two points $(1, 3)$ and $(2, 4.5)$, find the equation of the exponential function that passes through these two points.

Answer

$$f(x) = 2(1.5)^x$$

Q&A

Do two points always determine a unique exponential function?

Yes, provided the two points are either both above the x-axis or both below the x-axis and have different x-coordinates. But keep in mind that we also need to know that the graph is, in fact, an exponential function. Not every graph that looks exponential really is exponential. We need to know the graph is based on a model that shows the same percent growth with each unit increase in x, *which in many real world cases involves time.*

HOW TO

Given a graph or a table of an exponential function, write its equation.

1. First, identify two points on the graph or table. Choose the vertical intercept as one of the two points whenever possible. Try to choose points that are as far apart as possible to reduce round-off error.
2. Use one of the methods for finding an equation given two points above.

EXAMPLE 7: WRITING AN EXPONENTIAL FUNCTION GIVEN ITS GRAPH

Find an equation for the exponential function graphed in Figure 5.

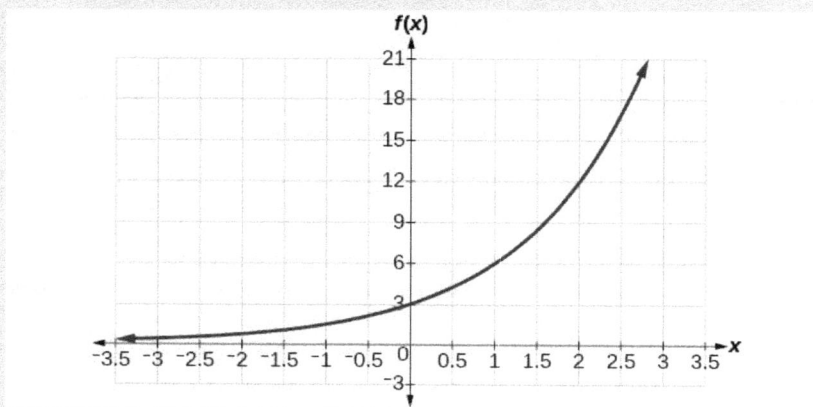

Figure 5.

Answer

We can choose the y-intercept of the graph, $(0, 3)$, as our first point. This gives us the initial value, $a = 3$. Next, choose a point on the curve some distance away from $(0, 3)$ that has integer coordinates. One such point is $(2, 12)$.

$f(x) = ab^x$	Write the general form of an exponential equation.
$f(x) = 3b^x$	Substitute the initial value 3 for a.
$12 = 3b^2$	Substitute in 12 for y and 2 for x.
$4 = b^2$	Divide by 3.
$b = 2$	Take the square root.

Because we restrict ourselves to positive values of b, we will use $b = 2$. Substitute a and b into the standard form to yield the equation $f(x) = 3(2)^x$.

TRY IT #7

Find an equation for the exponential function graphed in Figure 6.

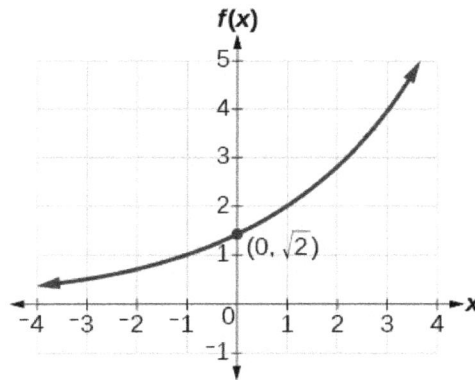

Figure 6.

Answer

$f(x) = \sqrt[2]{2}\left(\sqrt[2]{2}\right)^x$. **Answer**s may vary due to round-off error. The answer should be very close to $f(x) = 1.4142(1.4142)^x$.

EXAMPLE 8: WRITING AN EXPONENTIAL FUNCTION GIVEN A TABLE OF VALUES

Find an equation for the exponential function given by Table 4.

Table 4

x	1	3	6	8	12
$f(x)$	4.8	6.912	11.943936	17.19926784	35.664401793

Answer

First we choose any two points to use in finding our equation. Let's use (1, 4.8) and (3, 6.912).

For this example, we will use ratios to find the equation.

$$\frac{6.912}{4.8} = \frac{ab^3}{ab^1}$$

Next, we simplify and solve.

$$1.44 = b^{3-1}$$

$$1.44 = b^2$$

$$1.44^{\frac{1}{2}} = \left(b^2\right)^{\frac{1}{2}}$$

$$b = 1.2$$

Finally, we find a by plugging in 1.2 for b and 1 for x and 4.8 for $f\left(x\right)$ in $f\left(x\right) = ab^x$ to get

$$4.8 = a\left(1.2\right)^1 \text{ or } a = 4.$$

The final equation is $f\left(x\right) = 4\left(1.2\right)^x$.

Applying the Compound-Interest Formula

Savings instruments in which earnings are continually reinvested, such as mutual funds and retirement accounts, use **compound interest**. The term *compounding* refers to interest earned not only on the original value, but on the accumulated value of the account.

The **annual percentage rate (APR)** of an account, also called the **nominal rate**, is the yearly interest rate earned by an investment account. The term *nominal* is used when the compounding occurs a number of times other than once per year. In fact, when interest is compounded more than once a year, the effective interest rate ends up being *greater* than the nominal rate! This is a powerful tool for investing.

We can calculate the compound interest using the compound interest formula, which is an exponential function of the variables time t, principal P, APR r, and number of compounding periods in a year n :

$$A\left(t\right) = P\left(1 + \tfrac{r}{n}\right)^{nt}.$$

For example, observe Table 5, which shows the result of investing $1,000 at 10% for one year. Notice how the value of the account increases as the compounding frequency increases.

Table 5

Frequency	Value after 1 year
Annually	$1100
Semiannually	$1102.50
Quarterly	$1103.81
Monthly	$1104.71
Daily	$1105.16

THE COMPOUND INTEREST FORMULA

Compound interest can be calculated using the formula

$$A\left(t\right) = P\left(1 + \tfrac{r}{n}\right)^{nt}$$

where

- $A\left(t\right)$ is the account value,
- t is measured in years,
- P is the starting amount of the account, often called the principal, or more generally present value,
- r is the annual percentage rate (APR) expressed as a decimal, and
- n is the number of compounding periods in one year.

EXAMPLE 9: CALCULATING COMPOUND INTEREST

If we invest $3,000 in an investment account paying 3% interest compounded quarterly, how much will the account be worth in 10 years?

Answer

Because we are starting with $3,000, $P = 3000.$ Our interest rate is 3%, so $r = 0.03.$ Because we are compounding quarterly, we are compounding 4 times per year, so $n = 4.$ We want to know the value of the account in 10 years, so we are looking for $A\left(10\right)$, the value when $t = 10.$

$$A\left(t\right) = P\left(1 + \frac{r}{n}\right)^{nt} \qquad \text{Use the compound interest formula.}$$

$$A\left(10\right) = 3000\left(1 + \frac{0.03}{4}\right)^{4 \cdot 10} \qquad \text{Substitute using given values.}$$

$$\approx 4045.05.$$

The account will be worth about $4,045.05 in 10 years.

TRY IT #8

An initial investment of $100,000 at 12% interest is compounded weekly (use 52 weeks in a year). What will the investment be worth in 30 years?

Answer

about $3,644,675.88

EXAMPLE 10: USING THE COMPOUND INTEREST FORMULA TO SOLVE FOR THE PRINCIPAL

A 529 Plan is a college-savings plan that allows relatives to invest money to pay for a child's future college tuition; the account grows tax-free. Lily wants to set up a 529 account for her new granddaughter and wants the account to grow to $40,000 over 18 years. She believes the account will earn 6% compounded semi-annually (twice a year). To the nearest dollar, how much will Lily need to invest in the account now?

Answer

The nominal interest rate is 6%, so $r = 0.06$. Interest is compounded twice a year, so $n = 2$.

We want to find the initial investment, P, needed so that the value of the account will be worth $40,000 in 18 years. Substitute the given values into the compound interest formula, and solve for P.

$$A(t) = P\left(1 + \frac{r}{n}\right)^{nt} \qquad \text{Use the compound interest formula.}$$

$$40,000 = P\left(1 + \frac{0.06}{2}\right)^{2(18)} \qquad \text{Substitute using given values } A, r, n, \text{ and } t.$$

$$40,000 = P(1.03)^{36} \qquad \text{Simplify.}$$

$$\frac{40,000}{(1.03)^{36}} = P \qquad \text{Isolate } P.$$

$$P \approx 13,801 \qquad \text{Divide and round to the nearest dollar.}$$

Lily will need to invest $13,801 to have $40,000 in 18 years.

TRY IT #9

Refer to Example 10. To the nearest dollar, how much would Lily need to invest if the account is compounded quarterly?

Answer

$13,693

Evaluating Functions with Base e

As we saw earlier, the amount earned on an account increases as the compounding frequency increases. Table 6 shows that the increase from annual to semi-annual compounding is larger than the increase from monthly to daily compounding. This might lead us to ask whether this pattern will continue.

Examine the value of $1 invested at 100% interest for 1 year, compounded at various frequencies, listed in Table 6.

Table 6

Frequency	$A(n) = \left(1 + \frac{1}{n}\right)^n$	Value
Annually	$\left(1 + \frac{1}{1}\right)^1$	$2
Semiannually	$\left(1 + \frac{1}{2}\right)^2$	$2.25
Quarterly	$\left(1 + \frac{1}{4}\right)^4$	$2.441406
Monthly	$\left(1 + \frac{1}{12}\right)^{12}$	$2.613035
Daily	$\left(1 + \frac{1}{365}\right)^{365}$	$2.714567
Hourly	$\left(1 + \frac{1}{8760}\right)^{8760}$	$2.718127
Once per minute	$\left(1 + \frac{1}{525600}\right)^{525600}$	$2.718279
Once per second	$\left(1 + \frac{1}{31536000}\right)^{31536000}$	$2.718282

These values appear to be approaching a limit as n increases without bound. In fact, as n gets larger and larger, the expression $\left(1 + \frac{1}{n}\right)^n$ approaches a number used so frequently in mathematics that it has its own name: the letter e. This value is an irrational number, which means that its decimal expansion goes on forever without repeating. Its approximation to six decimal places is shown below.

EULER'S NUMBER

The letter e represents the irrational number $\left(1 + \frac{1}{n}\right)^n$, as n increases without bound.

The letter e is used as a base for many real-world exponential models. To work with base e, we use the approximation, $e \approx 2.718282$. The constant was named by the Swiss mathematician Leonhard Euler (1707–1783) who first investigated and discovered many of its properties.

EXAMPLE 11: USING A CALCULATOR TO FIND POWERS OF *E*

Calculate $e^{3.14}$. Round to five decimal places.

Answer

On a calculator, press the button labeled $\left[e^x\right]$. The window shows $\left[e^\wedge \ (\ \right]$. Type 3.14 and then close parenthesis, $[\ )\ ]$. Press [ENTER]. Rounding to 5 decimal places, $e^{3.14} \approx 23.10387$.

Caution: Many scientific calculators have an "Exp" button, which is used to enter numbers in scientific notation. It is not used to find powers of e.

TRY IT #10

Use a calculator to find $e^{-0.5}$. Round to five decimal places.

Answer

$$e^{-0.5} \approx 0.60653$$

Investigating Continuous Growth

So far we have worked with rational bases for exponential functions. For many real-world phenomena, however, e is used as the base for exponential functions. Exponential models that use e as the base are called ***continuous growth or decay models***. We see these models in finance, computer science, and most of the sciences, such as physics, toxicology, and fluid dynamics.

THE CONTINUOUS GROWTH/DECAY FORMULA

For all real numbers t, and all positive numbers a and k, continuous growth or decay is represented by the formula

$$A(t) = ae^{kt}$$

where

- a is the initial value,
- k is the **continuous rate** per unit time, and
- t is the elapsed time.

If $k > 0$, then the formula represents continuous growth. If $k < 0$, then the formula represents continuous decay.

HOW TO

Given the initial value, continuous rate of growth or decay, and time, solve a continuous growth or decay function.

1. Use the information in the problem to determine a, the initial value of the function.
2. Use the information in the problem to determine the growth rate k.

   ○ If the problem refers to continuous growth, then $k > 0$.

° If the problem refers to continuous decay, then $k < 0$.

3. Use the information in the problem to determine the time t.
4. Substitute the given information into the continuous growth formula and simplify.

EXAMPLE 12: CALCULATING CONTINUOUS GROWTH

A person invested \$1,000 in an account earning a nominal 10% per year compounded continuously. How much was in the account at the end of one year?

Answer

The rate given is continuous so we use the formula $A\left(t\right) = ae^{kt}$. Since the account is growing in value, this is a continuous compounding problem with growth rate $k = 0.10$. The initial investment was \$1,000, so $a = 1000$. To find the value after $t = 1$ year:

$$A\left(t\right) = ae^{kt} \qquad \text{Use the continuous compounding formula.}$$
$$= 1000(e)^{0.1} \qquad \text{Substitute known values for } a, k, \text{ and } t.$$
$$\approx 1105.17 \qquad \text{Use a calculator to approximate.}$$

The account is worth \$1,105.17 after one year.

TRY IT #11

A person invests \$100,000 at a nominal 12% interest per year compounded continuously. What will be the value of the investment in 30 years?

Answer

\$3,659,823.44

EXAMPLE 13: CALCULATING CONTINUOUS DECAY

Radon-222 decays at a continuous rate of 17.3% per day. How much will 100 mg of Radon-222 decay to in 3 days?

Answer

Since the substance is decaying, the rate, 17.3%, is negative. So, $k = -0.173$. The initial amount of radon-222 was 100mg, so $a = 100$. We use the continuous decay formula to find the value after $t = 3$ days:

$$A(t) = ae^{kt}$$
$$= 100e^{-0.173(3)}$$
$$\approx 59.5115$$

Use the continuous decay formula.
Substitute known values for $a, k,$ and t.
Use a calculator to approximate.

So 59.5115 mg of radon-222 will remain after 3 days.

TRY IT #12

Using the data in Example 13, how much radon-222 will remain after one year?

Answer

3.77E-26 (This is calculator notation for the number written as 3.7710^{-26} in scientific notation. While the output of an exponential function is never zero, this number is so close to zero that for all practical purposes we can accept zero as the answer.)

MEDIA

Access these online resources for additional instruction and practice with exponential functions.

- Exponential Growth Function
- Compound Interest

Key Equations

definition of exponential growth or decay	$f(x) = ab^x$, where $a > 0, b > 0, b \neq 1$
compound interest formula	$A(t) = P\left(1 + \frac{r}{n}\right)^{nt}$, where $A(t)$ is the account value at time t t is the number of years P is the initial investment, often called the principal r is the annual percentage rate (APR), or nominal rate n is the number of compounding periods in one year
continuous growth formula	$A(t) = ae^{kt}$, where t is the number of unit time periods of growth a is the starting amount (in the continuous compounding formula a may be replaced with P, the principal) e is the mathematical constant, $e \approx 2.718282$

KEY CONCEPTS

- An exponential function is defined as a function with a positive constant other than 1 raised to a variable exponent.
- A function is evaluated by solving at a specific value.
- An exponential model can be found when the growth rate and initial value are known.
- An exponential model can be found when the two data points from the model are known.
- An exponential model can be found using two data points from a graph or a table of the model.
- The value of an account at any time t can be calculated using the compound interest formula when the principal, annual interest rate, and compounding periods are known.
- The initial investment of an account can be found using the compound interest formula when the value of the account, annual interest rate, compounding periods, and life span of the account are known.
- The number e is a mathematical constant often used as the base of real world exponential growth and decay models. Its decimal approximation is $e \approx 2.718282.$
- Continuous growth or decay models are exponential models that use e as the base. Continuous growth and decay models can be found when the initial value and continuous growth or decay rate are known.

GLOSSARY

annual percentage rate (APR)
the yearly interest rate earned by an investment account, also called *nominal rate*

compound interest
interest earned on the total balance, not just the principal

exponential growth
a model that grows by a rate proportional to the amount present

nominal rate
the yearly interest rate earned by an investment account, also called *annual percentage rate*

2.3 Graphs of Exponential Functions

As we discussed in the previous section, exponential functions are used for many real-world applications such as finance, forensics, computer science, and most of the life sciences. Working with an equation that describes a real-world situation gives us a method for making predictions. Most of the time, however, the equation itself is not enough. We learn a lot about things by seeing their pictorial representations, and that is exactly why graphing exponential equations is a powerful tool. It gives us another layer of insight for predicting future events.

Graphing Exponential Functions

Before we begin graphing, it is helpful to review the behavior of exponential growth. Recall the table of values for a function of the form $f\left(x\right) = b^x$ whose base is greater than one. We'll use the function $f\left(x\right) = 2^x$. Observe how the output values in Table 1 change as the input increases by 1.

Table 1

x	-3	-2	-1	0	1	2	3
$f\left(x\right) = 2^x$	$\frac{1}{8}$	$\frac{1}{4}$	$\frac{1}{2}$	1	2	4	8

Each output value is the product of the previous output and the base, 2. We call the base 2 the **constant ratio** or **growth factor**. In fact, for any exponential function with the form $f\left(x\right) = ab^x$, b is the constant ratio of the function. This means that each time we increase the input by 1, we multiply the output by b. Notice from the table that the output values are positive for all values of x.

End Behavior of $f\left(x\right) = ab^x$

Often we want to know what happens to the output value of $f\left(x\right)$ as x moves far to the left or far to the right. This is known as the **end behavior** or **long term behavior** of the function. We will continue to study the function $f\left(x\right) = 2^x$ and determine its end behavior.

Begin by looking at two tables of values. Table 2 shows function values as x moves far to the left. We choose x values of -10, -100 and -250 and evaluate $f\left(x\right)$ at these values so we can observe what is happening to the output values to the far left. Note that -250 is not considered long term behavior for most functions but for an exponential function it is about the limit of what our current technology can compute. In Table 3, we move far toward the right choosing our x values to be 10, 100, and 250 and evaluate the function so we can observe what happens to the output when x gets large.

Table 2

x	-10	-100	-250
$f\left(x\right)=2^{x}$	9.77E-4	7.89E-31	5.53E-76

Table 3

x	10	100	250
$f\left(x\right)=2^{x}$	1024	1.27E30	1.81E75

Recognize that scientific notation is being used in Table 2 and Table 3 for the output values. In Table 2, we observe that as x decreases or becomes more and more negative, the output values get closer and closer to zero from above. We capture this idea using arrow notation and write as $x \to -\infty, f\left(x\right) \to 0$. This is read, "As x decreases without bound, f of x goes to zero." When we are studying end behavior and we observe that the output is getting closer and closer to a value, we say that there is a horizontal asymptote. In this example, $y = 0$ is the horizontal asymptote on the left hand side.

Further, Table 3 shows that as x increases or becomes larger and larger, the output values also become larger and larger or increase without bound. We write as $x \to \infty, f\left(x\right) \to \infty$. Since these output values increase without bound, there is not a horizontal asymptote in this direction.

DEFINITION

A **horizontal asymptote** of a graph is a horizontal line $y = b$ where the graph approaches the line as the inputs increase or decrease without bound. We write as $x \to \infty$ or $x \to -\infty, f\left(x\right) \to b$.

Figure 1 shows the exponential growth function $f\left(x\right) = 2^{x}$.

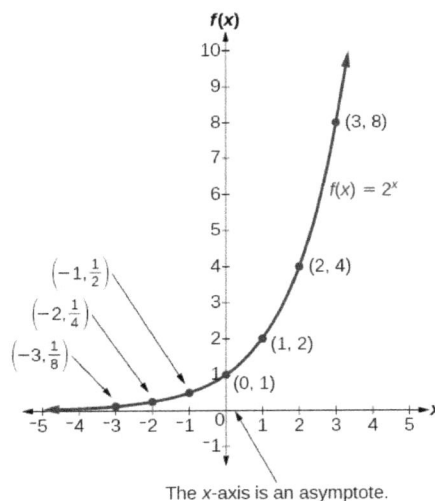

The x-axis is an asymptote.

Figure 1: Notice that the graph gets close to the x-axis, but never touches it.

We observe in the graph above that as x becomes more negative, the graph is getting closer to the x-axis but never touches it demonstrating the horizontal asymptote of $y = 0$. Other characteristics of the graph can also be observed. The domain of $f(x) = 2^x$ is all real numbers, and the range is $(0, \infty)$. Notice that the function is also increasing and concave up on $(-\infty, \infty)$.

Exponential Decay Graphically

To get a sense of the behavior of exponential decay, we can create a table of values for a function of the form $f(x) = b^x$ whose base is between zero and one. We'll use the function $g(x) = \left(\frac{1}{2}\right)^x$. Observe how the output values in Table 4 change as the input increases by 1.

Table 4

x	-3	-2	-1	0	1	2	3
$g(x) = \left(\frac{1}{2}\right)^x$	8	4	2	1	$\frac{1}{2}$	$\frac{1}{4}$	$\frac{1}{8}$

Again, notice that each time the input is increased by 2, the output is multiplied by the base, or constant ratio $b = \frac{1}{2}$.

To look at the end behavior of the exponential decay function, we again create tables with input values to the far left and right.

Table 5

x	-10	-100	-250
$g(x) = \left(\frac{1}{2}\right)^x$	1024	1.27E30	1.81E75

Table 6

x	10	100	250
$g(x) = \left(\frac{1}{2}\right)^x$	9.77E-4	7.89E-31	5.53E-76

Notice from the tables above that:

- the output values are positive for all values of x.
- as x decreases, the output values grow without bound so as $x \to -\infty, g(x) \to \infty$.
- as x increases without bound, the output values approach zero from above so as $x \to \infty, g(x) \to 0$. The horizontal asymptote is $y = 0$ on the right hand side.

Figure 2 shows the exponential decay function, $g(x) = \left(\frac{1}{2}\right)^x$.

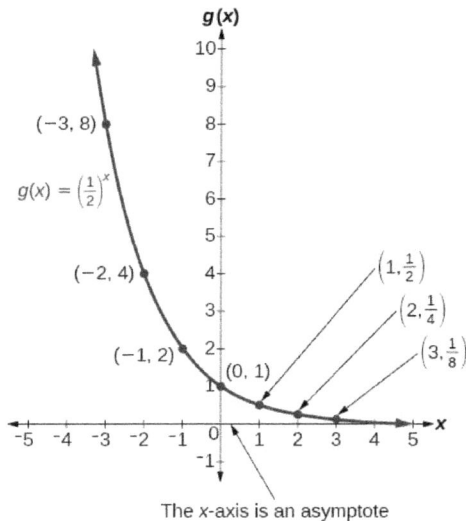

The x-axis is an asymptote

Figure 2 The graph shows that as x gets larger, the output gets close to zero.

Again we observe the end behavior and see that as x increases, the graph approaches the x-axis and there is a horizontal asymptote of $y = 0$. Other characteristics can also be observed from the graph. The domain of $g(x) = \left(\frac{1}{2}\right)^{x}$ is all real numbers, and the range is $(0, \infty)$. Notice that the function is decreasing and concave up on $(-\infty, \infty)$.

Characteristics of the Graph of the Function $f(x) = b^x$

An exponential function with the form $f(x) = b^x, b > 0, b \neq 1$, has these characteristics:

- one-to-one function
- horizontal asymptote: $y = 0$ on one side
- domain: $(-\infty, \infty)$
- range: $(0, \infty)$
- x-intercept: none
- y-intercept: $(0, 1)$
- increasing if $b > 1$
- decreasing if $b < 1$
- concave up

Figure 3 compares the graphs of exponential growth and decay functions.

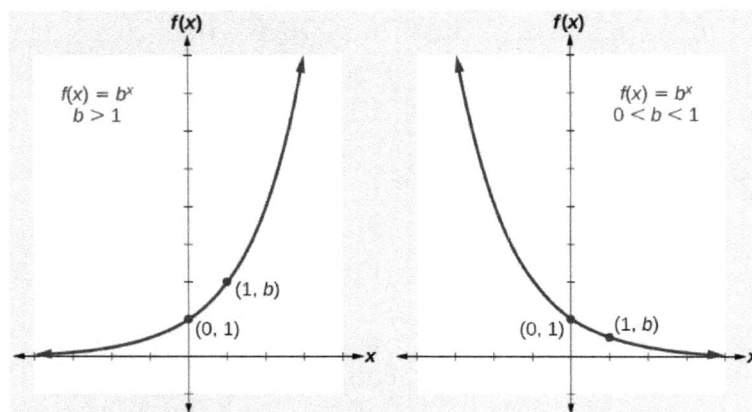

Figure 3

HOW TO

Given an exponential function of the form $f\left(x\right) = b^x$, graph the function.

1. Create a table of points.
2. Plot at least 3 points from the table, including the vertical intercept $\left(0, 1\right)$.
3. Draw a smooth curve through the points.
4. State the domain, $\left(-\infty, \infty\right)$, the range, $\left(0, \infty\right)$, and on which side the horizontal asymptote, $y = 0$ occurs.

EXAMPLE 1: SKETCHING THE GRAPH OF AN EXPONENTIAL FUNCTION OF THE FORM $F(X) = B^X$

Sketch a graph of $f\left(x\right) = 0.25^x$. State the domain, range, and horizontal asymptote.

Answer

Before graphing, identify the behavior and create a table of points for the graph.

- Since $b = 0.25$ is between zero and one, we know the function is decreasing. The left end behavior of the graph will increase without bound, and the right end behavior will approach the horizontal asymptote $y = 0$.
- Create a table of points as in Table 7.

Table 7

x	-3	-2	-1	0	1	2	3
$f\left(x\right) = 0.25^x$	64	16	4	1	0.25	0.0625	0.015625

- Plot the y-intercept, $\left(0, 1\right)$, along with two other points. We can use $\left(-1, 4\right)$ and $\left(1, 0.25\right)$.

Draw a smooth curve connecting the points as in Figure 4.

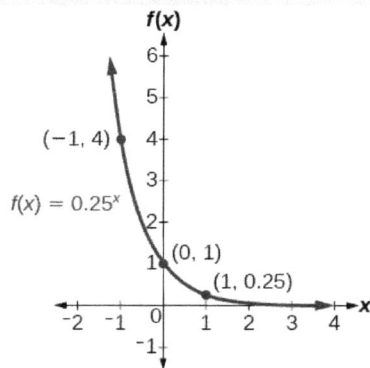

Figure 4

The domain is $(-\infty, \infty)$; the range is $(0, \infty)$; the horizontal asymptote is $y = 0$ on the right side.

TRY IT #1

Sketch the graph of $f(x) = 4^x$. State the domain, range, and horizontal asymptote.

Answer

The domain is $(-\infty, \infty)$; the range is $(0, \infty)$; the horizontal asymptote is $y = 0$ on the left side.

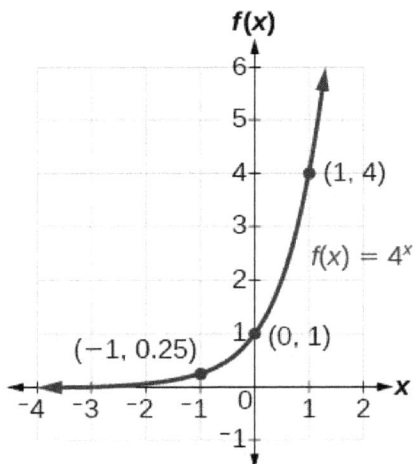

Graphing Transformations of Exponential Functions

Transformations of exponential graphs behave similarly to those of other functions. Just as with our toolkit functions, we can apply the four types of transformations—shifts, reflections, stretches, and compressions—to the exponential function

$f\left(x\right)=b^{x}$ without loss of shape. For instance, just as the quadratic function maintains its parabolic shape when shifted, reflected, stretched, or compressed, the exponential function also maintains its general shape regardless of the transformations applied.

NOTE: In this section we will be using different notation for horizontal and vertical shifts. Recall that in section 1.6, we considered functions in the form $g\left(x\right)=af\left(b\left(x-h\right)\right)+k$. In this notation, k indicated the vertical shift, h indicated the horizontal shift, a indicated the vertical stretch/compression, and b indicated the horizontal stretch/compression. When we study exponential functions, we have already designated k to indicate continuous growth. Therefore, we will modify our notation and use c to represent horizontal shifts and d to represent vertical shifts. Vertical stretches/compressions will still be represented by a. The variable b will represent the base of the exponential function and not represent a horizontal stretch or compression.

Graphing a Vertical Shift

The first transformation occurs when we add a constant d to the exponential function $f\left(x\right)=b^{x}$, giving us a vertical shift d units in the same direction as the sign. For example, if we begin by graphing the function, $f\left(x\right)=2^{x}$, we can then graph two vertical shifts alongside it, using $d=3$ and $d=-3$: the upward shift, $g\left(x\right)=f\left(x\right)+3=2^{x}+3$ and the downward shift, $h\left(x\right)=f\left(x\right)-3=2^{x}-3$. Both vertical shifts are shown in Figure 5.

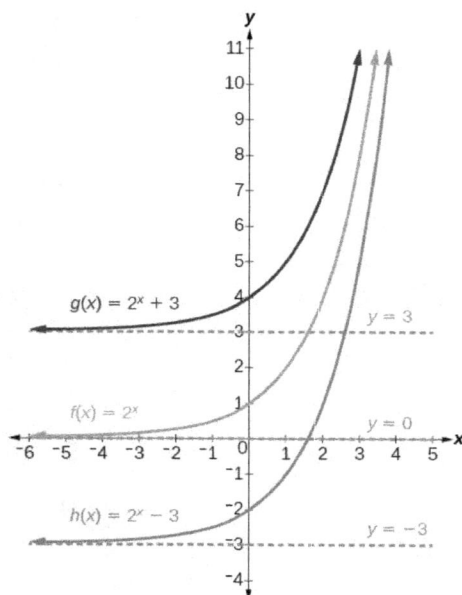

Figure 5

Observe the results of shifting $f\left(x\right)=2^{x}$ vertically:

- The domain, $\left(-\infty,\infty\right)$ remains unchanged.
- When the function is shifted up 3 units to $g\left(x\right)=2^{x}+3$:
  - The y-intercept shifts up 3 units to $\left(0,4\right)$.
  - The horizontal asymptote shifts up 3 units to $y=3$ on the left side.
  - The range becomes $\left(3,\infty\right)$.
- When the function is shifted down 3 units to $h\left(x\right)=2^{x}-3$:
  - The y-intercept shifts down 3 units to $\left(0,-2\right)$.
  - The horizontal asymptote also shifts down 3 units to $y=-3$ on the left side.
  - The range becomes $\left(-3,\infty\right)$.

Graphing a Horizontal Shift

The next transformation occurs when we subtract a constant C from the input of the exponential function $f(x) = b^x$, giving us a horizontal shift C units in the direction of the sign of C. The equation is given by $f(x - c) = b^{x-c}$. For example, if we begin by graphing the function $f(x) = 2^x$, we can then graph two horizontal shifts alongside it, using

$c = -3$: the shift left, $g(x) = f(x+3) = 2^{x+3}$, and using

$c = 3$: the shift right, $h(x) = f(x-3) = 2^{x-3}$.

Both horizontal shifts are shown in Figure 6.

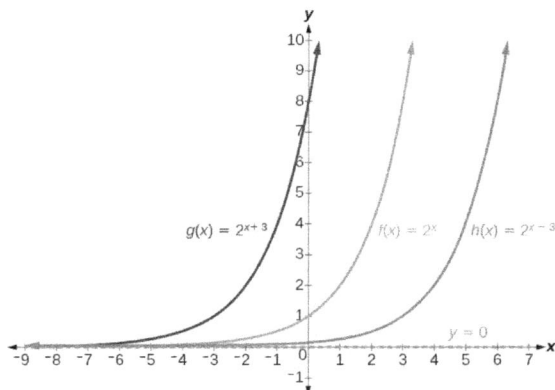

Figure 6

Observe the results of shifting $f(x) = 2^x$ horizontally:

- The domain, $(-\infty, \infty)$, remains unchanged.
- The horizontal asymptote, $y = 0$, remains unchanged.
- The vertical intercept shifts such that:
  - When the function is shifted left 3 units to $g(x) = 2^{x+3}$, the vertical intercept becomes $(0, 8)$. This is because $2^{x+3} = 2^x 2^3 = (8) 2^x$ using the rules of exponents, so the initial value of the function is 8.
  - When the function is shifted right 3 units to $h(x) = 2^{x-3}$, the vertical intercept becomes $\left(0, \frac{1}{8}\right)$. Again, see that $2^{x-3} = 2^x 2^{-3} = \left(\frac{1}{8}\right) 2^x$, so the initial value of the function is $\frac{1}{8}$.

Shifts of the Function y = b^x

For any constants C and d, the function $f(x) = b^{x-c} + d$ shifts the exponential function $y = b^x$

- vertically d units, in the direction of the sign of d, and
- horizontally C units, in the direction of the sign of C.
- The vertical intercept becomes $(0, b^{-c} + d)$.
- The horizontal asymptote becomes $y = d$ on the same side.
- The range becomes (d, ∞).
- The domain, $(-\infty, \infty)$, remains unchanged.

Given an exponential function with the form $f\left(x\right) = b^{x-c} + d$, graph the translation.

1. Draw the horizontal asymptote $y = d.$
2. Identify c and d. Shift the graph of $y = b^x$ right c units if c is positive, and left c units if c is negative.
3. Shift the graph of $y = b^x$ up d units if d is positive, and down d units if d is negative.
4. State the domain, $\left(-\infty, \infty\right)$, the range, $\left(d, \infty\right)$, and the horizontal asymptote $y = d.$

EXAMPLE 2: GRAPHING A SHIFT OF AN EXPONENTIAL FUNCTION

Graph $f\left(x\right) = 2^{x+1} - 3.$ State the domain, range, and horizontal asymptote.

Answer

We have an exponential equation of the form $f\left(x\right) = b^{x-c} + d$, with $b = 2, c = -1$, and $d = -3.$

Draw the horizontal asymptote $y = d$, so draw $y = -3.$ Shift the graph of $y = 2^x$ left 1 unit and down 3 units.

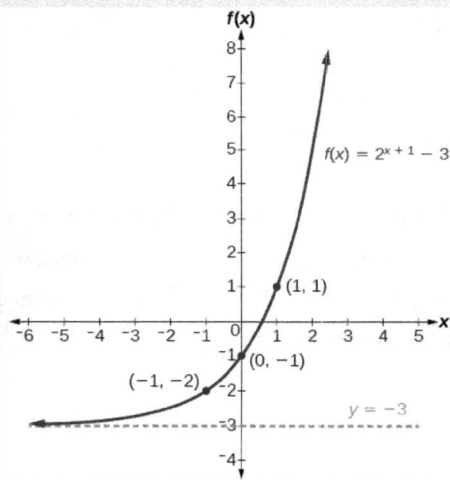

Figure 7

The domain is $\left(-\infty, \infty\right)$; the range is $\left(-3, \infty\right)$; the horizontal asymptote is $y = -3$ on the left side.

TRY IT #2

Graph $f\left(x\right) = 2^{x-1} + 3.$ State domain, range, and horizontal asymptote.

Answer

The domain is $(-\infty, \infty)$; the range is $(3, \infty)$; the horizontal asymptote is $y = 3$ on the left side.

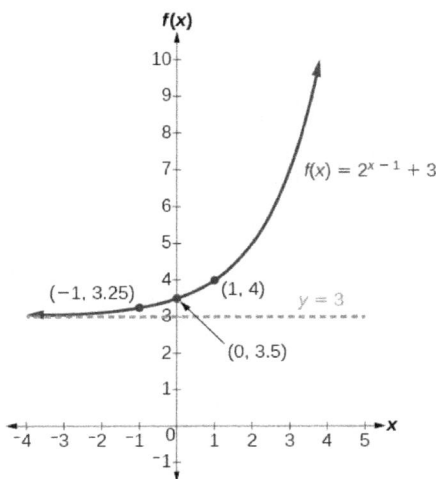

Graphing a Vertical Stretch or Compression

While horizontal and vertical shifts involve adding constants to the input or to the function itself, a stretch or compression occurs when we multiply the exponential function $f(x) = b^x$ by a constant $|a| > 0$. For example, if we begin by graphing the function $f(x) = 2^x$, we can then graph the vertical stretch, using $a = 3$, to get $g(x) = 3f(x) = 3(2)^x$ as shown on the left in Figure 8a, and the vertical compression, using $a = \frac{1}{3}$, to get $h(x) = \frac{1}{3}f(x) = \frac{1}{3}(2)^x$ as shown on the right in Figure 8b.

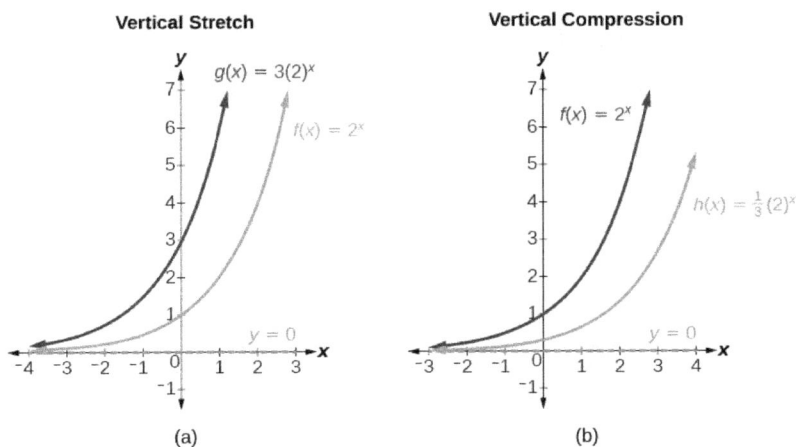

Figure 8. (a) $g(x) = 3(2)^x$ stretches the graph of $f(x) = 2^x$ vertically by a factor of 3. (b) $h(x) = \frac{1}{3}(2)^x$ compresses the graph of $f(x) = 2^x$ vertically by a factor of $\frac{1}{3}$.

Stretches and Compressions of the Function $y = b^x$

For any factor $a > 0$, the function $f(x) = a(b)^x$

- stretches $y = b^x$ vertically by a factor of a if $|a| > 1$.
- compresses $y = b^x$ vertically by a factor of a if $|a| < 1$.
- has a vertical intercept of $(0, a)$.
- has a horizontal asymptote at $y = 0$, a range of $(0, \infty)$, and a domain of $(-\infty, \infty)$, which are unchanged from $y = b^x$.

EXAMPLE 3: GRAPHING THE STRETCH OF AN EXPONENTIAL FUNCTION

Sketch a graph of $f(x) = 4\left(\frac{1}{2}\right)^x$. State the domain, range, and horizontal asymptote.

Answer

Before graphing, identify the behavior and key points on the graph.

- Since $b = \frac{1}{2}$ is between zero and one, the left end behavior of the graph will increase without bound as x decreases without bound, and the right end behavior will approach the x-axis as x increases without bound.
- Since $a = 4$, the graph of $y = \left(\frac{1}{2}\right)^x$ will be stretched by a factor of 4.
- Create a table of points as shown in Table 8.

Table 8

x	-3	-2	-1	0	1	2	3
$f(x) = 4\left(\frac{1}{2}\right)^x$	32	16	8	4	2	1	0.5

- Plot the vertical intercept, $(0, 4)$, along with two other points. We can use $(-1, 8)$ and $(1, 2)$.

Draw a smooth curve connecting the points, as shown in Figure 9.

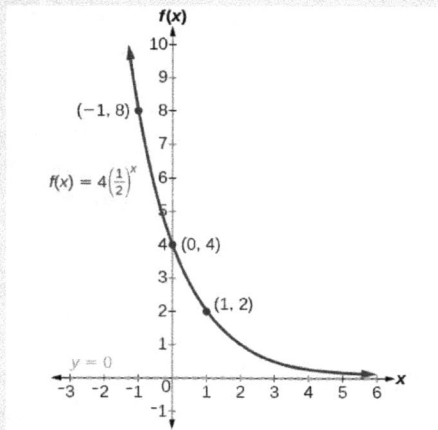

Figure 9

The domain is $(-\infty, \infty)$; the range is $(0, \infty)$; the horizontal asymptote is $y = 0$ on the right side.

TRY IT #3

Sketch the graph of $f\left(x\right)=\frac{1}{2}\left(4\right)^{x}$. State the domain, range, and horizontal asymptote.

Answer

The domain is $\left(-\infty,\infty\right)$; the range is $\left(0,\infty\right)$; the horizontal asymptote is $y=0$ on the left side.

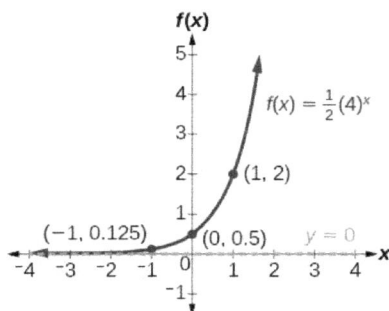

Graphing Reflections

In addition to shifting, compressing, and stretching a graph, we can also reflect an exponential function about the x-axis or the y-axis. When we multiply the exponential function $f\left(x\right)=b^{x}$ by -1, we get a vertical reflection about the x-axis. When we multiply the input by -1, we get a reflection about the y-axis. For example, if we begin by graphing the function $f\left(x\right)=2^{x}$, we can then graph the two reflections alongside it. The reflection about the x-axis, $g\left(x\right)=-2^{x}$, is shown on the left side of Figure 10a, and the reflection about the y-axis $h\left(x\right)=2^{-x}$, is shown on the right side of Figure 10b.

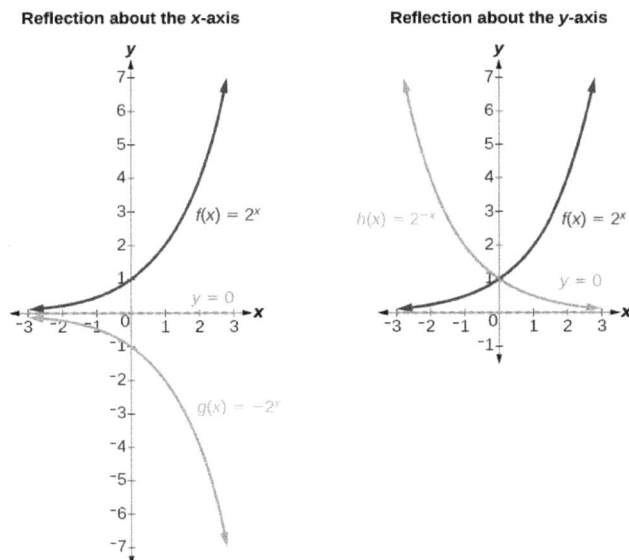

Figure 10 (a) $g\left(x\right)=-2^{x}$ reflects the graph of $f\left(x\right)=2^{x}$ about the x-axis. (b) $g\left(x\right)=2^{-x}$ reflects the graph of $f\left(x\right)=2^{x}$ about the y-axis.

Reflections of the Function $y = f(x) = b^x$

The function $g\left(x\right) = -b^x$

- reflects the function $y = b^x$ over the x-axis.
- has a vertical intercept of $(0, -1)$.
- has a range of $(-\infty, 0)$.
- has a horizontal asymptote at $y = 0$ and domain of $(-\infty, \infty)$, which are unchanged from the function $y = b^x$.

The function $h\left(x\right) = b^{-x}$

- reflects the function $y = b^x$ over the y-axis.
- has a vertical intercept of $(0, 1)$, a horizontal asymptote at $y = 0$, a range of $(0, \infty)$, and a domain of $(-\infty, \infty)$, which are unchanged from the function $y = b^x$.

EXAMPLE 4: WRITING AND GRAPHING THE REFLECTION OF AN EXPONENTIAL FUNCTION

Find and graph the equation for a function, $g\left(x\right)$, that reflects $f\left(x\right) = \left(\frac{1}{4}\right)^x$ over the x-axis. State its domain, range, and horizontal asymptote.

Answer

Since we want to reflect the function $f\left(x\right) = \left(\frac{1}{4}\right)^x$ about the x-axis, we multiply $f\left(x\right)$ by -1 to get, $g\left(x\right) = -\left(\frac{1}{4}\right)^x$. Next we create a table of points as in Table 9.

Table 9

x	-3	-2	-1	0	1	2	3
$g\left(x\right) = -\left(\frac{1}{4}\right)^x$	-64	-16	-4	-1	-0.25	-0.0625	-0.0156

Plot the y-intercept, $(0, -1)$, along with two other points. We can use $(-1, -4)$ and $(1, -0.25)$.

Draw a smooth curve connecting the points:

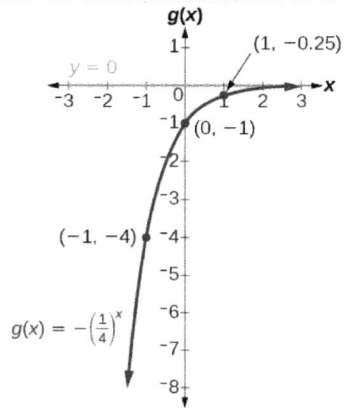

Figure 11.

The domain is $(-\infty, \infty)$; the range is $(-\infty, 0)$; the horizontal asymptote is $y = 0$ on the right side.

TRY IT #4

Find and graph the equation for a function, $g(x)$, that reflects $f(x) = 1.25^x$ over the y-axis. State its domain, range, and horizontal asymptote.

Answer

The function is $g(x) = 1.25^{-x} = 0.8^x$. The domain is $(-\infty, \infty)$; the range is $(0, \infty)$; the horizontal asymptote is $y = 0$ on the right side.

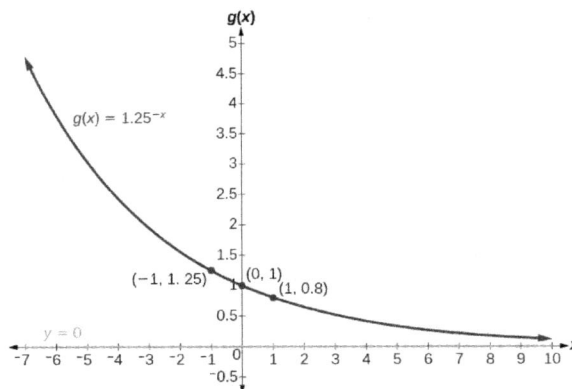

Summarizing Translations of the Exponential Function

Now that we have worked with each type of translation for the exponential function, we can summarize them in Table 10 to arrive at the general equation for translating exponential functions.

Table 10

Translations of the Function $y = b^x$	
Translation	**Form**
Shift • Horizontally c units to the left or right • Vertically d units up or down	$f(x) = b^{x-c} + d$
Vertical Stretch and Compression • Stretch if $\lvert a \rvert > 1$ • Compression if $0 < \lvert a \rvert < 1$	$f(x) = ab^x$
Reflect about the x-axis	$f(x) = -b^x$
Reflect about the y-axis	$f(x) = b^{-x} = \left(\frac{1}{b}\right)^x$
General equation for all translations	$f(x) = ab^{x-c} + d$

Q&A

Why isn't there a discussion on horizontal stretches and compressions?

Recall the exponential rule $b^{mn} = \left(b^m\right)^n$. Essentially a horizontal compression would be a change in the base of the function. For example, $b^{3x} = \left(b^3\right)^x$. The original base is b with a horizontal compression by a factor of $\frac{1}{3}$, but we can also simply consider this as a function with the new base b^3.

Think of $f(x) = 2^{3x}$. We can think of this as a horizontal compression by a factor of $\frac{1}{3}$ of the function $y = 2^x$. The point (1,2) will be compressed to the point $\left(\frac{1}{3}, 2\right)$. Notice that if we used the function $g(x) = \left(2^3\right)^x$ or $g(x) = 8^x$, we would also see the point $\left(\frac{1}{3}, 2\right)$. This helps us see that we can achieve the same results as horizontal compressions by rewriting the function with a different base.

EXAMPLE 5: WRITING A FUNCTION FROM A DESCRIPTION

Write the equation for the function described below. Give the horizontal asymptote, the domain, and the range.

• $f(x) = e^x$ is vertically stretched by a factor of 2, reflected across the y-axis, and then shifted up 4 units.

Answer

We want to find an equation of the general form $g(x) = ab^{x-c} + d$. We use the description provided to find a, b, c, and d.

- We are given the function $f\left(x\right) = e^x$, so $b = e$.
- The function is stretched by a factor of 2, so $a = 2$.
- The function is reflected about the y-axis. We replace x with $-x$ to get: e^{-x}.
- The graph is shifted vertically 4 units, so $d = 4$.

Substituting in the general form we get,

$$g\left(x\right) = ab^{x-c} + d$$
$$= 2e^{-x-0} + 4$$
$$= 2e^{-x} + 4$$

The domain is $\left(-\infty, \infty\right)$; the range is $\left(4, \infty\right)$; the horizontal asymptote is $y = 4$ on the right side.

TRY IT #5

Write the equation for function described below. Give the horizontal asymptote, the domain, and the range.

- $f\left(x\right) = e^x$ is compressed vertically by a factor of $\frac{1}{3}$, reflected across the x-axis and then shifted down 2 units.

Answer

$g\left(x\right) = -\frac{1}{3}e^x - 2$; the domain is $\left(-\infty, \infty\right)$; the range is $\left(-\infty, 2\right)$; the horizontal asymptote is $y = 2$ on the right side.

Approximating Solutions to an Exponential Equation with the Calculator

Sometimes we want to find out when an exponential function will have a particular output value. The next sections will focus on being able to do this algebraically. Currently, we can use technology to determine what input will give a particular output for the transformed exponential function.

HOW TO

Given an equation of the form $y = ab^{x-c} + d$, use a graphing calculator to approximate the solution.

- Press [Y=]. Enter the given exponential equation in the line headed "Y₁=".
- Enter the given value for y in the line headed "Y₂=".
- Press [**WINDOW**]. Adjust the y-axis so that it includes the value entered for "Y₂=".
- Press [**GRAPH**] to observe the graph of the exponential function along with the line for the specified value of y.
- To find the value of x, we compute the point of intersection. Press [**2ND**] then [**CALC**]. Select "intersect" and press [**ENTER**] three times. The point of intersection gives the value of x for the indicated output value of the function.

EXAMPLE 6: APPROXIMATING THE SOLUTION OF AN EXPONENTIAL EQUATION

Solve $42 = 1.2(5)^x + 2.8$ graphically. Round to the nearest thousandth.

Answer

Press [Y=] and enter $1.2(5)^x + 2.8$ next to **Y₁=**. Then enter 42 next to **Y2=**. For a window, use the values –3 to 3 for x and –5 to 55 for y. Press [GRAPH]. The graphs should intersect somewhere near $x = 2$.

For a better approximation, press [2ND] then [CALC]. Select [5: intersect] and press [ENTER] three times. The x-coordinate of the point of intersection is displayed as 2.1661943. (Your answer may be slightly different if you use a different window or use a different value for **Guess?**) To the nearest thousandth, $x \approx 2.166$.

TRY IT #6

Solve $4 = 7.85(1.15)^x - 2.27$ graphically. Round to the nearest thousandth.

Answer

$x \approx -1.608$

Access this online resource for additional instruction and practice with graphing exponential functions.

- Graph Exponential Functions

Key Equations

General Form for the Translation of the Function $y = b^x$	$f(x) = ab^{x-c} + d$

KEY CONCEPTS

- The graph of the function $f(x) = b^x$ has a y-intercept at $(0, 1)$, domain $(-\infty, \infty)$, range $(0, \infty)$, and horizontal asymptote $y = 0$.
- End behavior describes what happens to the output if you go very far to the left or right.

  - If $b > 1$, the function is increasing. The left end behavior of the graph will approach the horizontal asymptote $y = 0$, and the right end behavior will increase without bound.
  - If $0 < b < 1$, the function is decreasing. The left end behavior of the graph will increase without

bound, and the right end behavior will approach the horizontal asymptote $y = 0$.

- The equation $f(x) = b^x + d$ represents a vertical shift of the exponential function $y = b^x$.
- The equation $f(x) = b^{x-c}$ represents a horizontal shift of the exponential function $y = b^x$.
- The equation $f(x) = ab^x$, where $a > 0$, represents a vertical stretch if $|a| > 1$ or compression if $0 < |a| < 1$ of the exponential function $y = b^x$.
- When the exponential function $y = b^x$ is multiplied by -1, the result, $g(x) = -b^x$, is a reflection about the x-axis. When the input is multiplied by -1, the result, $h(x) = b^{-x}$, is a reflection about the y-axis.
- All translations of the exponential function can be summarized by the general equation
$$f(x) = ab^{x-c} + d.$$
- Using the general equation $f(x) = ab^{x-c} + d$, we can write the equation of a function given its description.
- Approximate solutions of the equation $y = b^{x-c} + d$ can be found using a graphing calculator.

2.4 Logarithmic Functions

Figure 1 Devastation of March 11, 2011 earthquake in Honshu, Japan.
(credit: Daniel Pierce)

In 2010, a major earthquake struck Haiti, destroying or damaging over 285,000 homes[1]. One year later, another, stronger earthquake devastated Honshu, Japan, destroying or damaging over 332,000 buildings,[2] like those shown in Figure 1. Even though both caused substantial damage, the earthquake in 2011 was 100 times stronger than the earthquake in Haiti. How do we know? The magnitudes of earthquakes are measured on a scale known as the Richter Scale. The Haitian earthquake registered a 7.0 on the Richter Scale[3] whereas the Japanese earthquake registered a 9.0[4]

The Richter Scale is a base-ten logarithmic scale. In other words, an earthquake of magnitude 8 is not twice as great as an earthquake of magnitude 4. It is $10^{8-4} = 10^4 = 10,000$ times as great! In this section, we will investigate the nature of the Richter Scale and the base-ten function upon which it depends.

---

1. http://earthquake.usgs.gov/earthquakes/eqinthenews/2010/us2010rja6/#summary. Accessed 3/4/2013.
2. http://earthquake.usgs.gov/earthquakes/eqinthenews/2011/usc0001xgp/#summary. Accessed 3/4/2013.
3. http://earthquake.usgs.gov/earthquakes/eqinthenews/2010/us2010rja6/. Accessed 3/4/2013.
4. http://earthquake.usgs.gov/earthquakes/eqinthenews/2011/usc0001xgp/#details. Accessed 3/4/2013.

Converting from Logarithmic to Exponential Form

In order to analyze the magnitude of earthquakes or compare the magnitudes of two different earthquakes, we need to be able to convert between logarithmic and exponential form. For example, suppose the amount of energy released from one earthquake were 500 times greater than the amount of energy released from another. We want to calculate the difference in magnitude. The equation that represents this problem is $10^x = 500$, where x represents the difference in magnitudes on the Richter Scale. How would we solve for x?

We have only learned a graphical method for approximating solutions of exponential equations. None of the algebraic tools discussed so far is sufficient to solve $10^x = 500$. We know that $10^2 = 100$ and $10^3 = 1000$, so it is clear that x must be some value between 2 and 3, since $y = 10^x$ is increasing. We can examine a graph, as in Figure 2, to better estimate the solution.

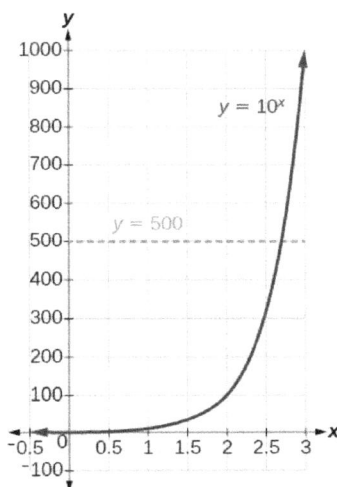

Figure 2

Estimating from a graph, however, is imprecise. To find an algebraic solution, we must introduce a new function. Observe that the graph in Figure 2 passes the horizontal line test. The exponential function $y = b^x$ is one-to-one, so its inverse is also a function. We use a logarithmic function of the form $y = \log_b(x)$ to describe the inverse. The base b **logarithm** of a number is the exponent by which we must raise b to get that number.

We read a logarithmic expression as, "The logarithm with base b of x is equal to y," or, simplified, "log base b of x is y." We can also say, "b raised to the power of y is x," because logs are exponents. For example, the base 2 logarithm of 32 is 5, because 5 is the exponent we must apply to 2 to get 32. Since $2^5 = 32$, we can write $\log_2 32 = 5$. We read this as "log base 2 of 32 is 5."

We can express the relationship between logarithmic form and its corresponding exponential form as follows:

$$\log_b(x) = y \qquad b^y = x, \quad b > 0, b \neq 1$$

Note that the base b is always positive.

$$\log_b(x) = y \qquad \text{Think} \atop b \text{ to the } y = x$$

Because logarithm is a function, it is most correctly written as $\log_b(x)$, using parentheses to denote function evaluation, just as we would with $f(x)$. However, when the input is a single variable or number, it is common to see the parentheses dropped and the expression written without parentheses, as $\log_b x$. Note that many calculators require parentheses around the x.

We can illustrate the notation of logarithms as follows:

$$\log_b (c) = a \;\; \text{means} \;\; b^a = c$$

to

Notice that, comparing the logarithm function and the exponential function, the input and the output are switched.

DEFINITION

A **logarithm** base b of a positive number x satisfies the following definition.

For $x > 0, \; b > 0, \; b \neq 1, \; y = \log_b (x)$ is equivalent to $b^y = x$.

- We read $\log_b (x)$ as, "the logarithm with base b of x" or the "log base b of x."
- The logarithm y is the exponent to which b must be raised to get x.

Also, since the logarithmic and exponential functions are inverses of each other, the domain and range of the exponential function are interchanged for the logarithmic function. Therefore,

- the domain of the logarithm function with base b is $(0, \infty)$, and
- the range of the logarithm function with base b is $(-\infty, \infty)$.

Q&A

Can we take the logarithm of a negative number?

No. We are working with functions of real numbers. Because the base of an exponential function is always a positive real number, no power of that base can ever be a negative real number. We can never take the logarithm of a negative real number. Also, we cannot take the logarithm of zero. Calculators may output a log of a negative number when in complex mode, but the log of a negative number is not a real number.

HOW TO

Given an equation in logarithmic form $\log_b (x) = y$, **convert it to exponential form.**

1. Examine the equation $y = \log_b (x)$ and identify $b, \; y$, and x.
2. Rewrite $\log_b (x) = y$ as $b^y = x$.

EXAMPLE 1: CONVERTING FROM LOGARITHMIC FORM TO EXPONENTIAL FORM

Write the following logarithmic equations in exponential form.

a. $\log_6\left(\sqrt{6}\right) = \frac{1}{2}$
b. $\log_{10}(100) = 2$

Answer

First, identify the values of b, y, and x in $y = \log_b(x)$. Then, write the equation in the form $b^y = x$.

a. $\log_6\left(\sqrt{6}\right) = \frac{1}{2}$
Here, $b = 6$, $y = \frac{1}{2}$, and $x = \sqrt{6}$. Therefore, the equation $\log_6\left(\sqrt{6}\right) = \frac{1}{2}$ is equivalent to $6^{\frac{1}{2}} = \sqrt{6}$.

b. $\log_{10}(100) = 2$
Here, $b = 10$, $y = 2$, and $x = 100$. Therefore, the equation $\log_{10}(100) = 2$ is equivalent to $10^2 = 100$.

TRY IT #1

Write the following logarithmic equations in exponential form.

a. $\log_{10}(1,000,000) = 6$
b. $\log_5(25) = 2$

Answer

a. $\log_{10}(1,000,000) = 6$ is equivalent to $10^6 = 1,000,000$.
b. $\log_5(25) = 2$ is equivalent to $5^2 = 25$.

Converting from Exponential to Logarithmic Form

To convert from exponents to logarithms, we follow the same steps in reverse. We identify the base b, exponent x, and output y in $b^x = y$. Then we write $x = \log_b(y)$.

EXAMPLE 2: CONVERTING FROM EXPONENTIAL FORM TO

LOGARITHMIC FORM

Write the following exponential equations in logarithmic form.

a. $2^3 = 8$
b. $10^4 = 10000$
c. $10^{-4} = \frac{1}{10,000}$

Answer

First, identify the values of $b, \ y$, and x in $b^x = y$. Then, write the equation in the form $x = \log_b (y)$.

a. $2^3 = 8$
   Here, $b = 2$, $x = 3$, and $y = 8$. Therefore, the equation $2^3 = 8$ is equivalent to $\log_2 (8) = 3$.

b. $10^4 = 10000$
   Here, $b = 10$, $x = 4$, and $y = 10000$. Therefore, the equation $10^4 = 10000$ is equivalent to $\log_{10} (10000) = 4$.

c. $10^{-4} = \frac{1}{10,000}$
   Here, $b = 10$, $x = -4$, and $y = \frac{1}{10,000}$. Therefore, the equation $10^{-4} = \frac{1}{10,000}$ is equivalent to $\log_{10} \left(\frac{1}{10,000} \right) = -4$.

TRY IT #2

Write the following exponential equations in logarithmic form.

a. $1.1^2 = 1.21$
b. $5^3 = 125$
c. $10^{-1} = \frac{1}{10}$

Answer

a. $1.1^2 = 1.21$ is equivalent to $\log_{1.1} (1.21) = 2$.
b. $5^3 = 125$ is equivalent to $\log_5 (125) = 3$.
c. $10^{-1} = \frac{1}{10}$ is equivalent to $\log_{10} \left(\frac{1}{10} \right) = -1$.

Evaluating Logarithms

Knowing the squares, cubes, and roots of numbers allows us to evaluate many logarithms mentally. For example, consider $\log_2{(8)}$. We ask, "To what exponent must 2 be raised in order to get 8?" Because we already know $2^3 = 8$, it follows that $\log_2{(8)} = 3.$

Now consider solving $\log_7{(49)}$ and $\log_3{(27)}$ mentally.

- We ask, "To what exponent must 7 be raised in order to get 49?" We know $7^2 = 49.$ Therefore, $\log_7{(49)} = 2.$
- We ask, "To what exponent must 3 be raised in order to get 27?" We know $3^3 = 27.$ Therefore, $\log_3{(27)} = 3.$

Even some seemingly more complicated logarithms can be evaluated without a calculator. For example, let's evaluate $\log_{\frac{2}{3}}\left(\frac{4}{9}\right)$ mentally.

- We ask, "To what exponent must $\frac{2}{3}$ be raised in order to get $\left(\frac{4}{9}\right)$?" We know $2^2 = 4$ and $3^2 = 9$, so $\left(\frac{2}{3}\right)^2 = \frac{4}{9}.$ Therefore, $\log_{\frac{2}{3}}\left(\frac{4}{9}\right) = 2.$

HOW TO

Given a logarithm of the form $y = \log_b{(x)}$, **evaluate it mentally.**

1. Rewrite the argument x as a power of $b : b^y = x.$
2. Use previous knowledge of powers of b identify y by asking, "To what exponent should b be raised in order to get $x?$"

EXAMPLE 3: SOLVING LOGARITHMS MENTALLY

Solve $y = \log_4{(64)}$ without using a calculator.

Answer

First we rewrite the logarithm in exponential form: $4^y = 64.$ Next, we ask, "To what exponent must 4 be raised in order to get 64?"

We know $4^3 = 64.$ Therefore, $\log_4{(64)} = 3.$

TRY IT #3

Solve $y = \log_{121}{(11)}$ without using a calculator.

Answer

$$\log_{121}(11) = \tfrac{1}{2} \text{ (recalling that } \sqrt{121} = (121)^{\frac{1}{2}} = 11 \text{)}$$

EXAMPLE 4: EVALUATING THE LOGARITHM OF A RECIPROCAL

Evaluate $y = \log_{10}\left(\frac{1}{100}\right)$ without using a calculator.

Answer

First we rewrite the logarithm in exponential form: $10^y = \frac{1}{100}$. Next, we ask, "To what exponent must 10 be raised in order to get $\frac{1}{100}$?"

We know $10^2 = 100$, but what must we do to get the reciprocal, $\frac{1}{100}$? Recall from working with exponents that $b^{-a} = \frac{1}{b^a}$. We use this information to write

$$10^{-2} = \frac{1}{10^2}$$
$$= \frac{1}{100}$$

Therefore, $\log_{10}\left(\frac{1}{100}\right) = -2.$

TRY IT #4

Evaluate $y = \log_2\left(\frac{1}{32}\right)$ without using a calculator.

Answer

$$\log_2\left(\tfrac{1}{32}\right) = -5$$

Using Common Logarithms

Sometimes we may see a logarithm written without a base. In this case, we assume that the base is 10. In other words, the expression $\log(x)$ means $\log_{10}(x)$. We call a base-10 logarithm a **common logarithm**. Common logarithms are used to measure the Richter Scale mentioned at the beginning of the section. Scales for measuring the brightness of stars and the pH of acids and bases also use common logarithms.

DEFINITION

A **common logarithm** is a logarithm with base 10. We write $\log_{10}(x)$ simply as $\log(x)$. The common logarithm of a positive number x satisfies the following definition.

For $x > 0$, $y = \log(x)$ is equivalent to $10^y = x$.

- We read $\log(x)$ as, "the logarithm with base 10 of x" or "log base 10 of x."
- The logarithm y is the exponent to which 10 must be raised to get x.

HOW TO

Given a common logarithm of the form $y = \log(x)$, evaluate it mentally.

1. Rewrite the argument x as a power of 10: $10^y = x$.
2. Use previous knowledge of powers of 10 to identify y by asking, "To what exponent must 10 be raised in order to get x?"

EXAMPLE 5: FINDING THE VALUE OF A COMMON LOGARITHM MENTALLY

Evaluate $y = \log(1000)$ without using a calculator.

Answer

First we rewrite the logarithm in exponential form: $10^y = 1000$. Next, we ask, "To what exponent must 10 be raised in order to get 1000?" We know $10^3 = 1000$. Therefore, $\log(1000) = 3$.

TRY IT #5

Evaluate $y = \log(1,000,000)$.

Answer

$\log(1,000,000) = 6$

EXAMPLE 6: FINDING THE VALUE OF A COMMON LOGARITHM USING A CALCULATOR

Evaluate $y = \log(321)$ to four decimal places using a calculator.

Answer

- Press [**LOG**].
- Enter 321, followed by [**)**].
- Press [**ENTER**].

Rounding to four decimal places, $\log(321) \approx 2.5065$.

Analysis

Note that $10^2 = 100$ and that $10^3 = 1000$. Since 321 is between 100 and 1000, we know that $\log(321)$ must be between $\log(100)$ and $\log(1000)$. This gives us the following:

$$100 \; < \; 321 \; < \; 1000$$
$$2 \; < \; 2.5065 \; < \; 3$$

TRY IT #6

Evaluate $y = \log(123)$ to four decimal places using a calculator.

Answer

$\log(123) \approx 2.0899$

EXAMPLE 7: REWRITING AND SOLVING A REAL-WORLD EXPONENTIAL MODEL

The amount of energy released from one earthquake was 500 times greater than the amount of energy released from another. The equation $10^x = 500$ represents this situation, where x is the difference in magnitudes on the Richter Scale. To the nearest thousandth, what was the difference in magnitudes?

Answer

We begin by rewriting the exponential equation in logarithmic form.

$$10^x = 500$$

$$\log(500) = x \qquad \text{Use the definition of the common logarithm.}$$

Next we evaluate the logarithm using a calculator:

- Press **[LOG]**.
- Enter 500, followed by **[ )]**.
- Press **[ENTER]**.
- To the nearest thousandth, $\log(500) \approx 2.699$.

The difference in magnitudes was about 2.699.

TRY IT #7

The amount of energy released from one earthquake was $8{,}500$ times greater than the amount of energy released from another. The equation $10^x = 8500$ represents this situation, where x is the difference in magnitudes on the Richter Scale. To the nearest thousandth, what was the difference in magnitudes?

Answer

The difference in magnitudes was about 3.929.

Using Natural Logarithms

The most frequently used base for logarithms is e. Base e logarithms are important in calculus and some scientific applications; they are called **natural logarithms**. The base e logarithm, $\log_e(x)$, has its own notation, $\ln(x)$.

Most values of $\ln(x)$ can be found only using a calculator. The major exception is that, because the logarithm of 1 is always 0 in any base, $\ln(1) = 0$. For other natural logarithms, we can use the **[LN]** key that can be found on most scientific calculators. We can also find the natural logarithm of any power of e using the inverse property of logarithms.

A **natural logarithm** is a logarithm with base e. We write $\log_e(x)$ simply as $\ln(x)$. The natural logarithm of a positive number x satisfies the following definition.

For $x > 0$, $y = \ln(x)$ is equivalent to $e^y = x$.

- We read $\ln(x)$ as, "the logarithm with base e of x" or "the natural logarithm of x."
- The logarithm y is the exponent to which e must be raised to get x.

Given a natural logarithm with the form $y = \ln(x)$, evaluate it using a calculator.

1. Press [**LN**].
2. Enter the value given for x, followed by [**)**].
3. Press [**ENTER**].

EXAMPLE 8: EVALUATING A NATURAL LOGARITHM USING A CALCULATOR

Evaluate $y = \ln(500)$ to four decimal places using a calculator.

Answer

- Press [**LN**].
- Enter 500, followed by [**)**].
- Press [**ENTER**].

Rounding to four decimal places, $\ln(500) \approx 6.2146$.

Evaluate: a. $\ln(-500)$. b. $\ln(8)$.

Answer

a. It is not possible to take the logarithm of a negative number in the set of real numbers.

b. Approximately 2.0794.

Properties of Logarithms

In applications, equation solving and advance mathematics, it is often easier to work with simplified logarithmic expressions. We will next study the properties of logarithms which allow logarithmic expressions to be written in multiple ways so that they can be interpreted from multiple view points.

The logarithmic and exponential functions with the same base "undo" each other. This means that logarithms have similar properties to exponents. Some important properties of logarithms are given here. First, the following properties are easy to prove.

$$\log_b 1 = 0$$
$$\log_b b = 1$$

For example, $\log_5 1 = 0$ since $5^0 = 1$, and $\log_5 5 = 1$ since $5^1 = 5$.

Next, we have the **inverse property**.

$$\log_b (b^x) = x$$
$$b^{\log_b x} = x, \ \ x > 0$$

Since the functions $y = e^x$ and $y = \ln(x)$ are inverse functions, we know that the composition of the functions produce the identity over the appropriate domain. Therefore, $\ln(e^x) = x$ for all x and $e^{\ln(x)} = x$ for $x > 0$. Similarly, $\log(10^x) = x$ for all x and $10^{\log(x)} = x$ for $x > 0$. For example, to evaluate $\log(100)$, we can rewrite the logarithm as $\log_{10}(10^2)$, and then apply the inverse property $\log_b (b^x) = x$ to get $\log_{10}(10^2) = 2$. To evaluate $e^{\ln(7)}$, we can rewrite the logarithm as $e^{\log_e 7}$, and then apply the inverse property $b^{\log_b x} = x$ to get $e^{\log_e 7} = 7$.

Product Rule of Logarithms

Recall that we use the *product rule of exponents* to combine the product of exponents by adding: $x^a x^b = x^{a+b}$. We have a similar property for logarithms, called the **product rule for logarithms**, which says that the logarithm of a product is equal to a sum of logarithms. Because logs are exponents, and we multiply like bases, we can add the exponents. We will use the inverse property to derive the product rule below.

Given any real number x and positive real numbers M, N, and b, where $b \neq 1$, we will show
$$\log_b (MN) = \log_b (M) + \log_b (N).$$

Let $m = \log_b (M)$ and $n = \log_b (N)$. In exponential form, these equations are $b^m = M$ and $b^n = N$. It follows that

$$
\begin{aligned}
\log_b (MN) &= \log_b (b^m b^n) && \text{Substitute for } M \text{ and } N. \\
&= \log_b (b^{m+n}) && \text{Apply the product rule for exponents.} \\
&= m + n && \text{Apply the inverse property of logs.} \\
&= \log_b (M) + \log_b (N) && \text{Substitute for } m \text{ and } n.
\end{aligned}
$$

Note that repeated applications of the product rule for logarithms allow us to simplify the logarithm of the product of any number of factors. For example, consider $\log_b (wxyz)$. Using the product rule for logarithms, we can rewrite this logarithm of a product as the sum of logarithms of its factors:

$$\log_b (wxyz) = \log_b (w) + \log_b (x) + \log_b (y) + \log_b (z)$$

HOW TO

Given the logarithm of a product, use the product rule of logarithms to write an equivalent sum of logarithms.

1. Factor the argument completely, expressing each whole number factor as a product of prime numbers.
2. Write the equivalent expression by summing the logarithms of each factor.

EXAMPLE 9: USING THE PRODUCT RULE FOR LOGARITHMS

Expand $\log_3 (30x (3x + 4))$.

Answer

The input expression is the product of the three factors $30, x$ and $3x + 4$. Using the product rule of logarithms we write the equivalent equation by summing the logarithms of each factor.

$$\log_3 (30x (3x + 4)) = \log_3 (30) + \log_3 (x) + \log_3 (3x + 4)$$

TRY IT #9

Expand $\log_b (10k)$.

Answer

$\log_b (10) + \log_b (k)$

Quotient Rule for Logarithms

For quotients, we have a similar rule for logarithms. Recall that we use the *quotient rule of exponents* to combine the quotient of exponents by subtracting: $\frac{x^a}{x^b} = x^{a-b}$. The **quotient rule for logarithms** says that the logarithm of a quotient is equal to a difference of logarithms. Just as with the product rule, we can use the inverse property to derive the quotient rule.

Given any real number x and positive real numbers M, N, and b, where $b \neq 1$, we will show

$$\log_b \left(\tfrac{M}{N}\right) = \log_b (M) - \log_b (N).$$

Let $m = \log_b(M)$ and $n = \log_b(N)$. In exponential form, these equations are $b^m = M$ and $b^n = N$. It follows that

$$\begin{aligned} \log_b\left(\frac{M}{N}\right) &= \log_b\left(\frac{b^m}{b^n}\right) & \text{Substitute for } M \text{ and } N. \\ &= \log_b\left(b^{m-n}\right) & \text{Apply the quotient rule for exponents.} \\ &= m - n & \text{Apply the inverse property of logs.} \\ &= \log_b(M) - \log_b(N) & \text{Substitute for } m \text{ and } n. \end{aligned}$$

For example, to expand $\log\left(\frac{2x^2+6x}{3x+9}\right)$, we must first express the quotient in lowest terms. Factoring and canceling we get,

$$\begin{aligned} \log\left(\frac{2x^2+6x}{3x+9}\right) &= \log\left(\frac{2x\,(x+3)}{3\,(x+3)}\right) & \text{Factor.} \\ &= \log\left(\frac{2x}{3}\right) & \text{Cancel the common factors.} \end{aligned}$$

When the common factors are canceled, keep in mind that $x = -3$ is not in the domain of the original function and should be noted in the work as the expression is simplified.

Next we apply the quotient rule by subtracting the logarithm of the denominator from the logarithm of the numerator. Then we apply the product rule to the first term.

$$\begin{aligned} \log\left(\frac{2x}{3}\right) &= \log(2x) - \log(3) \\ &= \log(2) + \log(x) - \log(3) \end{aligned}$$

Finally we notice that $x = -3$ is not in the domain of the simplified expression either.

HOW TO

Given the logarithm of a quotient, use the quotient rule of logarithms to write an equivalent difference of logarithms.

1. Express the argument in lowest terms by factoring the numerator and denominator and canceling common terms.
2. Write the equivalent expression by subtracting the logarithm of the denominator from the logarithm of the numerator.
3. Check to see that each term is fully expanded. If not, apply the product rule for logarithms to expand completely.

EXAMPLE 10: USING THE QUOTIENT RULE FOR LOGARITHMS

Expand $\log_2\left(\frac{15x(x-1)}{(3x+4)(2-x)}\right)$.

Answer

First we note that the quotient is factored and in lowest terms, so we apply the quotient rule.

$$\log_2 \left(\frac{15x(x-1)}{(3x+4)(2-x)}\right) = \log_2\left(15x\,(x-1)\right) - \log_2\left((3x+4)\,(2-x)\right)$$

Notice that the resulting terms are logarithms of products. To expand completely, we apply the product rule.

$$\log_2\left(15x\,(x-1)\right) - \log_2\left((3x+4)\,(2-x)\right)$$
$$= \left[\log_2\left(15\right) + \log_2\left(x\right) + \log_2\left(x-1\right)\right] - \left[\log_2\left(3x+4\right) + \log_2\left(2-x\right)\right]$$
$$= \log_2\left(15\right) + \log_2\left(x\right) + \log_2\left(x-1\right) - \log_2\left(3x+4\right) - \log_2\left(2-x\right)$$

Analysis

There are exceptions to consider in this and many other examples. First, because denominators must never be zero, this expression is not defined for $x = -\frac{4}{3}$ and $x = 2$. Also, since the argument of a logarithm must be positive, we note as we observe the expanded logarithm, that $x > 0$, $x > 1$, $x > -\frac{4}{3}$, and $x < 2$. Combining these conditions is beyond the scope of this section, and we will not consider them here.

TRY IT #10

Expand $\log\left(\frac{7x^2+21x}{7x(x-1)(x-2)}\right)$.

Answer

$$\log\left(x+3\right) - \log\left(x-1\right) - \log\left(x-2\right)$$

Power Rule for Logarithms

We've explored the product rule and the quotient rule, but how can we take the logarithm of a power, such as x^2? One method is as follows:

$$\begin{aligned}\log_b\left(x^2\right) &= \log_b\left(x \cdot x\right)\\ &= \log_b\left(x\right) + \log_b\left(x\right)\\ &= 2\log_b\left(x\right)\end{aligned}$$

Notice that we used the product rule for logarithms to find a solution for the example above. By doing so, we have derived the **power rule for logarithms**, which says that the log of a power is equal to the exponent times the log of the base. Keep in mind that, although the input to a logarithm may not be written as a power, we may be able to change it to a power. For example,

$$100 = 10^2 \qquad \sqrt{3} = 3^{\frac{1}{2}} \qquad \tfrac{1}{e} = e^{-1}$$

The power rule for logarithms can be used to simplify the logarithm of a power by rewriting it as the product of the exponent times the logarithm of the base.

$$\log_b\left(M^n\right) = n\log_b\left(M\right)$$

HOW TO

Given the logarithm of a power, use the power rule of logarithms to write an equivalent product of a factor and a logarithm.

1. Express the argument as a power, if needed.
2. Write the equivalent expression by multiplying the exponent times the logarithm of the base.

EXAMPLE 11: EXPANDING A LOGARITHM WITH POWERS

Expand $\log_2 \left(x^5 \right)$.

Answer

The argument is already written as a power, so we identify the exponent, 5, and the base, x, and rewrite the equivalent expression by multiplying the exponent times the logarithm of the base.

$$\log_2 \left(x^5 \right) = 5\log_2 \left(x \right)$$

TRY IT #11

Expand $\ln \left(x^2 \right)$.

Answer

$2\ln \left(x \right)$

EXAMPLE 12: REWRITING AN EXPRESSION AS A POWER BEFORE USING THE POWER RULE

Expand $\log_3 \left(25 \right)$ using the power rule for logs.

Answer

Expressing the argument as a power, we get $\log_3 \left(25 \right) = \log_3 \left(5^2 \right)$.

Next we identify the exponent, 2, and the base, 5, and rewrite the equivalent expression by multiplying the exponent times the logarithm of the base.

$$\log_3\left(5^2\right) = 2\log_3\left(5\right)$$

TRY IT #12

Expand $\ln\left(\frac{1}{x^2}\right)$.

Answer

$-2\ln\left(x\right)$

Access this online resource for additional instruction and practice with logarithms.

- Introduction to Logarithms

Key Equations

Definition of the logarithmic function	For $x > 0, b > 0, b \neq 1,$ $y = \log_b\left(x\right)$ if and only if $b^y = x$.
Definition of the common logarithm	For $x > 0, y = \log\left(x\right)$ if and only if $10^y = x$.
Definition of the natural logarithm	For $x > 0, y = \ln\left(x\right)$ if and only if $e^y = x$.
Properties of logarithms	$\log_b\left(b^x\right) = x$ $b^{\log_b x} = x, \ x > 0$ $\log_b\left(MN\right) = \log_b\left(M\right) + \log_b\left(N\right)$ for $b > 0$ $\log_b\left(\frac{M}{N}\right) = \log_b\left(M\right) - \log_b\left(N\right)$ $\log_b\left(M^n\right) = n\log_b\left(M\right)$

KEY CONCEPTS

- The inverse of an exponential function is a logarithmic function, and the inverse of a logarithmic function is an exponential function.
- Logarithmic equations can be written in an equivalent exponential form, using the definition of a logarithm.

- Exponential equations can be written in their equivalent logarithmic form using the definition of a logarithm.
- Logarithmic functions with base b can be evaluated mentally using previous knowledge of powers of b.
- Common logarithms can be evaluated mentally using previous knowledge of powers of 10.
- When common logarithms cannot be evaluated mentally, a calculator can be used.
- Real-world exponential problems with base 10 can be rewritten as a common logarithm and then evaluated using a calculator.
- Natural logarithms can be evaluated using a calculator.
- Properties of logarithms can be used to simplify expressions and expand them into sums and differences.

GLOSSARY

common logarithm
the exponent to which 10 must be raised to get x; $\log_{10}(x)$ is written simply as $\log(x)$.

logarithm
the exponent to which b must be raised to get x; written $y = \log_b(x)$.

natural logarithm
the exponent to which the number e must be raised to get x; $\log_e(x)$ is written as $\ln(x)$.

2.5 Exponential and Logarithmic Equations

LEARNING OBJECTIVES

In this section, you will:

- Use logarithms to solve exponential equations.
- Use the definition of a logarithm to solve logarithmic equations.
- Solve applied problems involving exponential and logarithmic equations.

Figure 1. Wild rabbits in Australia. The rabbit population grew so quickly in Australia that the event became known as the "rabbit plague." (credit: Richard Taylor, Flickr)

In 1859, an Australian landowner named Thomas Austin released 24 rabbits into the wild for hunting. Because Australia had few predators and ample food, the rabbit population exploded. In fewer than ten years, the rabbit population numbered in the millions.

Uncontrolled population growth, as in the wild rabbits in Australia, can be modeled with exponential functions. Equations resulting from those exponential functions can be solved to analyze and make predictions about exponential growth. In this section, we will learn techniques for solving exponential equations.

Solving Exponential Equations Using Logarithms

Many times we need to solve equations of the form $n = ab^x$ where n is a real number. To do this we will divide both sides by a to get $\frac{n}{a} = b^x$ and then take the logarithm of both sides giving the equation $\log\left(\frac{n}{a}\right) = \log\left(b^x\right)$. Next, we use the power rule for logarithms to get $\log\left(\frac{n}{a}\right) = x\log\left(b\right)$. Finally, divide both sides by $\log\left(b\right)$ to get $x = \frac{\log\left(\frac{n}{a}\right)}{\log(b)}$. Note that any base for the logarithm can be used but base 10 and base e are most commonly used.

HOW TO

Given an exponential equation in the form $n = ab^x$, solve for the unknown.

1. Divide both sides by a or the initial condition.
2. Apply the logarithm of both sides of the equation.

   - If one of the terms in the equation has base 10, use the common logarithm.
   - If none of the terms in the equation has base 10, use the natural logarithm.

3. Use the rules of logarithms to solve for the unknown.

EXAMPLE 1: SOLVING A BASIC EXPONENTIAL EQUATION

Solve $5 = 3\,(2)^x$.

Answer

Begin by dividing both sides by 3 to get

$$5/3 = (2)^x.$$

Then, take the natural logarithm of both sides to get the equation

$$\ln(5/3) = \ln(2^x).$$

Use the power rule of logarithms $\ln(b^x) = x\ln(b)$ to get

$$\ln(5/3) = x\ln(2).$$

Finally, divide both sides by $\ln(2)$ to get

$$x = \frac{\ln(5/3)}{\ln(2)} \approx 0.7370.$$

TRY IT #1

Solve $7 = 15\,(4)^x$.

Answer

$$x = \frac{\ln(7/15)}{\ln(4)} \approx -0.5498.$$

EXAMPLE 2: SOLVING AN EXPONENTIAL EQUATION WITH AN ALGEBRAIC EXPRESSION IN THE EXPONENT

Solve $15 = 3\left(0.5\right)^{x+1}$.

Answer

Since the algebraic expression is in the exponent, this equation will be solved using the same steps as above but the x+1 should be kept in parenthesis until we are ready to simplify at the end. Begin by dividing both sides by 3 to get

$$5 = \left(0.5\right)^{(x+1)}.$$

Then, take the natural logarithm of both sides to get the equation

$$\ln\left(5\right) = \ln\left(0.5^{(x+1)}\right).$$

Use the power rule of logarithms $\ln\left(b^x\right) = x\ln\left(b\right)$ so

$$\ln\left(5\right) = (x+1)\ln\left(0.5\right).$$

Divide both sides of the equation by $\ln\left(0.5\right)$ to get the equation

$$(x+1) = \frac{\ln(5)}{\ln(0.5)}.$$

Note that this is just a linear equation so we subtract 1 from both sides and the solution is

$$x = \frac{\ln(5)}{\ln(0.5)} - 1 \approx -3.3219.$$

TRY IT #2

Solve $15 = 2\left(7\right)^{x^2+1}$.

Answer

$$x = \pm\sqrt{\frac{\ln(7.5)}{\ln(7)} - 1} \approx \pm 0.1883$$

Equations Containing e

One common type of exponential equations uses base e. This constant occurs again and again in nature, in mathematics, in science, in engineering, and in finance. When we have an equation with a base e on either side, we can use the natural logarithm to solve it.

HOW TO

Given an equation of the form $y = ae^{kt}$, **solve for** t.

1. Divide both sides of the equation by a.
2. Apply the natural logarithm of both sides of the equation.
3. Divide both sides of the equation by k.

EXAMPLE 3: SOLVE AN EQUATION WITH CONTINUOUS GROWTH

Solve $100 = 20e^{2t}$.

Answer

For this problem, use the natural logarithm since the equation contains base e.

$$100 = 20e^{2t}$$

$$5 = e^{2t} \qquad \text{Divide both sides by 20.}$$

$$\ln(5) = 2t \qquad \text{Take the natural logarithm of both sides.}$$

$$\text{Use the fact that } \ln(e^x) = x.$$

$$t = \frac{\ln(5)}{2} \qquad \text{Divide by the coefficient of } t.$$

Analysis

Using laws of logs, we can also write this answer in the form $t = \ln\sqrt{5}$. If we want a decimal approximation of the answer, we use a calculator.

TRY IT #3

Solve $3e^{0.5t} = 11$.

Answer

$t = 2\ln\left(\frac{11}{3}\right)$ or $\ln\left(\left(\frac{11}{3}\right)^2\right)$

Do all exponential equations have a solution? If not, how can we tell if there is a solution during the problem-solving process?

No. Recall that the range of an exponential function is always positive. While solving the equation, we may obtain an expression that is undefined.

EXAMPLE 4: SOLVING AN EQUATION WITH POSITIVE AND NEGATIVE POWERS

Solve $3^{x+1} = -2$.

Answer

This equation has no solution. There is no real value of x that will make the equation a true statement because any power of a positive number is positive.

Figure 2 shows that the two graphs do not cross so the left side is never equal to the right side. Thus the equation has no solution.

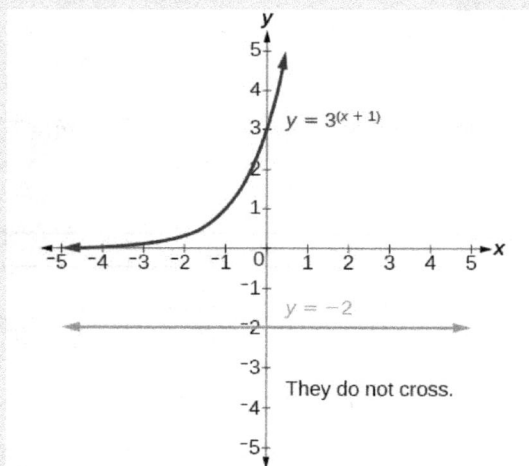

Figure 2

TRY IT #4

Solve $2^x = -100$.

Answer

The equation has no solution.

Sometimes there are exponential functions on both sides of the equation such as $ab^x = cd^x$. The process is similar to solving $n = ab^x$ but we will need to be careful to use the product rule of logarithms before applying the power rule of logarithms.

HOW TO

Given an equation with exponential expression on each side, solve for the unknown.

1. Divide both sides by one of the initial conditions.
2. Apply the logarithm of both sides of the equation.

   ○ If one of the terms in the equation has base 10, use the common logarithm.
   ○ If none of the terms in the equation has base 10, use the natural logarithm.

3. Use the product and power rules of logarithms and then solve for the unknown.

EXAMPLE 5: SOLVING AN EQUATION CONTAINING POWERS OF DIFFERENT BASES

Solve $6\left(5\right)^{x+2} = 2\left(4\right)^{x}$.

Answer

$$6\,(5)^{x+2} = 2\,(4)^x$$

$$3\,(5)^{x+2} = (4)^x \qquad \text{Divide both sides by 2.}$$

$$\ln\left(3\,(5)^{x+2}\right) = \ln\left(4^x\right) \qquad \text{Take the natural logarithm of both sides.}$$

$$\ln\,(3) + \ln\left(5^{x+2}\right) = \ln\left(4^x\right) \qquad \text{Use the product rule of logarithms.}$$

$$\ln\,(3) + (x+2)\ln\,(5) = x\ln\,(4) \qquad \text{Use the power law of logarithms.}$$

$$\ln\,(3) + x\ln\,(5) + 2\ln\,(5) = x\ln\,(4) \qquad \text{Use the distributive law.}$$

$$x\ln\,(5) - x\ln\,(4) = -2\ln\,(5) - \ln\,(3) \qquad \text{Move terms containing x on one side,}$$
$$\text{and terms without x on the other.}$$

$$x\,(\ln\,(5) - \ln\,(4)) = -2\ln\,(5) - \ln\,(3) \qquad \text{On the left hand side, factor out an } x.$$

$$x\ln\left(\frac{5}{4}\right) = \ln\left(\frac{1}{75}\right) \qquad \text{Use the laws of logarithms.}$$

$$x = \frac{\ln\left(\frac{1}{75}\right)}{\ln\left(\frac{5}{4}\right)} \qquad \text{Divide by the coefficient of } x.$$

Analysis

Notice that the product rule of logarithms was used before the power rule because the power rule cannot be applied to ab^x; a is not raised to the x power. Also, note that when the power rule is used on $\ln\left(5^{x+2}\right)$ the $(x+2)$ has parenthesis so that the $\ln\,(5)$ gets properly distributed.

TRY IT #5

Solve $2^x = 3^{x+1}$.

Answer

$$x = \frac{\ln(3)}{\ln\left(\frac{2}{3}\right)}$$

Q&A

Does every equation of the form $y = Ae^{kt}$ have a solution?

No. There is a solution when $k \neq 0$, and when y and A are either both 0 or neither 0, and they have the same sign. An example of an equation with this form that has no solution is $2 = -3e^t$.

EXAMPLE 6: SOLVING AN EQUATION THAT REQUIRES ALGEBRA FIRST

Solve

1. $4e^{2x} + 5 = 12.$
2. $e^{2t} - 3 = -4e^{2t}$

Answer

$$4e^{2x} + 5 = 12$$

$4e^{2x} = 7$ Combine like terms.

$e^{2x} = \dfrac{7}{4}$ Divide by the coefficient.

$2x = \ln\left(\dfrac{7}{4}\right)$ Take the natural logarithm of both sides.

$x = \dfrac{1}{2}\ln\left(\dfrac{7}{4}\right)$ Solve for x.

1.

$$e^{2t} - 3 = -4e^{2t}$$

$5e^{2t} - 3 = 0$ Add $4e^{2t}$ to both sides.

$5e^{2t} = 3$ Add 3 to both sides.

$e^{2t} = \dfrac{3}{5}$ Divide both sides by 5.

$2t = \ln\left(\dfrac{3}{5}\right)$ Take the natural logarithm of both sides.

$t = \dfrac{1}{2}\ln\left(\dfrac{3}{5}\right)$ Solve for x.

2.

TRY IT #6

Solve $3 + e^{2t} = 7e^{2t}$.

Answer

$t = \ln\left(\frac{1}{\sqrt{2}}\right) = -\frac{1}{2}\ln(2)$

Extraneous Solutions

Sometimes the methods used to solve an equation introduce an extraneous solution, which is a solution that is correct algebraically but does not satisfy the conditions of the original equation. One such situation arises in solving when the logarithm is taken on both sides of the equation. In such cases, remember that the argument of the logarithm must be positive. If the number we are evaluating in a logarithm function is negative, there is no output.

These extraneous solutions frequently occur when the exponential function is a quadratic form. Recall that quadratic equations can be solved by factoring and setting each factor equal to zero, or by the quadratic equation. W will look for the pattern of the quadratic and then choose which technique can most easily be used.

EXAMPLE 7: SOLVING EXPONENTIAL FUNCTIONS IN QUADRATIC FORM

Solve $e^{2x} - e^x = 56$.

Answer

We first re-write the equation using the fact that $e^{2x} = (e^x)^2$. Notice that this equation has the form e^x squared minus e^x equals 56. This is a quadratic equation which can be solved by factoring.

$$e^{2x} - e^x = 56$$

$$e^{2x} - e^x - 56 = 0 \qquad \text{Get one side of the equation equal to zero.}$$

$$(e^x + 7)(e^x - 8) = 0 \qquad \text{Factor by the FOIL method.}$$

$$e^x + 7 = 0 \text{ or } e^x - 8 = 0 \qquad \text{If a product is zero, then one factor must be zero.}$$

$$e^x = -7 \text{ or } e^x = 8 \qquad \text{Isolate the exponentials.}$$

$$e^x = 8 \qquad \text{Reject the equation that has no solution.}$$

$$x = \ln(8) \qquad \text{Write as a logarmithm.}$$

Analysis

When we plan to use factoring to solve a problem, we always get zero on one side of the equation, because zero has the unique property that when a product is zero, one or both of the factors must be zero. We reject the equation $e^x = -7$ because a positive number never equals a negative number. The solution $\ln(-7)$ is not a real number, and in the real number system this solution is rejected as an extraneous solution.

TRY IT #7

Solve $e^{2x} = e^x + 2$.

Answer

$x = \ln 2$.

Using the Definition of a Logarithm to Solve Logarithmic Equations

We have already seen that the logarithmic equation $\log_b (x) = y$ is equivalent to the exponential equation $b^y = x$, for $x > 0$. We can use this fact, along with the rules of logarithms, to solve logarithmic equations where the argument is an algebraic expression.

EXAMPLE 8: SOLVING A LOGARITHMIC EQUATION

Solve the equation $\log_2 (2) + \log_2 (3x - 5) = 3$.

Answer

To solve this equation, we can use rules of logarithms to rewrite the left side in compact form and then apply the definition of logs to solve for x :

$$\log_2 (2) + \log_2 (3x - 5) = 3$$

$$\log_2 (2(3x - 5)) = 3 \qquad \text{Apply the product rule of logarithms.}$$

$$\log_2 (6x - 10) = 3 \qquad \text{Distribute.}$$

$$2^3 = 6x - 10 \qquad \text{Apply the definition of a logarithm.}$$

$$8 = 6x - 10 \qquad \text{Calculate } 2^3.$$

$$18 = 6x \qquad \text{Add 10 to both sides.}$$

$$x = 3 \qquad \text{Divide by 6.}$$

EXAMPLE 9: USING ALGEBRA TO SOLVE A LOGARITHMIC EQUATION

Solve $2\ln (x) + 3 = 7$.

Answer

$$2\ln (x) + 3 = 7$$

$$2\ln (x) = 4 \qquad \text{Subtract 3.}$$

$$\ln (x) = 2 \qquad \text{Divide by 2.}$$

$$x = e^2 \qquad \text{Rewrite in exponential form.}$$

TRY IT #8

Solve $6 + \ln (x) = 10$.

Answer

$$x = e^4$$

EXAMPLE 10: USING ALGEBRA BEFORE AND AFTER USING THE DEFINITION OF THE NATURAL LOGARITHM

Solve $2\ln(6x) = 7$.

Answer

$$2\ln(6x) = 7$$

$$\ln(6x) = \frac{7}{2} \qquad \text{Divide by 2.}$$

$$6x = e^{\left(\frac{7}{2}\right)} \qquad \text{Use the definition of the natural logarithm.}$$

$$x = \frac{1}{6}e^{\left(\frac{7}{2}\right)} \qquad \text{Divide by 6.}$$

TRY IT #9

Solve $2\ln(x+1) = 10$.

Answer

$$x = e^5 - 1$$

EXAMPLE 11: USING A GRAPH TO UNDERSTAND THE SOLUTION TO A LOGARITHMIC EQUATION

Solve $\ln(x) = 3$.

Answer

To solve $\ln(x) = 3$, write the equation in exponential form using the definition of the natural logarithm to get $x = e^3$.

Figure 3 represents the graph of the equation. On the graph, the x-coordinate of the point at which the two graphs intersect is close to 20. In other words $e^3 \approx 20$. A calculator gives a better approximation: $e^3 \approx 20.0855$.

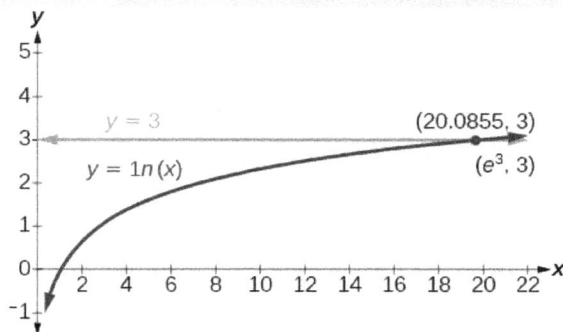

Figure 3. The graphs of $y = \ln(x)$ and $y = 3$ cross at the point $\left(e^3, 3\right)$, which is approximately (20.0855, 3).

TRY IT #10

Use a graphing calculator to estimate the approximate solution to the logarithmic equation $2^x = 1000$ to 2 decimal places.

Answer

$x \approx 9.97$

Solving Applied Problems Using Exponential Equations

In previous sections, we learned the properties and rules for both exponential and logarithmic functions. We have seen that any exponential function can be written as a logarithmic function and vice versa. We have used exponents to solve logarithmic equations and logarithms to solve exponential equations. We are now ready to combine our skills to solve equations that model real-world situations, whether the unknown is in an exponent or in the argument of a logarithm.

EXAMPLE 12: REAL WORLD APPLICATION; DEPRECIATION

In 2018, a car was purchased for $32,000 and depreciates 20% per year. When will the car be worth $8,000?

Answer

We begin by writing a formula for the situation. Since a noncontinuous rate is given, we use the formula $f(t) = ab^t$. Let $t = 0$ represent 2019. The initial condition is $a = 32000$, the rate is $r = -0.2$, and the growth factor is $b = 1 + r = 1 - 0.2 = 0.8$. Therefore, the formula is

$$f(t) = 32000\,(0.8)^t.$$

Next, we need to know when the function's output will be 8,000. Set the formula equal to 8000 and solve.

$$8000 = 32000\,(0.8)^t$$

$$0.25 = 0.8^t \qquad\qquad\text{Divide both sides by 32000.}$$

$$\ln(0.25) = \ln\left(0.8^t\right) \qquad\text{Take the natural logarithm of both sides.}$$

$$\ln(0.25) = t\ln(0.8) \qquad\quad\text{Use the power rule for logarithms.}$$

$$t = \frac{\ln(0.25)}{\ln(0.8)} \approx 6.21 \qquad\text{Divide both sides by } \ln(0.8).$$

It will take approximately 6.2 years for the car to depreciate to $8,000. In 2024, the car will be worth $8,000.

EXAMPLE 13: REAL WORLD APPLICATION; BLOOD ALCOHOL CONTENT

A person's blood alcohol content (BAC) is a measure of how much alcohol is in the bloodstream. When a person stops drinking, over time the BAC will decay exponentially. For a particular individual, the formula $f(t) = 0.1e^{-0.0067t}$ models their BAC over time, t, measured in minutes. When will their BAC be 0.04?

Answer

We are asked to find the input when the output is 0.04 so solve the equation $0.04 = 0.1e^{-0.0067t}$.

$$0.4 = e^{-0.0067t} \qquad\qquad\text{Divide both sides by 0.1.}$$

$$\ln(0.4) = -0.0067t \qquad\quad\text{Take the natural logarithm and simplify.}$$

$$t = \frac{\ln(0.4)}{-0.0067} \qquad\qquad\text{Divide both sides by } -0.0067.$$

$$\approx 136.8$$

It will take approximately 137 minutes for the BAC to drop to 0.04 for this individual.

EXAMPLE 14: REAL WORLD APPLICATION; TUITION

College A is charging $40,000 tuition in 2019 and it is increasing at a continuous rate of 7% per year. College B charges $45,000 in 2019, but it is increasing at 4% per year. When will college A cost more than college B?

Answer

First, we determine the model representing each of the tuition amounts. College A has a continuous rate so use $A(t) = ae^{kt}$ with $a = 40000$ and $k = 0.07$. Therefore, the model for college A is $A(t) = 40000e^{0.07t}$. College B has a noncontinuous rate so use $B(t) = ab^t$ where $a = 45000$ and $b = 1 + r = 1 + 0.04 = 1.04$. The model for college B is $B(t) = 45000\,(1.04)^t$.

We need to know when these two models are equal so we set the equations equal to each other and solve.

$$40000e^{0.07t} = 45000\,(1.04)^t$$

$$\frac{8}{9}e^{0.07t} = 1.04^t \qquad \text{Divide both sides by 45000 and simplify.}$$

$$\ln\left(\frac{8}{9}e^{0.07t}\right) = \ln\left(1.04^t\right) \qquad \text{Take the natural logarithm of both sides.}$$

$$\ln\left(\frac{8}{9}\right) + 0.07t = \ln\left(1.04^t\right) \qquad \text{Use the product rule and simplify.}$$

$$\ln\left(\frac{8}{9}\right) + 0.07t = t\ln\,(1.04) \qquad \text{Use the power rule on the right side.}$$

$$t\,(0.07 - \ln\,(1.04)) = -\ln\left(\frac{8}{9}\right) \qquad \text{Collect like terms and factor } t.$$

$$t = \frac{-\ln\left(\frac{8}{9}\right)}{0.07 - \ln\,(1.04)} \qquad \text{Divide by the coefficient of } t.$$

$$\approx 3.83.$$

College B will have a higher tuition in approximately 4 years.

TRY IT #11

A town's population is 14,000 people in 2017 and is increasing at a rate of 2.1% each year. When will the town's population reach 18,000 people?

Answer

In approximately 12.1 years or in 2029, the population will be 18,000 people.

Conversions Between Continuous and Noncontinuous Growth Rates

Recall that there are two possible formulas that can be used to represent an exponential function: $f(x) = ab^x = a(b)^x$ for noncontinuous growth and $f(x) = ae^{kx} = a\left(e^k\right)^x$ for continuous growth. When comparing the two forms, we see that $b = e^k$. Further, since $b = 1 + r$ where r is the noncontinuous growth rate, we have that $1 + r = e^k$.

EXAMPLE 15: CONTINUOUS GROWTH TO NONCONTINUOUS GROWTH

Given a continuous growth rate of 12%, find the noncontinuous rate.

Answer

The continuous growth rate k is given as $k = 0.12.$ We solve the equation $1 + r = e^k$ with this value of k plugged in.

$$1 + r = e^{0.12}$$
$$r = e^{0.12} - 1 \approx 0.1275.$$

The noncontinuous rate is approximately 12.75%. Notice that since both rates are modeling the same growth, the noncontinuous rate must be slightly higher than the continuous rate, because the continuous rate allows for growth on the growth immediately.

EXAMPLE 16: NONCONTINUOUS GROWTH TO CONTINUOUS GROWTH

Given a noncontinuous growth rate of 15%, find the continuous rate.

Answer

The noncontinuous growth rate r is given as $r = 0.15.$ Solve the equation $1 + r = e^k$ with this value of r plugged in.

$$1 + 0.15 = e^k$$
$$k = \ln{(1.15)} \approx 0.1398.$$

We took the natural logarithm of both sides to solve the equation.

The continuous rate is approximately 13.98%.

TRY IT #12

a. Given a continuous decreasing rate of 5%, find the noncontinuous rate.

b. Given a noncontinuous increasing rate of 7%, find the continuous rate.

Answer

a. The noncontinuous rate of decrease is approximately 4.877% when the continuous rate of decrease is 5%.

b. The continuous rate of increase is approximately 6.766%, when the noncontinuous rate is 7%.

Solving Logarithmic Applications

Exponential growth and decay often involve very large or very small numbers. It is common to use a logarithmic scale when measurements result in extremely large values or extremely small values. The Richter Scale, pH and decibels are examples of such scales.

To describe these numbers, we often use **orders of magnitude**. The order of magnitude is the power of ten, when the number is expressed in scientific notation, with one digit to the left of the decimal. For example, the distance to the nearest star, Proxima Centauri, measured in kilometers, is 40,113,497,200,000 kilometers. Expressed in scientific notation, this is $4.01134972 \times 10^{13}$. So, we could describe this number as having order of magnitude of 13.

The magnitude (size) of an earthquake is measured on a scale known as the Richter Scale. The Richter Scale is a base-ten logarithmic scale. In other words, an earthquake of magnitude 6 is not twice as great as an earthquake of magnitude 3. It is $10^{6-3} = 10^3 = 1,000$ times as great!

The Richter scale strength of an earthquake, M, is given by $M = \log\left(\frac{W}{W_0}\right)$, where W is the strength of the seismic waves of an earthquake and W_0 is the strength of normally occurring earthquakes. Minor earthquakes occur regularly allowing W_0 to be determined.

EXAMPLE 17: THE RICHTER SCALE

An earthquake Richter Scale strength of 5 is considered a moderate strength earthquake. How much stronger was the 2010 magnitude 5.5 earthquake that occurred between Ontario and Quebec compared to a standard earthquake?

Answer

First we note that $M = 5.5$ and substitute this into the Richter Scale equation. $5.5 = \log\left(\frac{W}{W_0}\right)$ written as an exponential equation is $10^{5.5} = \frac{W}{W_0}$. This can be written as $W = 10^{5.5}W_0$ or $W \approx 316228 W_0$.

The earthquake was 316,228 times stronger than a standard earthquake.

In chemistry, pH is used as a measure of the acidity or alkalinity of a substance. The pH scale runs from 0 to 14. Substances with a pH less than 7 are considered *acidic*, and substances with a pH greater than 7 are said to be *alkaline*. Our bodies, for instance, must maintain a pH close to 7.35 in order for enzymes to work properly. To get a feel for what is acidic and what is alkaline, consider the following pH levels of some common substances:

- Battery acid: 0.8
- Stomach acid: 2.7
- Orange juice: 3.3
- Pure water: 7 (at 25° C)
- Human blood: 7.35
- Fresh coconut: 7.8
- Sodium hydroxide (lye): 14

To determine whether a solution is acidic or alkaline, we find its pH, which is a measure of the number of active positive hydrogen ions in the solution. The pH is defined by the following formula, where $[H^+]$ is the *concentration of hydrogen ions* in the solution measured in moles per liter.

$$pH = -\log\left([H^+]\right) = \log\left(\frac{1}{[H^+]}\right)$$

EXAMPLE 18: PH

If the concentration of hydrogen ions in a liquid is doubled, what is the effect on pH?

Answer

Suppose c is the original concentration of hydrogen ions, and p is the original pH of the liquid. Then $p = -\log(c)$. If the concentration is doubled, the new concentration is $2c$. Then the pH of the new liquid is

$$pH = -\log(2c).$$

Using the product rule of logarithms,

$$pH = -\left(\log(2) + \log(c)\right) = -\log(2) - \log(c).$$

Since $p = -log(c)$, the new pH is

$$pH = p - \log(2) \approx p - 0.301.$$

When the concentration of hydrogen ions is doubled, the pH decreases by about 0.301.

EXAMPLE 19: PH

Orange juice has a pH of approximately 3.3. Determine the hydrogen ion concentration.

Answer

Since the pH of orange juice is 3.3, we substitute 3.3 into the pH formula and get $3.3 = -\log\left([H^+]\right)$. Rewriting this as an exponential function, $[H^+] = 10^{-3.3} \approx 5.012 \times 10^{-4}$. Therefore the concentration is $[H^+] \approx 5 \times 10^{-4} = 0.0005$, or 0.0005 moles per liter.

A *decibel* (dB) is a measure of how loud a sound is when compared to a reference value. A commonly used reference value is the sound intensity of the softest sound a human can typically hear; usually that of a child. We will call this value I_0. Since there is a very wide range of sounds that humans can hear, the logarithmic scale is used. The formula for decibels is

$$\text{Sound level in decibels} = 10\log\left(\frac{I}{I_0}\right),$$

where I is the sound intensity of the sound being measured.

EXAMPLE 20: DECIBELS

A vacuum cleaner sound level measures at 75dB and a balloon popping measures 125dB. A balloon popping is how many times more intense than the sound intensity of the vacuum cleaner?

Answer

Let I_B be the sound intensity of the balloon popping and I_V be the sound intensity of the vacuum cleaner. We then look at the difference.

$$125 \text{ dB } - 75 \text{ dB } = 10\log\left(\frac{I_B}{I_0}\right) - 10\log\left(\frac{I_V}{I_0}\right)$$

$$50 \text{ dB } = 10\log\left(\frac{\frac{I_B}{I_0}}{\frac{I_V}{I_0}}\right) \qquad \text{Use the quotient rule for logarithms.}$$

$$5 \text{ dB } = \log\left(\frac{I_B}{I_V}\right) \qquad \text{Simplify.}$$

$$10^5 = \frac{I_B}{I_V} \qquad \text{Write as an exponential equation.}$$

$$100000 I_V = I_B \qquad \text{Multiply both sides by } I_V.$$

The balloon's sound intensity is 100,000 times more than the vacuum cleaner's sound intensity.

Access these online resources for additional instruction and practice with exponential and logarithmic equations.

- Solving Logarithmic Equations
- Solving Exponential Equations with Logarithms

KEY CONCEPTS

- An exponential equation can be solved by taking the logarithm of each side.
- We can solve exponential equations with base e, by applying the natural logarithm of both sides because exponential and logarithmic functions are inverses of each other.
- After solving an exponential equation, check each solution in the original equation to find and eliminate any extraneous solutions.
- When given an equation of the form $\log_b(S) = c$, where S is an algebraic expression, we can use the definition of a logarithm to rewrite the equation as the equivalent exponential equation $b^c = S$, and solve for the unknown.
- We can also use graphing to solve equations with the form $\log_b(S) = c$. We graph both equations $y = \log_b(S)$ and $y = c$ on the same coordinate plane and identify the solution as the x-value of the intersecting point.

- Combining the skills learned in this and previous sections, we can solve equations that model real world situations, whether the unknown is in an exponent or in the argument of a logarithm.

GLOSSARY

extraneous solution
a solution introduced while solving an equation that does not satisfy the conditions of the original equation

2.6 Graphs of Logarithmic Functions

In Section 2.3, Graphs of Exponential Functions, we saw how creating a graphical representation of an exponential model gives us another layer of insight for predicting future events. How do logarithmic graphs give us insight into situations? Because every logarithmic function is the inverse function of an exponential function, we can think of every output on a logarithmic graph as the input for the corresponding inverse exponential equation.

To illustrate, suppose we invest $2500 in an account that offers an annual interest rate of 5%, compounded continuously. We already know that the balance in our account for any year t can be found with the equation $A = 2500e^{0.05t}$.

But what if we wanted to know the year given any balance? We would need to create a corresponding new function by using logarithms to solve for the year; thus we would need to create a logarithmic model for this situation. By graphing the model, we can see the output (year) for any input (account balance). For instance, what if we wanted to know how many years it would take for our initial investment to double? Figure 1 shows this point on the logarithmic graph.

Logarithmic Model Showing Years as a Function of the Balance in the Account

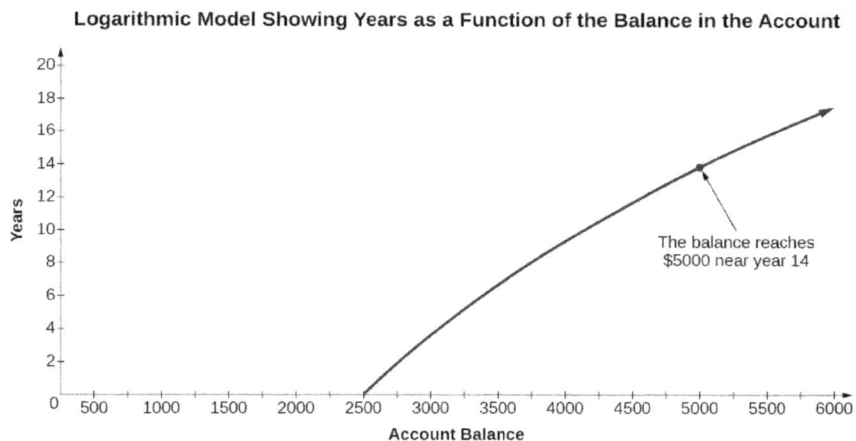

The balance reaches $5000 near year 14

Figure 1.

In this section, we will discuss the values for which a logarithmic function is defined, and then turn our attention to graphing the family of logarithmic functions.

Finding the Domain of a Logarithmic Function

Before working with graphs, we will take a look at the domain (the set of input values) for which the logarithmic function is defined.

Recall that the exponential function is defined as $f(x) = b^x$ for any real number x and constant $b > 0, b \neq 1$, where

- The domain of $f(x)$ is $(-\infty, \infty)$.
- The range of $f(x)$ is $(0, \infty)$.

In the last section we learned that the logarithmic function $f(x) = \log_b(x)$ is the inverse of the exponential function with base b. So, as inverse functions:

- The domain of $f(x) = \log_b(x)$ is the range of $f^{-1}(x) = b^x : (0, \infty)$.
- The range of $f(x) = \log_b(x)$ is the domain of $f^{-1}(x) = b^x : (-\infty, \infty)$.

Transformations of the logarithmic function $y = \log_b(x)$ behave similarly to those of other functions. Just as with toolkit and exponential functions, we can apply the four types of transformations—shifts, stretches, compressions, and reflections—to the original function without loss of shape.

In Section 2.3, Graphs of Exponential Functions we saw that certain transformations can change the *range* of $y = b^x$. Similarly, applying transformations to the function $y = \log_b(x)$ can change the *domain*. When finding the domain of a logarithmic function, therefore, it is important to remember that we can only take the logarithm *of positive real numbers*. That is, the value of the input expression of the logarithmic function must be greater than zero.

For example, consider $f(x) = \log(2x - 3)$. This function is defined for any values of x such that the input expression, in this case $2x - 3$, is greater than zero. To find the domain, we set up an inequality and solve for x :

$$2x - 3 > 0 \qquad \text{Show the input expression is greater than zero.}$$
$$2x > 3 \qquad \text{Add 3.}$$
$$x > 1.5 \qquad \text{Divide by 2.}$$

In interval notation, the domain of $f(x) = \log(2x - 3)$ is $(1.5, \infty)$.

HOW TO

Given a logarithmic function, identify the domain.

1. Set up an inequality showing the input expression greater than zero.
2. Solve for x.
3. Write the domain in interval notation.

EXAMPLE 1: IDENTIFYING THE DOMAIN OF A LOGARITHMIC SHIFT

What is the domain of $f(x) = \log_2(x + 3)$?

Answer

The logarithmic function is defined only when the input expression is positive, so this function is defined when $x + 3 > 0$. Solving this inequality,

$$x + 3 > 0 \qquad \text{The input expression must be positive.}$$
$$x > -3 \qquad \text{Subtract 3.}$$

The domain of $f(x) = \log_2(x+3)$ is $(-3, \infty)$.

TRY IT #1

What is the domain of $f(x) = \log_5(x-2) + 1$?

Answer

$(2, \infty)$

EXAMPLE 2: IDENTIFYING THE DOMAIN OF A LOGARITHMIC SHIFT AND REFLECTION

What is the domain of $f(x) = \log(5 - 2x)$?

Answer

The logarithmic function is defined only when the input expression is positive, so this function is defined when $5 - 2x > 0$. Solving this inequality,

$$5 - 2x > 0 \qquad \text{The input expression must be positive.}$$
$$-2x > -5 \qquad \text{Subtract 5.}$$
$$x < \frac{5}{2} \qquad \text{Divide by } -2 \text{ and switch the inequality.}$$

The domain of $f(x) = \log(5 - 2x)$ is $\left(-\infty, \frac{5}{2}\right)$.

TRY IT #2

What is the domain of $f(x) = \log(x-5) + 2$?

Answer

$(5, \infty)$

Graphing Logarithmic Functions

Now that we have a feel for the set of values for which a logarithmic function is defined, we move on to graphing logarithmic functions. The family of logarithmic functions includes the function $y = \log_b(x)$ along with all its transformations: shifts, stretches, compressions, and reflections.

We begin with the function $y = \log_b(x)$. Because every logarithmic function of this form is the inverse of an exponential function with base b, their graphs will be reflections of each other across the line $y = x$. To illustrate this, we can observe the relationship between the input and output values of $y = 2^x$ and its equivalent $x = \log_2(y)$ in Table 1.

Table 1

x	-3	-2	-1	0	1	2	3
$2^x = y$	$\frac{1}{8}$	$\frac{1}{4}$	$\frac{1}{2}$	1	2	4	8

$y = 2^x$	$\frac{1}{8}$	$\frac{1}{4}$	$\frac{1}{2}$	1	2	4	8
$x = \log_2(y)$	-3	-2	-1	0	1	2	3

Using the inputs and outputs from Table 1, we can build another table to observe the relationship between points on the graphs of the inverse functions $f(x) = 2^x$ and $g(x) = \log_2(x)$. See Table 2.

Table 2

$f(x) = 2^x$	$\left(-3, \frac{1}{8}\right)$	$\left(-2, \frac{1}{4}\right)$	$\left(-1, \frac{1}{2}\right)$	$(0, 1)$	$(1, 2)$	$(2, 4)$	$(3, 8)$
$g(x) = \log_2(x)$	$\left(\frac{1}{8}, -3\right)$	$\left(\frac{1}{4}, -2\right)$	$\left(\frac{1}{2}, -1\right)$	$(1, 0)$	$(2, 1)$	$(4, 2)$	$(8, 3)$

As we'd expect, the x– and y-coordinates are reversed for the inverse functions. Figure 2 shows the graph of f and g.

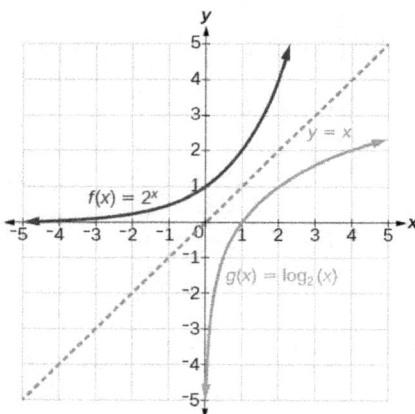

Figure 2. Notice that the graphs of $f(x) = 2^x$ and $g(x) = \log_2(x)$ are reflections about the line $y = x$.

Observe the following from the graph:

- $f(x) = 2^x$ has a y-intercept at $(0, 1)$ and $g(x) = \log_2(x)$ has an x– intercept at $(1, 0)$.
- The domain of $f(x) = 2^x$, $(-\infty, \infty)$, is the same as the range of $g(x) = \log_2(x)$.

- The range of $f(x) = 2^x$, $(0, \infty)$, is the same as the domain of $g(x) = \log_2(x)$.

Recall that for exponential growth functions like $f(x) = 2^x$, we observed that as x decreased without bound, $f(x)$ gets closer and closer to zero from above. We concluded that exponential growth functions of the form $f(x) = ab^x$ have a horizontal asymptote of $y = 0$ on one side. We created tables supporting this idea numerically. Using the inverse relationship, we will study what happens in the logarithmic function as the input gets closer and closer to zero from the right of zero or through values of x slightly larger than zero.

To create a table of values to explore this idea, we choose input values that get closer and closer to zero from the right side and then evaluate $g(x) = \log_2(x)$. We choose values slightly larger than zero because the logarithm is defined only for positive values and we want to observe what happens near the boundary of the domain. See Table 3.

Table 3

x	0.1	0.01	0.001	0.0001
$g(x) = \log_2(x)$	-3.322	-6.644	-9.966	-13.288

As x gets closer and closer to zero from the right (or from the positive side), the function values decrease without bound (go toward minus infinity). Referring back to Figure 2, we observe that the graph of $g(x) = \log_2(x)$ decreases without bound or decreases rapidly as x approaches zero from the right.

We use **arrow notation** to express these ideas. See Table 4.

Table 4: Arrow Notation

Symbol	Meaning
$x \to a^-$	x approaches a from the left ($x < a$ but values are increasing to get closer and closer to a)
$x \to a^+$	x approaches a from the right ($x > a$ but values are decreasing to get closer and closer to a)
$f(x) \to \infty$	the output goes toward infinity (the output increases without bound)
$f(x) \to -\infty$	the output goes toward negative infinity (the output decreases without bound)

For the function $g(x) = \log_2(x)$, we write in arrow notation, as $x \to 0^+$, $g(x) \to -\infty$. This behavior demonstrates a **vertical asymptote**, which is a vertical line where the graph decreases rapidly as the input values get closer and closer to 0 from the right hand side.

DEFINITION

A **vertical asymptote** of a graph is a vertical line $x = a$ where the function's output tends toward positive or negative infinity as the inputs approach a. We write

$$\text{as } x \to a^-, f(x) \to \infty, \text{ or as } x \to a^+, f(x) \to \infty, \text{ or}$$
$$\text{as } x \to a^-, f(x) \to -\infty, \text{ or as } x \to a^+, f(x) \to -\infty.$$

Note that vertical asymptotes may exist as $x = a$ is approached from one side or the other. In the case of the logarithmic function, the domain will only exist on one side of the asymptote so the asymptote will be one-sided. When we study other families of functions, we will see examples where the function increases and/or decreases rapidly on both sides of the vertical asymptote.

Note that when technology is used to graph logarithmic functions, it often appears that there is a vertical intercept rather than a vertical asymptote. Creating a table of values demonstrates that, in fact, there is a vertical asymptote.

CHARACTERISTICS OF THE GRAPH OF THE FUNCTION, $F(X) = LOG_b(X)$

For any real number x and constant $b > 0$, $b \neq 1$, we can see the following characteristics in the graph of $f(x) = \log_b(x)$:

- one-to-one function
- vertical asymptote: $x = 0$
- domain: $(0, \infty)$
- range: $(-\infty, \infty)$
- horizontal intercept: $(1, 0)$ and key point $(b, 1)$
- vertical intercept: none
- increasing if $b > 1$ and decreasing if $0 < b < 1$

See Figure 3.

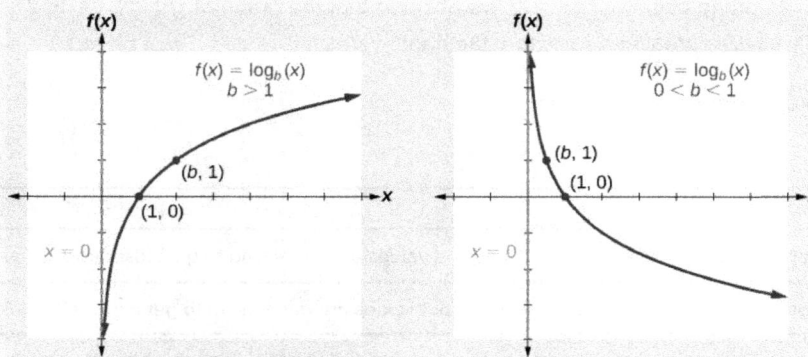

Figure 3.

HOW TO

Given a logarithmic function with the form $f(x) = \log_b(x)$, **graph the function.**

1. Draw and label the vertical asymptote, $x = 0$.
2. Plot the horizontal intercept, $(1, 0)$.
3. Plot the key point $(b, 1)$.
4. Draw a smooth curve through the points.
5. State the domain, $(0, \infty)$, the range, $(-\infty, \infty)$, and the vertical asymptote, $x = 0$.

EXAMPLE 3: GRAPHING A LOGARITHMIC FUNCTION WITH THE FORM $F(X) = LOG_b(X)$.

Graph $f(x) = \log_5(x)$. State the domain, range, and asymptote.

Answer

Before graphing, identify the behavior and key points for the graph.

- Since $b = 5$ is greater than one, we know the function is increasing. The left end of the graph will approach the vertical asymptote $x = 0$ from the positive side, and the right end behavior will increase slowly without bound.
- The horizontal intercept is $(1, 0)$.
- The key point $(5, 1)$ is on the graph.
- We draw and label the asymptote, plot and label the points, and draw a smooth curve through the points (see Figure 4).

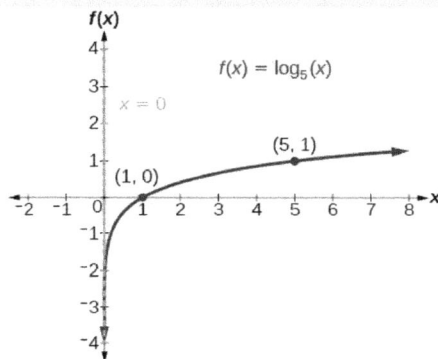

Figure 4.

The domain is $(0, \infty)$, the range is $(-\infty, \infty)$, and there is a vertical asymptote from the positive side of $x = 0$.

TRY IT #3

Graph $f(x) = \log_{\frac{1}{5}}(x)$. State the domain, range, and vertical asymptote.

Answer

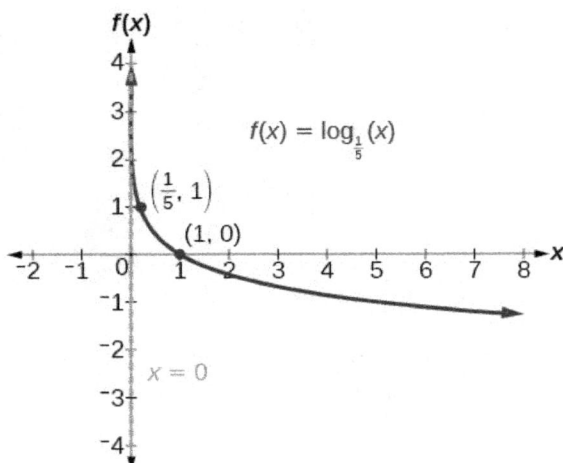

The domain is $(0, \infty)$, the range is $(-\infty, \infty)$, and there is a vertical asymptote from the positive side of $x = 0$.

For $b > 1$, the base of the logarithm effects how quickly the graph increases as x gets larger or decreases as x goes to zero. Figure 5 shows the graphs of three logarithmic functions $f\left(x\right) = \log_b\left(x\right)$ with the base $b = 2$, $b = e \approx 2.718$, and $b = 10$. Observe that the graphs compress vertically as the value of the base increases. The key points for the graphs are (2, 1), (e, 1) and (10, 1) respectively showing that base $b = 2$ reaches an output value of 1 more quickly than base e or 10.

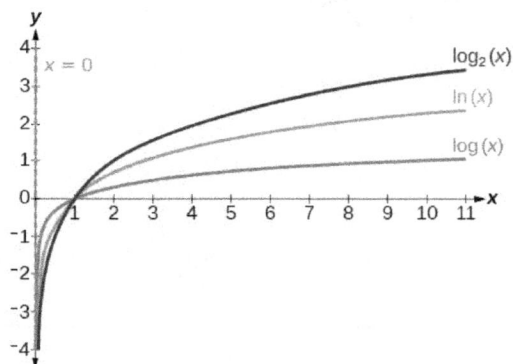

Figure 5. The graphs of three logarithmic functions with different bases, all greater than 1.

Graphing Transformations of Logarithmic Functions

As we mentioned in the beginning of the section, transformations of logarithmic graphs behave similarly to those of toolkit and exponential functions. We can shift, stretch, compress, and reflect the function $y = \log_b\left(x\right)$ without loss of shape.

Graphing a Horizontal Shift of y = log_b(x)

We will begin by looking at a horizontal shift of the function $y = \log_b\left(x\right)$. Consider the function $f\left(x\right) = \log_b\left(x - c\right)$, where c is a constant. If c is positive, then the horizontal shift is to the right and if c is negative, the horizontal shift is to the left. Note that, as we observed in earlier sections, because the general formula for the horizontal shift contains a minus sign, c will have the opposite sign of what you observe in the formula.

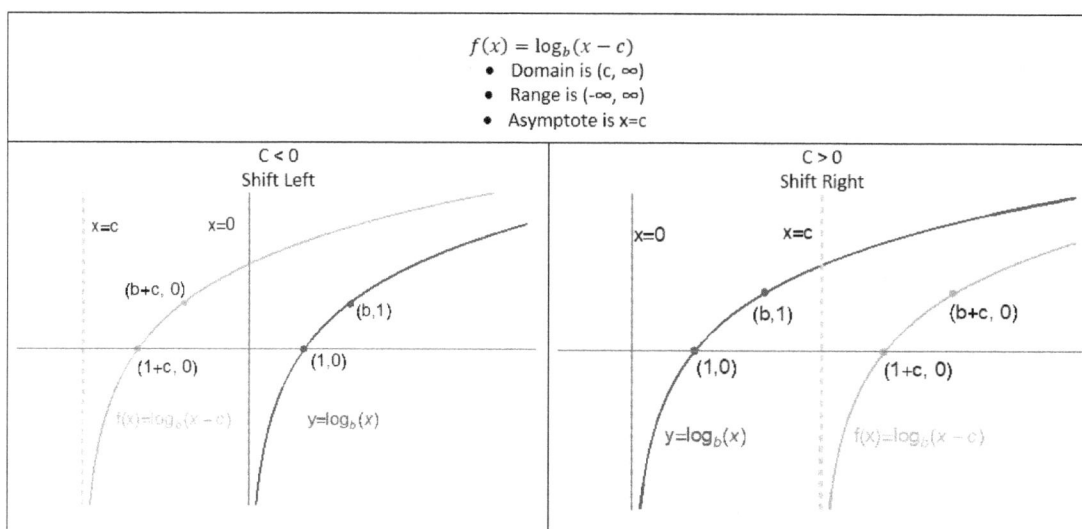

$$f(x) = \log_b(x - c)$$
- Domain is (c, ∞)
- Range is $(-\infty, \infty)$
- Asymptote is x=c

Figure 6

HORIZONTAL SHIFTS OF THE FUNCTION $Y = LOG_b(X)$

For any constant c, the function $f(x) = \log_b(x - c)$

- shifts the function $y = \log_b(x)$ right c units if $c > 0$.
- shifts the function $y = \log_b(x)$ left c units if $c < 0$.
- has the vertical asymptote $x = c$.
- has domain (c, ∞).
- has range $(-\infty, \infty)$.

HOW TO

Given a logarithmic function with the form $f(x) = \log_b(x - c)$, graph the translation.

1. Determine the value for c. It will have the opposite sign of what you see in the simplified formula.
2. Identify the horizontal shift:

   - If $c > 0$, shift the graph of $y = \log_b(x)$ right c units.
   - If $c < 0$, shift the graph of $y = \log_b(x)$ left c units.

3. Draw the vertical asymptote $x = c$.
4. Identify two or three key points from the function $y = \log_b(x)$. Find new coordinates for the shifted functions by adding c to the x coordinate.
5. Label the points.
6. The domain is (c, ∞), the range is $(-\infty, \infty)$, and the vertical asymptote is $x = c$.

Sketch the horizontal shift $f(x) = \log_3(x-2)$ alongside the function $y = \log_3(x)$. Include the key points and the vertical asymptote on the graph. State the domain, range, and vertical asymptote.

Answer

Since the function is $f(x) = \log_3(x-2)$, the input expression is $x - (2)$. Thus $c = 2$, so $c > 0$. This means we will shift the function $y = \log_3(x)$ right 2 units.

The vertical asymptote is $x = 2$.

Consider the two key points from the function $y = \log_3(x)$, $(1,0)$, and $(3,1)$. The new coordinates are found by adding 2 to the x coordinates.

Label the points $(3,0)$, and $(5,1)$.

The domain is $(2, \infty)$, the range is $(-\infty, \infty)$, and the vertical asymptote is $x = 2$.

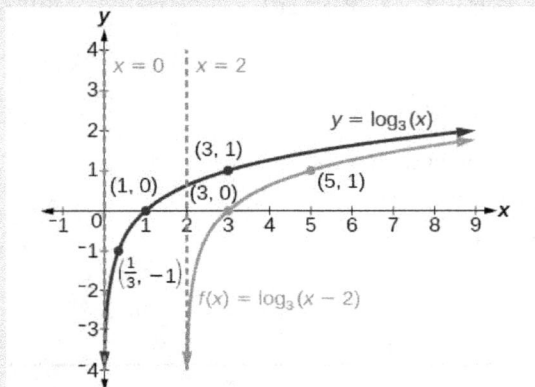

Figure 7.

TRY IT #4

Sketch a graph of $f(x) = \log_3(x+4)$ alongside the function $y = \log_3(x)$. Include the key points and the vertical asymptote on the graph. State the domain, range, and vertical asymptote.

Answer

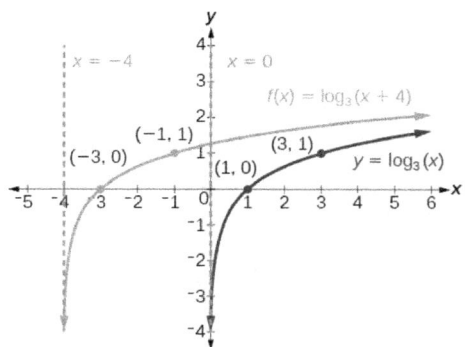

The domain is $(-4, \infty)$, the range $(-\infty, \infty)$, and the vertical asymptote $x = -4$.

Graphing a Vertical Shift of y = log_b(x)

When a constant d is added to the function $y = \log_b (x)$, the result is a vertical shift d units in the direction of the sign of d. To visualize vertical shifts, we can observe the general graph of the function $y = \log_b (x)$ alongside the shift, $f (x) = \log_b (x) + d$. See Figure 8.

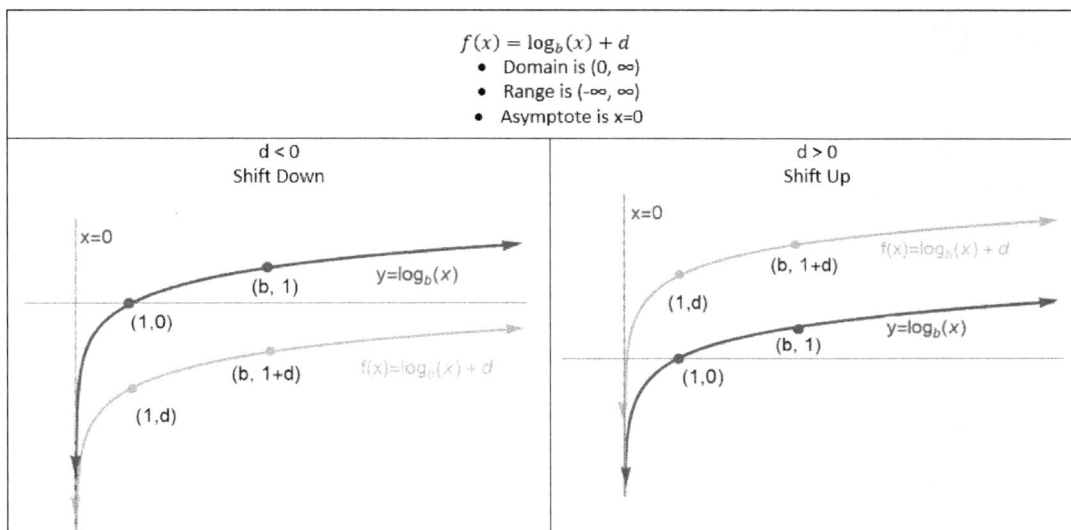

Figure 8

VERTICAL SHIFTS OF THE FUNCTION Y = LOG_b(X)

For any constant d, the function $f (x) = \log_b (x) + d$

- shifts the function $y = \log_b (x)$ up d units if $d > 0$.
- shifts the function $y = \log_b (x)$ down d units if $d < 0$.
- has the vertical asymptote $x = 0$.
- has domain $(0, \infty)$.
- has range $(-\infty, \infty)$.

HOW TO

Given a logarithmic function with the form $f(x) = \log_b(x) + d$, **graph the translation.**

1. Identify the vertical shift:

   ◦ If $d > 0$, shift the graph of $y = \log_b(x)$ up d units.
   ◦ If $d < 0$, shift the graph of $y = \log_b(x)$ down d units.

2. Draw the vertical asymptote $x = 0$.
3. Identify two or three key points from the function $y = \log_b(x)$. Find new coordinates for the shifted functions by adding d to the y coordinate.
4. Label the points.
5. The domain is $(0, \infty)$, the range is $(-\infty, \infty)$, and the vertical asymptote is $x = 0$.

EXAMPLE 5: GRAPHING A VERTICAL SHIFT OF THE FUNCTION $Y = LOG_b(X)$

Sketch a graph of $f(x) = \log_3(x) - 2$ alongside the function $y = \log_3(x)$. Include the key points and the vertical asymptote on the graph. State the domain, range, and vertical asymptote.

Answer

Since the function is $f(x) = \log_3(x) - 2$, we observe that $d = -2$. Thus $d < 0$. This means we will shift the function $y = \log_3(x)$ down 2 units.

The vertical asymptote is $x = 0$.

Consider the three key points from the function $y = \log_3(x)$, $\left(\frac{1}{3}, -1\right)$, $(1, 0)$, and $(3, 1)$. The new coordinates are found by subtracting 2 from the y coordinates. Label the points $\left(\frac{1}{3}, -3\right)$, $(1, -2)$, and $(3, -1)$.

The domain is $(0, \infty)$, the range is $(-\infty, \infty)$, and the vertical asymptote is $x = 0$.

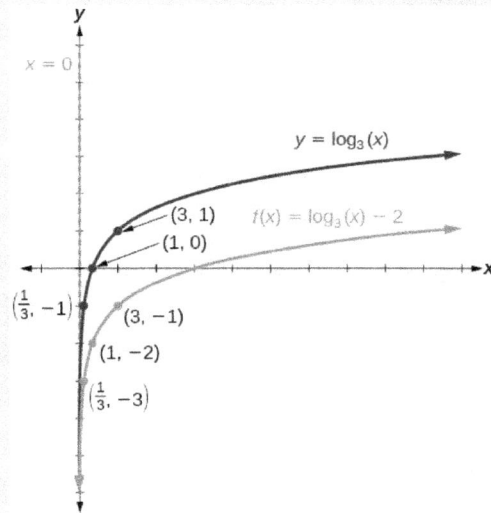

Figure 9.

TRY IT #5

Sketch a graph of $f(x) = \log_2(x) + 2$ alongside the function $y = \log_2(x)$. Include the key points and vertical asymptote on the graph. State the domain, range, and vertical asymptote.

Answer

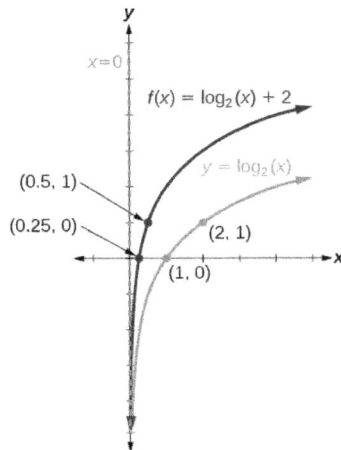

The domain is $(0, \infty)$, the range is $(-\infty, \infty)$, and the vertical asymptote is $x = 0$.

Graphing Vertical Stretches and Compressions of y = log_b(x)

When the function $y = \log_b(x)$ is multiplied by a constant $a > 0$, the result is a **vertical stretch** or **compression** of the original graph. To visualize stretches and compressions, we set $a > 0$ and observe the general graph of the function $y = \log_b(x)$ alongside the vertical stretch or compression, $f(x) = a\log_b(x)$. See Figure 10.

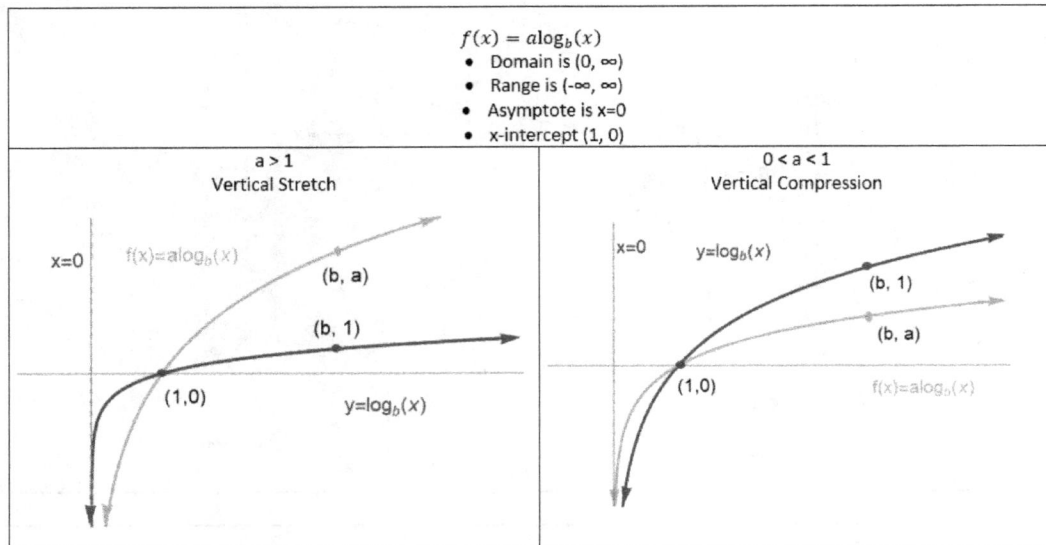

Figure 10

VERTICAL STRETCHES AND COMPRESSIONS OF THE FUNCTION Y = LOG_b(X)

For any constant $a > 1$, the function $f(x) = a\log_b(x)$

- stretches the function $y = \log_b(x)$ vertically by a factor of a if $a > 1$.
- compresses the function $y = \log_b(x)$ vertically by a factor of a if $0 < a < 1$.
- has the vertical asymptote $x = 0$.
- has the horizontal intercept $(1, 0)$.
- has domain $(0, \infty)$.
- has range $(-\infty, \infty)$.

HOW TO

Given a logarithmic function with the form $f(x) = a\log_b(x), a > 0$, **graph the translation.**

1. Identify the vertical stretch or compressions:

   ○ If $|a| > 1$, the graph of $y = \log_b(x)$ is stretched by a factor of a units.
   ○ If $|a| < 1$, the graph of $y = \log_b(x)$ is compressed by a factor of a units.

2. Draw the vertical asymptote $x = 0$.
3. Identify two or three key points from the function $y = \log_b (x)$. Find new coordinates for the stretched or compressed function by multiplying the y coordinates by a.
4. Label the points.
5. The domain is $(0, \infty)$, the range is $(-\infty, \infty)$, and the vertical asymptote is $x = 0$.

EXAMPLE 6: GRAPHING A STRETCH OR COMPRESSION OF THE FUNCTION Y = LOG_b(X)

Sketch a graph of $f(x) = 2\log_4 (x)$ alongside the function $y = \log_4 (x)$. Include the key points and the vertical asymptote on the graph. State the domain, range, and vertical asymptote.

Answer

Since the function is $f(x) = 2\log_4 (x)$, we will notice $a = 2$. This means we will stretch the function $y = \log_4 (x)$ by a factor of 2.

The vertical asymptote is $x = 0$.

Consider the two key points from the function $y = \log_4 (x)$, $(1, 0)$, and $(4, 1)$. The new coordinates are found by multiplying the y coordinates by 2. Label the points $(1, 0)$, and $(4, 2)$.

The domain is $(0, \infty)$, the range is $(-\infty, \infty)$, and the vertical asymptote is $x = 0$. See Figure 11.

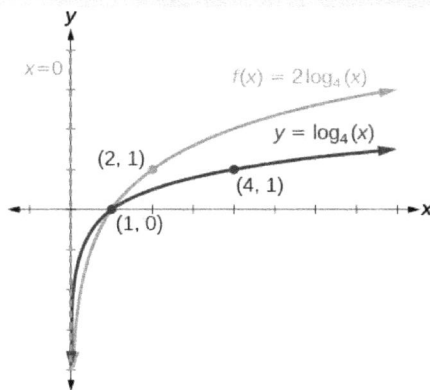

Figure 11.

TRY IT #6

Sketch a graph of $f(x) = \frac{1}{2} \log_4 (x)$ alongside the function $y = \log_4 (x)$. Include the key points and the vertical asymptote on the graph. State the domain, range, and vertical asymptote.

Answer

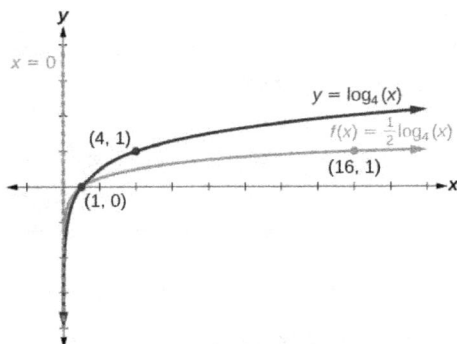

The domain is $(0, \infty)$, the range is $(-\infty, \infty)$, and the vertical asymptote is $x = 0$.

EXAMPLE 7: COMBINING A SHIFT AND A STRETCH

Sketch a graph of $f(x) = 5\log(x + 2)$. State the domain, range, and vertical asymptote.

Answer

First, we move the graph of $y = \log(x)$ left 2 units, then stretch the function vertically by a factor of 5, as in Figure 12. The vertical asymptote will be shifted to $x = -2$. The x-intercept will be $(-1, 0)$. The domain will be $(-2, \infty)$. Two points will help give the shape of the graph: $(-1, 0)$ and $(8, 5)$. We chose $x = 8$ as the x-coordinate of one point to graph because when $x = 8, x + 2 = 10$, the base of the common logarithm.

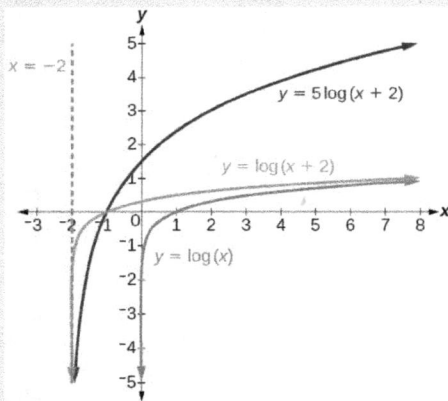

Figure 12.

The domain is $(-2, \infty)$, the range is $(-\infty, \infty)$, and the vertical asymptote is $x = -2$.

TRY IT #7

Sketch a graph of the function $f(x) = 3\log(x-2) + 1$. State the domain, range, and vertical asymptote. Identify at least 2 key points on $f(x) = 3\log(x-2) + 1$.

Answer

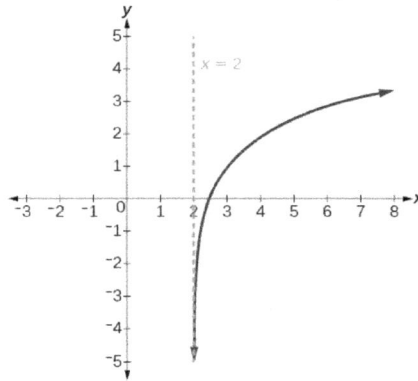

The domain is $(2, \infty)$, the range is $(-\infty, \infty)$, and the vertical asymptote is $x = 2$. Key points may include $(3, 1)$, and $(12, 4)$.

Graphing Reflections of $y = \log_b(x)$

When the function $y = \log_b(x)$ is multiplied by -1 we are negating the output so, the result is a **reflection** about the x-axis. When the *input* is multiplied by -1, the result is a reflection about the y-axis. To visualize reflections, we restrict $b > 1$, and observe the general graph of the function $y = \log_b(x)$ alongside the reflection about the x-axis, $g(x) = -\log_b(x)$ and the reflection about the y-axis, $h(x) = \log_b(-x)$.

Reflection about the x-axis $g(x) = -\log_b(x), \quad b > 1$	Reflection about the y-axis $h(x) = \log_b(-x), \quad b > 1$
• Domain $(0, \infty)$ • Range $(-\infty, \infty)$ • x-intercept $(1, 0)$ • Vertical asymptote x=0 • $(b, 1)$ reflects to $(b, -1)$	• Domain $(-\infty, 0)$ • Range $(-\infty, \infty)$ • x-intercept $(-1, 0)$ • Vertical asymptote x=0 • $(b, 1)$ reflects to $(-b, 1)$

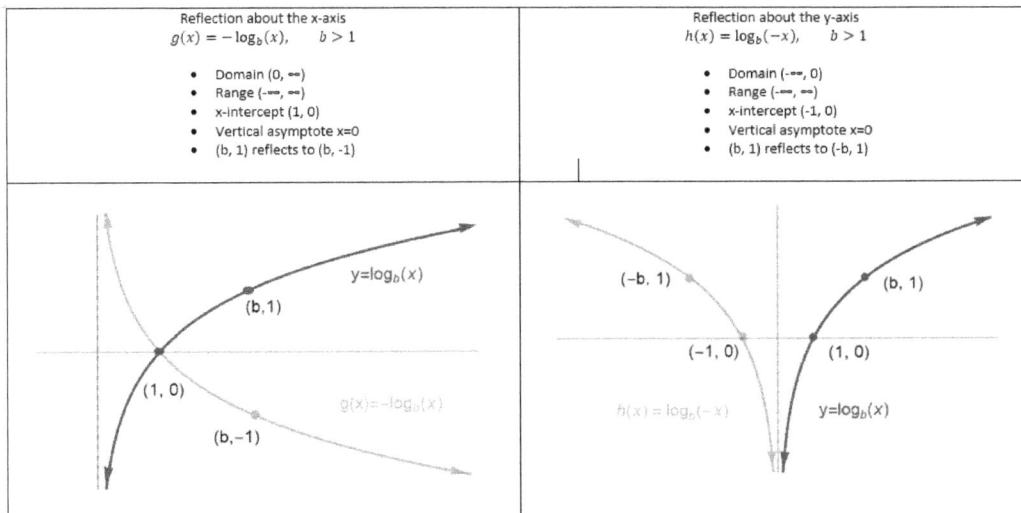

Figure 13

REFLECTIONS OF THE FUNCTION $Y = LOG_b(X)$

The function $f(x) = -\log_b(x)$

- reflects the function $y = \log_b(x)$ about the x-axis, and
- has domain, $(0, \infty)$, range, $(-\infty, \infty)$, and vertical asymptote, $x = 0$, which are unchanged from the original function.

The function $f(x) = \log_b(-x)$

- reflects the function $y = \log_b(x)$ about the y-axis,
- has domain $(-\infty, 0)$, and
- has range, $(-\infty, \infty)$, and vertical asymptote, $x = 0$, which are unchanged from the original function.

HOW TO

Given a logarithmic function with the function $y = \log_b(x)$, graph a translation.

If $f(x) = -\log_b(x)$	**If $f(x) = \log_b(-x)$**
1. Draw the vertical asymptote, $x = 0$.	1. Draw the vertical asymptote, $x = 0$.
2. Plot the x-intercept, $(1, 0)$.	2. Plot the x-intercept, $(-1, 0)$.
3. Reflect the graph of the function $f y = \log_b(x)$ about the x-axis. Key point $(b, \ 1)$ reflects to $(b, \ -1)$.	3. Reflect the graph of the function $y = \log_b(x)$ about the y-axis. Key point $(b, \ 1)$ reflects to $(-b, \ 1)$.
4. Draw a smooth curve through the points.	4. Draw a smooth curve through the points.
5. State the domain, $(0, \infty)$, the range, $(-\infty, \infty)$, and the vertical asymptote $x = 0$.	5. State the domain, $(-\infty, 0)$, the range, $(-\infty, \infty)$, and the vertical asymptote $x = 0$.

EXAMPLE 8: GRAPHING A REFLECTION OF A LOGARITHMIC FUNCTION

Sketch a graph of $f(x) = \log(-x)$ alongside the function $y = \log(-x)$. Include the key points and vertical asymptote on the graph. State the domain, range, and vertical asymptote.

Answer

Before graphing $f(x) = \log(-x)$, identify the behavior and key points for the graph.

- Since $b = 10$ is greater than one, we know that the function $y = \log(x)$ is increasing. Since the *input* value is multiplied by -1, f is a reflection of the graph of $y = \log(x)$ about the y-axis. Thus, $f(x) = \log(-x)$ will be decreasing as x moves from negative infinity to zero, and as x approaches zero from the left or through negative values, the graph will decrease without bound.
- The x-intercept is $(-1, 0)$.
- We draw and label the vertical asymptote, plot and label the points, and draw a smooth curve through the points.

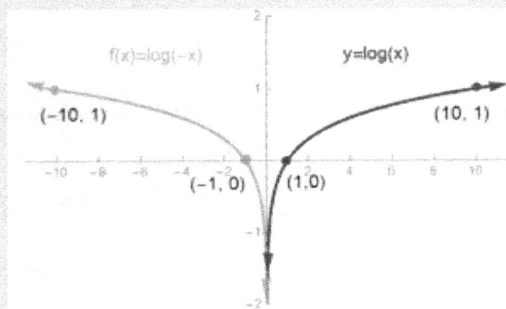

Figure 14

The domain is $(-\infty, 0)$, the range is $(-\infty, \infty)$, and the vertical asymptote is $x = 0$.

TRY IT #8

Graph $f(x) = -\log(-x)$. State the domain, range, and vertical asymptote.

Answer

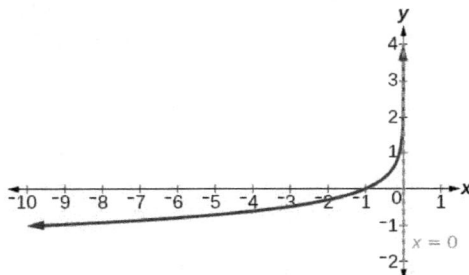

The domain is $(-\infty, 0)$, the range is $(-\infty, \infty)$, and the vertical asymptote is $x = 0$. Since there is both a horizontal and vertical reflection, both the x and y coordinates will be negated. Key points $(1, \ 0)$ and $(10, \ 1)$ on $f(x) = \log(x)$ will transform to $(-1, \ 0)$ and $(-10, \ -1)$ for the translated function.

Summarizing Translations of the Logarithmic Function

Now that we have worked with each type of translation for the logarithmic function, we can summarize each in Table 5 to arrive at the general equation for translating exponential functions.

Table 5

Translations of the Function $y = \log_b(x)$	
Translation	**Form**
Shift • Horizontally C units to the left or right	$f(x) = \log_b(x - c) + d$

Translations of the Function $y = \log_b(x)$	
Translation	**Form**
• Vertically d units up or down	
Stretch and Compress • Vertical stretch if $\|a\| > 1$ • Vertical compression if $\|a\| < 1$	$f(x) = a\log_b(x)$
Reflect about the x-axis	$f(x) = -\log_b(x)$
Reflect about the y-axis	$f(x) = \log_b(-x)$
General equation for all translations	$f(x) = a\log_b(x - c) + d$

EXAMPLE 9: FINDING THE VERTICAL ASYMPTOTE OF A LOGARITHM GRAPH

What is the vertical asymptote of $f(x) = -2\log_3(x + 4) + 5$?

Answer

The vertical asymptote is $x = -4$.

Analysis

The coefficient, the base, and the upward translation do not affect the vertical asymptote. The shift of the curve 4 units to the left shifts the vertical asymptote to $x = -4$.

TRY IT #9

What is the vertical asymptote of $f(x) = 3 + \ln(x - 1)$?

Answer

$x = 1$

EXAMPLE 10: FINDING THE EQUATION FROM A GRAPH

Find a possible equation for the common logarithmic function graphed in Figure 15.

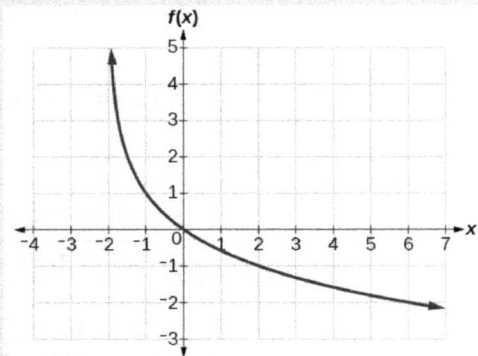

Figure 15.

Answer

This graph has a vertical asymptote at $x = -2$ and has been vertically reflected. We do not know yet the vertical shift or the vertical stretch. We know so far that the equation will have form:

$$f(x) = -a\log(x+2) + d.$$

It appears the graph passes through the points $(-1, 1)$ and $(2, -1)$. Substituting $(-1, 1)$,

$$
\begin{aligned}
1 &= -a\log(-1+2) + d && \text{Substitute } (-1, 1). \\
1 &= -a\log(1) + d && \text{Arithmetic.} \\
1 &= d && \log(1) = 0.
\end{aligned}
$$

Therefore, we now have $f(x) = -a\log(x+2) + 1$. Next, substituting into this equation the point $(2, -1)$ we have,

$$
\begin{aligned}
-1 &= -a\log(2+2) + 1 && \text{Plug in } (2, -1). \\
-2 &= -a\log(4) && \text{Arithmetic.} \\
a &= \frac{2}{\log(4)} && \text{Solve for } a.
\end{aligned}
$$

This gives us the equation $f(x) = -\frac{2}{\log(4)}\log(x+2) + 1$.

Analysis

We can verify this answer by comparing the function values in Table 6 with the points on the graph in Figure 15.

Table 6

x	−1	0	1	2	3
$f(x)$	1	0	−0.58496	−1	−1.3219

x	4	5	6	7	8
$f(x)$	-1.5850	-1.8074	-2	-2.1699	-2.3219

TRY IT #10

Give the equation of the natural logarithm graphed in Figure 16.

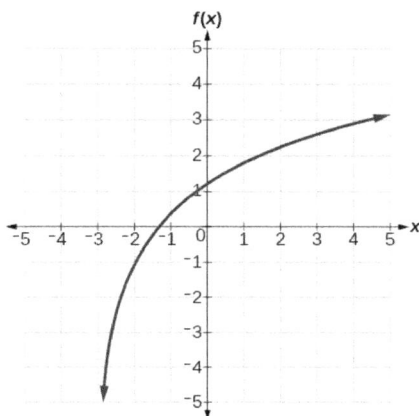

Figure 16.

Answer

$$f(x) = 2\ln(x+3) - 1$$

Q&A

Is it possible to tell the domain and range and describe the end behavior of a function just by looking at the graph?

Yes, if we know the function is a general logarithmic function. For example, look at the graph in Figure 16. The graph approaches $x = -3$ (or thereabouts) more and more closely, so $x = -3$ is, or is very close to, the vertical asymptote. It approaches from the right, so the domain is all points to the right, $\{x \mid x > -3\}$. The range, as with all general logarithmic functions, is all real numbers. And we can see the end behavior because the graph goes down as it goes left and up as it goes right. The end behavior is that as $x \to -3^+$, $f(x) \to -\infty$ and as $x \to \infty$, $f(x) \to \infty$.

Access these online resources for additional instruction and practice with graphing logarithms.

- Graph an Exponential Function and Logarithmic Function
- Match Graphs with Exponential and Logarithmic Functions

- Find the Domain of Logarithmic Functions

Key Equations

General Form for the Translation of the Logarithmic Function $y = \log_b(x)$	$f(x) = a\log_b(x - c) + d$

KEY CONCEPTS

- To find the domain of a logarithmic function, set up an inequality showing the input expression greater than zero, and solve for x.
- The graph of the function $f(x) = \log_b(x)$ has an x-intercept at $(1, 0)$, domain $(0, \infty)$, range $(-\infty, \infty)$, vertical asymptote $x = 0$, and
  - if $b > 1$, the function is increasing.
  - if $0 < b < 1$, the function is decreasing.

- The equation $f(x) = \log_b(x - c)$ shifts the function $y = \log_b(x)$ horizontally
  - right c units if $c > 0$.
  - left c units if $c < 0$.

- The equation $f(x) = \log_b(x) + d$ shifts the function $y = \log_b(x)$ vertically
  - up d units if $d > 0$.
  - down d units if $d < 0$.
- The equation $f(x) = a\log_b(x)$
  - stretches the function $y = \log_b(x)$ vertically by a factor of a if $|a| > 1$.
  - compresses the function $y = \log_b(x)$ vertically by a factor of a if $|a| < 1$.
- The equation $f(x) = -\log_b(x)$ represents a reflection of the function $y = \log_b(x)$ about the x-axis.
- The equation $f(x) = \log_b(-x)$ represents a reflection of the function $y = \log_b(x)$ about the y-axis.
- All translations of the logarithmic function can be summarized by the general equation $f(x) = a\log_b(x - c) + d.$
- Given an equation with the general form $f(x) = a\log_b(x - c) + d$, we can identify the vertical asymptote $x = c$ for the transformation.
- Using the general equation $f(x) = a\log_b(x - c) + d$, we can write the equation of a logarithmic function given its graph.

2.7 Exponential and Logarithmic Models

Figure 1. A nuclear research reactor inside the Neely Nuclear Research Center on the Georgia Institute of Technology campus (credit: Georgia Tech Research Institute)

We have already explored some basic applications of exponential and logarithmic functions. In this section, we explore some important applications in more depth.

Modeling Exponential Growth and Decay

In real-world applications, we need to model the behavior of a function. In mathematical modeling, we choose a familiar general function with properties that suggest that it will model the real-world phenomenon we wish to analyze. In the case of rapid growth, we may choose the exponential growth function:

$$y = A_0 e^{kt}$$

where A_0 is equal to the value at time zero, e is the natural base (Euler's constant), and k is a positive constant that determines the rate (percentage) of continuous growth. We also may use

$$y = A_0 b^t$$

where A_0 is equal to the value at time zero, and b is the growth factor which is greater than 1. We may use the **exponential growth** function in applications involving **doubling time**, the time it takes for a quantity to double. Such phenomena as wildlife populations, financial investments, biological samples, and natural resources may exhibit growth based on a doubling time.

On the other hand, if a quantity is falling rapidly toward zero, without ever reaching zero, then we should probably choose the **exponential decay** model. Again, we can use the form $y = A_0 e^{kt}$ where A_0 is the starting value, e is Euler's constant, and k is the (negative) continuous decay rate or we can use the form $y = A_0 b^t$ where A_0 is the starting value, and b is the decay factor between zero and one. We may use the exponential decay model when we are calculating **half-life**, or the time it takes for a substance to exponentially decay to half of its original quantity. We use half-life in applications involving radioactive isotopes.

In our choice of a function to serve as a mathematical model, we often use data points gathered by careful observation and measurement to construct points on a graph and hope we can recognize the shape of the graph. Exponential growth and decay graphs have a distinctive shape, as we can see in the graphs below. It is important to remember that, although parts of each of the two graphs seem to lie on the x-axis, they are really a tiny distance above the x-axis.

CHARACTERISTICS OF THE EXPONENTIAL FUNCTION, $y = A_0 e^{kt}$ AND $y = A_0 b^t$

An exponential function with the form $y = A_0 e^{kt}$ or $y = A_0 b^t$ has the following characteristics:

- one-to-one function
- horizontal asymptote: $y = 0$
- domain: $(-\infty, \infty)$
- range: $(0, \infty)$
- x intercept: none
- y-intercept: $(0, A_0)$
- increasing if $k > 0$ or $b > 1$ and decreasing if $k < 0$ or $0 < b < 1$

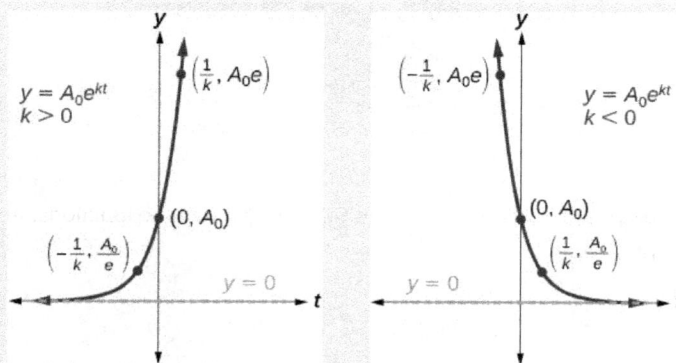

Figure 2. An exponential function models exponential growth when $k > 0$ and exponential decay when $k < 0$.

Expressing an Exponential Model in Base e

While powers and logarithms of any base can be used in modeling, the two most common bases are 10 and e. In science and mathematics, the base e is often preferred. We can use laws of exponents and laws of logarithms to change any base to base e.

HOW TO

Given a model with the form $y = ab^x$, change it to the form $y = A_0 e^{kx}$.

1. Since $b^x = e^{\ln(b^x)}$, rewrite $y = ab^x$ as $y = ae^{\ln(b^x)}$.
2. Use the power rule of logarithms to rewrite y as $y = ae^{x\ln(b)} = ae^{\ln(b)x}$.
3. Note that $a = A_0$ and $k = \ln(b)$ in the equation $y = A_0 e^{kx}$.

EXAMPLE 1: CHANGING TO THE NATURAL BASE

Change the function $y = 2.5(3.1)^x$ so that this same function is written in the form $y = A_0 e^{kx}$.

Answer

The formula is derived as follows

$$y = 2.5(3.1)^x$$
$$= 2.5e^{\ln(3.1^x)} \qquad \text{Insert exponential and its inverse.}$$
$$= 2.5e^{x\ln(3.1)} \qquad \text{Laws of logs.}$$
$$= 2.5e^{(\ln(3.1))x} \qquad \text{Commutative law of multiplication.}$$

In the above equation, we see that $A_0 = 2.5$, and $k = \ln(3.1) \approx 1.1314$.

TRY IT #1

Change the function $y = 3(0.5)^x$ to one having e as the base.

Answer

$$y = 3e^{(\ln(0.5))x} \approx 3e^{-.6931x}$$

Calculating Doubling Time

For growing quantities, we might want to find out how long it takes for a quantity to double. The time it takes for a quantity to double is called the **doubling time**. Let A_0 represent the initial quantity at time zero. We then want to find when the quantity is doubled or equal to $2A_0$. Using the exponential growth equation $f(t) = A_0 b^t$ or $f(t) = A_0 e^{kt}$, we solve the equation $2A_0 = A_0 b^t$ or the equation $2A_0 = A_0 e^{kt}$ for t to find the doubling time t, if the growth rate or growth factor is known. Similarly, if we know the doubling time but not the growth rate or factor, we can use $2A_0 = A_0 e^{kt}$ to solve for the continuous growth rate k or we can use $2A_0 = A_0 b^t$ to solve for the growth factor b.

EXAMPLE 2: GRAPHING EXPONENTIAL GROWTH

A population of bacteria doubles every hour. If the culture started with 10 bacteria, graph the population as a function of time.

Answer

When an amount grows at a fixed percent per unit time, the growth is exponential. To find A_0 we use the fact that A_0 is the amount at time zero, so $A_0 = 10$, and we can use the model $A = 10e^{kt}$. To find k, use the fact that after one hour $(t = 1)$ the population doubles from 10 to 20, and plug in the point $(1, \ 20)$ to the formula. The formula is derived as follows:

$$20 = 10e^{k \cdot 1}$$

$$2 = e^k \qquad \qquad \text{Divide by 10.}$$

$$\ln(2) = k \qquad \qquad \text{Take the natural logarithm of both sides.}$$

Therefore, $k = \ln(2) \approx 0.6931$. Thus the equation we want to graph is $y = 10e^{(\ln 2)t}$ or $y \approx 10e^{0.6931t}$. We can also write this as, $y = 10\left(e^{\ln 2}\right)^t = 10(2)^t$.

The graph is shown in Figure 3.

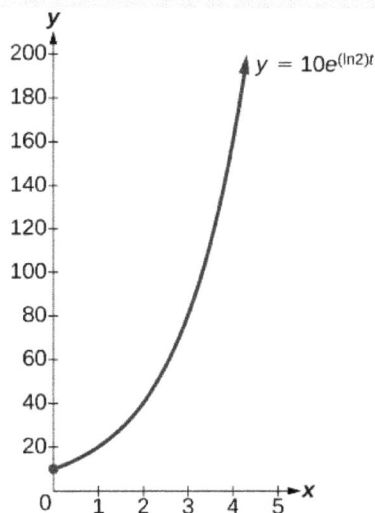

Figure 3. The graph of $y = 10e^{(\ln 2)t}$

Analysis

The population of bacteria after 10 hours is 10,240. We could describe this amount as being of the order of magnitude 10^4. The population of bacteria after twenty hours is 10,485,760 which is of the order of magnitude 10^7, so we could say that the population has increased by three order of magnitude in 10 hours.

EXAMPLE 3: FINDING A FUNCTION THAT DESCRIBES EXPONENTIAL GROWTH

According to Moore's Law, the doubling time for the number of transistors that can be put on a computer chip is approximately two years. Give a function that describes this behavior.

Answer

Let A_0 be the quantity of transistors that can currently be put on the computer chip. We will then work with the model $y = A_0 b^t$. We know that in two years, the number of transistors on the computer chip will double to be $2A_0$. We have the second point $(2, 2A_0)$ that can be plugged into our model to get:

$$2A_0 = A_0 b^2$$

$$2 = b^2 \qquad \text{Divide both sides by } A_0.$$

$$b = \sqrt{2} \approx 1.414 \qquad \text{Take the square root of both sides}$$

Substitute b into the formula to get the function $y = A_0 1.414^t$.

Half-Life

We now turn to **exponential decay**. One of the common terms associated with exponential decay, as stated above, is **half-life**, the length of time it takes an exponentially decaying quantity to decrease to half its original amount. Every radioactive isotope has a half-life, and the process describing the exponential decay of an isotope is called radioactive decay.

To find the half-life of a function describing exponential decay, solve one of the following equations:

$$\tfrac{1}{2}A_0 = A_o e^{kt} \text{ or } \tfrac{1}{2}A_0 = A_o b^t$$

The half-life depends only on the constant k or b and not on the starting quantity A_0.

HOW TO

Given the half-life, find the continuous decay rate.

1. Write $A = A_o e^{kt}$.
2. Replace A by $\tfrac{1}{2}A_0$ and replace t by the given half-life.
3. Solve to find k. Express k as an exact value (do not round).

Recall from Section 2.5, you can change the continuous rate to a noncontinuous rate if that is the desired rate.

EXAMPLE 4: FINDING THE FUNCTION THAT DESCRIBES RADIOACTIVE DECAY

The half-life of carbon-14 is 5,730 years. Express the amount of carbon-14 remaining as a function of time, t.

Answer

Let A_0 be the initial amount of carbon-14. We can work with the continuous or noncontinuous growth rate but for this example choose to work with $A = A_0 e^{kt}$. When $t = 5730$, there will be $\tfrac{1}{2}A_0$ of the carbon-14 remaining. Substitute $\left(5730, \tfrac{1}{2}A_0\right)$ into the equation and solve as follows.

$$A = A_0 e^{kt} \qquad \text{The continuous growth formula.}$$

$$0.5A_0 = A_0 e^{k \cdot 5730} \qquad \text{Substitute the half-life for } t \text{ and } 0.5A_0 \text{ for } A.$$

$$0.5 = e^{5730k} \qquad \text{Divide by } A_0.$$

$$\ln(0.5) = 5730k \qquad \text{Take the natural logarithm of both sides.}$$

$$k = \frac{\ln(0.5)}{5730} \qquad \text{Divide by the coefficient of } k.$$

$$A = A_0 e^{\left(\frac{\ln(0.5)}{5730}\right)t} \qquad \text{Substitute for } k \text{ in the continuous growth formula.}$$

The function that describes this continuous decay is $f(t) = A_0 e^{\left(\frac{\ln(0.5)}{5730}\right)t}$. We observe that the coefficient of t, $\frac{\ln(0.5)}{5730} \approx -1.209710^{-4}$ is negative, as expected in the case of exponential decay.

Evaluating $e^{\frac{\ln(0.5)}{5730}}$, we have a growth factor of $b \approx 0.999879$ and an equivalent equation of $f(t) \approx A_0 0.999879^t$.

TRY IT #3

The half-life of plutonium-244 is 80,000,000 years. Find function gives the amount of plutonium-244 remaining as a function of time, measured in years.

Answer

$$f(t) = A_0 e^{-0.0000000087t} \quad \text{or} \quad f(t) = A_0 0.9999999913^t$$

Table 1 lists the half-life for several of the more common radioactive substances.

Table 1

Substance	Use	Half-life
gallium-67	nuclear medicine	80 hours
cobalt-60	manufacturing	5.3 years
technetium-99m	nuclear medicine	6 hours
americium-241	construction	432 years
carbon-14	archeological dating	5,715 years
uranium-235	atomic power	703,800,000 years

We can see how widely the half-lives for these substances vary. Knowing the half-life of a substance allows us to calculate the amount remaining after a specified time.

EXAMPLE 5: USING THE FORMULA FOR RADIOACTIVE DECAY TO FIND THE QUANTITY OF A SUBSTANCE

How long will it take for ten percent of a 1000-gram sample of uranium-235 to decay?

Answer

We start with 1000 grams so we let $A_0 = 1000$. We will use the formula $f(t) = A_0 b^t$ so $f(t) = 1000 b^t$. The half life of uranium-235 is 703,800,000. Therefore, we will plug in the point (703800000, 500) to the equation to find the growth factor b.

The initial equation is $500 = 1000 b^{703800000}$ and we divide both sides by 1000 to get $0.5 = b^{703800000}$. Take the 703,800,000th root of both sides and we have $b \approx 0.999999999015036$ and the equation

$$f(t) = 1000 \left(0.999999999015036\right)^t.$$

Now that we have found the model to use, we can turn our attention to the question asked. Ten percent of 1000 grams is 100 grams. If 100 grams decay, the amount of uranium-235 remaining is 900 grams.

Solve $900 = 1000 \left(0.999999999015036\right)^t$ for t. We begin by dividing both sides by 1000 to get

$$0.9 = 0.999999999015036^t.$$

Next take the natural logarithm of both sides to get

$$\ln(0.9) = \ln\left(0.999999999015036^t\right).$$

Using the power property of logarithms the t comes out front to give

$$\ln(0.9) = t\ln(0.999999999015036).$$

Finally, dividing both sides by $\ln(0.999999999015036)$ gives

$$t = \frac{\ln(0.9)}{\ln(0.999999999015036)} \approx 106{,}968{,}248.$$

There will be 900 grams of uranium-235 after approximately 107 million years have passed.

TRY IT #4

How long will it take before twenty percent of our 1000-gram sample of uranium-235 has decayed?

Answer

$$t = 703{,}800{,}000\frac{\ln(0.8)}{\ln(0.5)} \text{ years} \approx 226{,}572{,}993 \text{ years}.$$

Choosing an Appropriate Model for Data

Now that we have discussed various mathematical models, we need to learn how to choose the appropriate model for the raw data we have. Many factors influence the choice of a mathematical model, among which are experience, scientific laws, and patterns in the data itself. Not all data can be described by elementary functions. Sometimes, a function is chosen that approximates the data over a given interval. For instance, suppose data were gathered on the number of homes bought in the

United States from the years 1960 to 2013. After plotting these data in a scatter plot, we notice that the shape of the data from the years 2000 to 2013 follow a logarithmic curve. We could restrict the interval from 2000 to 2010, apply regression analysis using a logarithmic model, and use it to predict the number of home buyers for the year 2015.

Three kinds of functions that are often useful in mathematical models are linear functions, exponential functions, and logarithmic functions. If the data lies on a straight line, or seems to lie approximately along a straight line, a linear model may be best. If the data is non-linear, we often consider an exponential or logarithmic model, though other models, such as quadratic models, may also be considered.

In choosing between an exponential model and a logarithmic model, we look at the way the data curves. This is called the concavity. If we draw a line between two data points, and all (or most) of the data between those two points lies above that line, we say the curve is concave down. We can think of it as a bowl that bends downward and therefore cannot hold water. If all (or most) of the data between those two points lies below the line, we say the curve is concave up. In this case, we can think of a bowl that bends upward and can therefore hold water. An exponential curve, whether rising or falling, whether representing growth or decay, is always concave up away from its horizontal asymptote. A logarithmic curve is always concave away from its vertical asymptote. In the case of positive data, which is the most common case, an exponential curve is always concave up, and a logarithmic curve always concave down.

After using the graph to help us choose a type of function to use as a model, we substitute points, and solve to find the parameters. We reduce round-off error by choosing points as far apart as possible.

EXAMPLE 6: CHOOSING A MATHEMATICAL MODEL

Does a linear, exponential, logarithmic, or quadratic model best fit the values listed in Table 2? Find the model, and use a graph to check your choice.

Table 2

x	1	2	3	4	5	6	7	8	9
y	0	1.386	2.197	2.773	3.219	3.584	3.892	4.159	4.394

Answer

First, plot the data on a graph as in Figure 4. For the purpose of graphing, round the data to two significant digits.

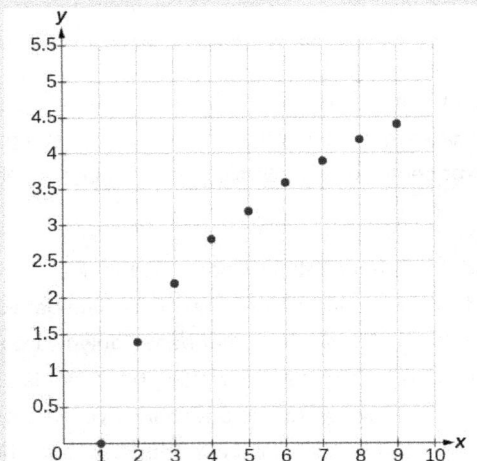

Figure 4.

Clearly, the points do not lie on a straight line, so we reject a linear model. If we draw a line between any two of the points, most or all of the points between those two points lie above the line, so the graph is concave down, suggesting a logarithmic model. We can try $y = a\ln(bx)$. Plugging in the first point, $(1,0)$, gives $0 = a\ln b$. We reject the case that $a = 0$ (if it were, all outputs would be 0), so we know $\ln(b) = 0$. Thus, $b = 1$ and $y = a\ln(x)$. Next, we can use the point $(9, 4.394)$ to solve for a:

$$y = a\ln(x)$$
$$4.394 = a\ln(9)$$
$$a = \frac{4.394}{\ln(9)}$$

Because $a = \frac{4.394}{\ln(9)} \approx 2$, an appropriate model for the data is $y = 2\ln(x)$.

To check the accuracy of the model, we graph the function together with the given points as in Figure 5.

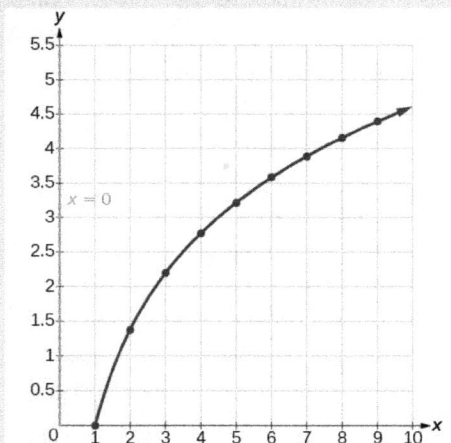

Figure 5. The graph of $y = 2\ln(x)$.

We can conclude that the model is a good fit to the data.

Compare Figure 5 to the graph of $y = \ln\left(x^2\right)$ shown in Figure 6.

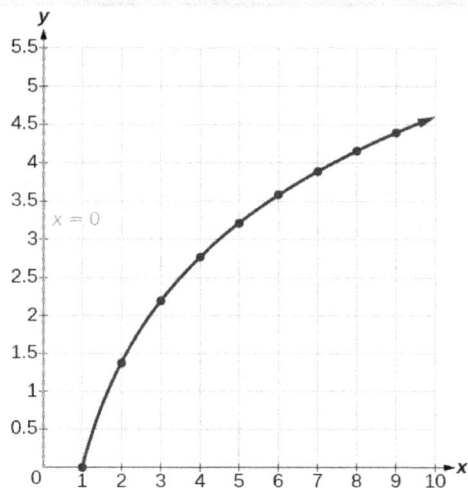

Figure 6. The graph of $y = \ln\left(x^2\right)$.

The graphs appear to be identical when $x > 0.$ A quick check confirms this conclusion: $y = \ln\left(x^2\right) = 2\ln\left(x\right)$ for $x > 0.$

However, if $x < 0$, the graph of $y = \ln\left(x^2\right)$ includes a "extra" branch, as shown in Figure 9. This occurs because, while $y = 2\ln\left(x\right)$ cannot have negative values in the domain (as such values would force the argument to be negative), the function $y = \ln\left(x^2\right)$ can have negative domain values.

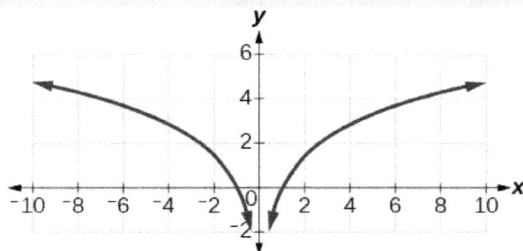

Figure 7.

TRY IT #5

Does a linear, exponential, or logarithmic model best fit the data in Table 3? Find the model.

Table 3

x	1	2	3	4	5	6	7	8	9
y	3.297	5.437	8.963	14.778	24.365	40.172	66.231	109.196	180.034

Using Newton's Law of Cooling (Optional)

Exponential decay can also be applied to temperature. When a hot object is left in surrounding air that is at a lower temperature, the object's temperature will decrease exponentially, leveling off as it approaches the surrounding air temperature. On a graph of the temperature function, the leveling off will correspond to a horizontal asymptote at the temperature of the surrounding air. Unless the room temperature is zero, this will correspond to a **vertical shift** of the generic **exponential decay** function. This translation leads to **Newton's Law of Cooling**, the scientific formula for temperature as a function of time as an object's temperature is equalized with the ambient temperature

$$T(t) = ae^{kt} + T_s.$$

NEWTON'S LAW OF COOLING

The temperature of an object, T, in surrounding air with temperature T_s will behave according to the formula

$$T(t) = Ae^{kt} + T_s$$

where

- t is time,
- A is the difference between the initial temperature of the object and the surroundings, and
- k is a constant, the continuous rate of cooling of the object.

HOW TO

Given a set of conditions, apply Newton's Law of Cooling.

1. Set T_s equal to the y-coordinate of the horizontal asymptote (usually the ambient temperature).
2. Substitute the given values into the continuous growth formula $T(t) = Ae^{kt} + T_s$ to find the parameters A and k.
3. Substitute in the desired time to find the temperature or the desired temperature to find the time.

EXAMPLE 7: USING NEWTON'S LAW OF COOLING

A cheesecake is taken out of the oven with an ideal internal temperature of $165F$, and is placed into a $35F$ refrigerator. After 10 minutes, the cheesecake has cooled to $150F$. If we must wait until the cheesecake has cooled to $70F$ before we eat it, how long will we have to wait?

Answer

Because the surrounding air temperature in the refrigerator is 35 degrees, the cheesecake's temperature will decay exponentially toward 35, we have the following equation

$$T(t) = Ae^{kt} + 35.$$

We know the initial temperature was 165, so $T(0) = 165$.

$$165 = Ae^{k0} + 35 \qquad \text{Substitute } (0, 165).$$
$$A = 130 \qquad \text{Solve for } A.$$

We were given another data point, $T(10) = 150$, which we can use to solve for k.

$$150 = 130e^{k10} + 35 \qquad \text{Substitute } (10, 150).$$
$$115 = 130e^{k10} \qquad \text{Subtract } 35.$$
$$\frac{115}{130} = e^{10k} \qquad \text{Divide by } 130.$$
$$\ln\left(\frac{115}{130}\right) = 10k \qquad \text{Take the natural log of both sides.}$$
$$k = \frac{\ln\left(\frac{115}{130}\right)}{10} \approx -0.0123 \qquad \text{Divide by the coefficient of } k.$$

This gives us the equation for the cooling of the cheesecake: $T(t) = 130e^{-0.0123t} + 35.$

Now we can solve for the time it will take for the temperature to cool to 70 degrees.

$$70 = 130e^{-0.0123t} + 35 \qquad \text{Substitute in 70 for } T(t).$$
$$35 = 130e^{-0.0123t} \qquad \text{Subtract } 35.$$
$$\frac{35}{130} = e^{-0.0123t} \qquad \text{Divide by } 130.$$
$$\ln\left(\frac{35}{130}\right) = -0.0123t \qquad \text{Take the natural logarithm.}$$
$$t = \frac{\ln\left(\frac{35}{130}\right)}{-0.0123} \approx 106.68 \qquad \text{Divide by the coefficient of } t.$$

It will take about 107 minutes, or one hour and 47 minutes, for the cheesecake to cool to $70F$.

TRY IT #6

A pitcher of water at 40 degrees Fahrenheit is placed into a 70 degree room. One hour later, the temperature has risen to 45 degrees. How long will it take for the temperature to rise to 60 degrees?

Answer

6.026 hours

Using Logistic Growth Models (Optional)

Exponential growth cannot continue forever. Exponential models, while they may be useful in the short term, tend to fall apart the longer they continue. Consider an aspiring writer who writes a single line on day one and plans to double the number of lines she writes each day for a month. By the end of the month, she must write over 17 billion lines, or one-half-billion pages. It is impractical, if not impossible, for anyone to write that much in such a short period of time. Eventually, an exponential model must begin to approach some limiting value, and then the growth is forced to slow. For this reason, it is often better to use a model with an upper bound instead of an **exponential growth** model, though the exponential growth model is still useful over a short term, before approaching the limiting value.

The **logistic growth model** is approximately exponential at first, but it has a reduced rate of growth as the output approaches the model's upper bound, called the **carrying capacity**. For constants a, b, and c, the logistic growth of a population over time x is represented by the model

$$f\left(x\right) = \frac{c}{1+ae^{-bx}}.$$

The graph in Figure 8 shows how the growth rate changes over time. The graph increases from left to right, but the growth rate only increases until it reaches its point of maximum growth rate, at which point the rate of increase decreases.

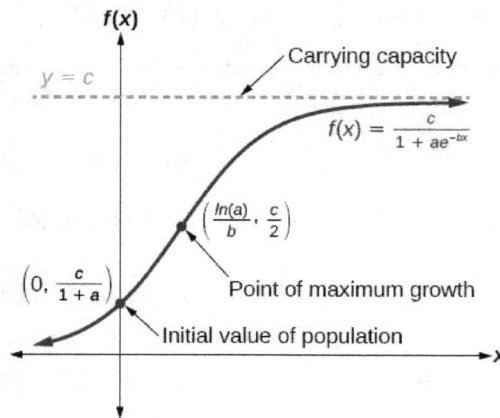

Figure 8.

LOGISTIC GROWTH

The logistic growth model is

$$f(x) = \frac{c}{1+ae^{-bx}}$$

where

- $\frac{c}{1+a}$ is the initial value
- c is the *carrying capacity*, or *limiting value,* and
- b is a constant determined by the rate of growth.

EXAMPLE 8: USING THE LOGISTIC-GROWTH MODEL

An influenza epidemic spreads through a population rapidly, at a rate that depends on two factors: The more people who have the flu, the more rapidly it spreads, and also the more uninfected people there are, the more rapidly it spreads. These two factors make the logistic model a good one to study the spread of communicable diseases. And, clearly, there is a maximum value for the number of people infected: the entire population.

For example, at time $t = 0$ there is one person in a community of 1,000 people who has the flu. So, in that community, at most 1,000 people can have the flu. Researchers find that for this particular strain of the flu, the logistic growth constant is $b = 0.6030.$ Estimate the number of people in this community who will have had this flu after ten days. Predict how many people in this community will have had this flu after a long period of time has passed.

Answer

We substitute the given data into the logistic growth model

$$f(x) = \frac{c}{1+ae^{-bx}}.$$

Because at most 1,000 people, the entire population of the community, can get the flu, we know the limiting value is $c = 1000.$ To find a, we use the formula that the number of cases at time $t = 0$ is $\frac{c}{1+a} = 1,$ from which it follows that $a = 999.$ This model predicts that, after ten days, the number of people who have had the flu is $f(10) = \frac{1000}{1+999e^{-0.6030\times10}} \approx 293.8.$ Because the actual number must be a whole number (a person has either had the flu or not) we round to 294. In the long term, the number of people who will contract the flu is the limiting value, $c = 1000.$

Analysis

Remember that, because we are dealing with a virus, we cannot predict with certainty the number of people infected. The model only approximates the number of people infected and will not give us exact or actual values.

The graph in Figure 9 gives a good picture of how this model fits the data.

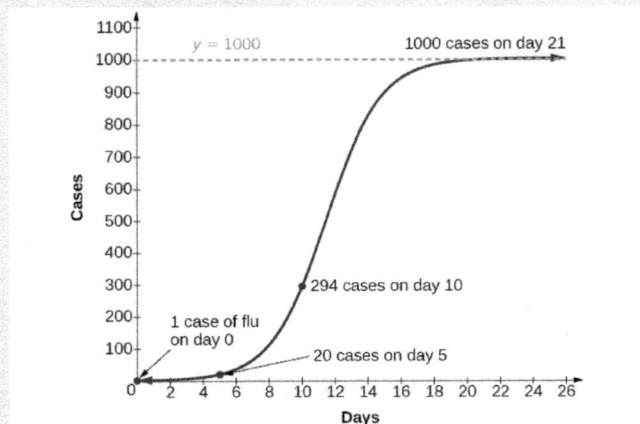

Figure 9. The graph of $f(x) = \dfrac{1000}{1 + 999e^{-0.6030x}}$

TRY IT #7

Using the model in Example 7, estimate the number of cases of flu on day 15.

Answer

895 cases on day 15

MEDIA:

Access these online resources for additional instruction and practice with exponential and logarithmic models.

- Logarithm Application – pH
- Exponential Model – Age Using Half-Life
- Newton's Law of Cooling
- Exponential Growth Given Doubling Time
- Exponential Growth – Find Initial Amount Given Doubling Time

---

KEY CONCEPTS

- The basic exponential function is $f(x) = ab^x$. If $b > 1$, we have exponential growth; if $0 < b < 1$, we have exponential decay.

- We can also write this formula in terms of continuous growth as $A = A_0 e^{kx}$, where A_0 is the starting value. If A_0 is positive, then we have exponential growth when $k > 0$ and exponential decay when $k < 0.$

- In general, we solve problems involving exponential growth or decay in two steps. First, we set up a model and use the model to find the parameters. Then we use the formula with these parameters to predict growth and decay.

- Given a substance's doubling time or half-time, we can find a function that represents its exponential growth or decay.

- We can use real-world data gathered over time to observe trends. Knowledge of linear, exponential, logarithmic, and logistic (Optional) graphs help us to develop models that best fit our data.

- Any exponential function with the form $y = ab^x$ can be rewritten as an equivalent exponential function with the form $y = A_0 e^{kx}$ where $k = \ln(b) .$

- (Optional) We can use Newton's Law of Cooling to find how long it will take for a cooling object to reach a desired temperature, or to find what temperature an object will be after a given time.

- (Optional) We can use logistic growth functions to model real-world situations where the rate of growth changes over time, such as population growth, spread of disease, and spread of rumors.

GLOSSARY

carrying capacity
in a logistic model, the limiting value of the output

doubling time
the time it takes for a quantity to double

half-life
the length of time it takes for a substance to exponentially decay to half of its original quantity

logistic growth model
a function of the form $f(x) = \frac{c}{1+ae^{-bx}}$ where $\frac{c}{1+a}$ is the initial value, c is the carrying capacity, or limiting value, and b is a constant determined by the rate of growth

Newton's Law of Cooling
the scientific formula for temperature as a function of time as an object's temperature is equalized with the ambient temperature

order of magnitude
the power of ten, when a number is expressed in scientific notation, with one non-zero digit to the left of the decimal

TRIGONOMETRIC FUNCTIONS

3.1 Trigonometric Functions of an Acute Angle

Consider a right triangle $\triangle$ ABC, with the right angle at C and with lengths a, b, and c, as in the Figure 1 below. For the acute angle A, call the leg BC its **opposite side**, and call the leg AC its **adjacent side**. Recall that the hypotenuse of the triangle is always opposite the right angle. In the triangle below, this is the side AB. The ratios of sides of a right triangle occur often enough in practical applications to warrant their own names, so we define the **six trigonometric functions** of A as follows:

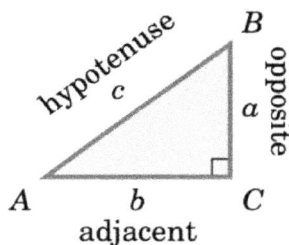

Figure 1: Sides of a right triangle with respect to angle A.

Table 1 The six trigonometric functions of A

Name of function	Abbreviation	Definition	
sine (A)	sin (A)	$= \dfrac{\text{opposite side}}{\text{hypotenuse}}$	$= \dfrac{a}{c}$
cosine (A)	cos (A)	$= \dfrac{\text{adjacent side}}{\text{hypotenuse}}$	$= \dfrac{b}{c}$
tangent (A)	tan (A)	$= \dfrac{\text{opposite side}}{\text{adjacent side}}$	$= \dfrac{a}{b}$
cosecant (A)	csc (A)	$= \dfrac{\text{hypotenuse}}{\text{opposite side}}$	$= \dfrac{c}{a}$
secant (A)	sec (A)	$= \dfrac{\text{hypotenuse}}{\text{adjacent side}}$	$= \dfrac{c}{b}$
cotangent (A)	cot (A)	$= \dfrac{\text{adjacent side}}{\text{opposite side}}$	$= \dfrac{b}{a}$

We will usually use the abbreviated names of the functions. Notice from Table 1 that the pairs sin(A) and csc(A), cos(A) and sec(A), and tan(A) and cot(A) are reciprocals:

$\csc(A) = \frac{1}{\sin(A)}$	$\sec(A) = \frac{1}{\cos(A)}$	$\cot(A) = \frac{1}{\tan(A)}$
$\sin(A) = \frac{1}{\csc(A)}$	$\cos(A) = \frac{1}{\sec(A)}$	$\tan(A) = \frac{1}{\cot(A)}$

EXAMPLE 1: FINDING TRIGONOMETRIC FUNCTIONS GIVEN SIDES

For the right triangle △ ABC shown in Figure 2 below, find the values of all six trigonometric functions of the acute angles A and B.

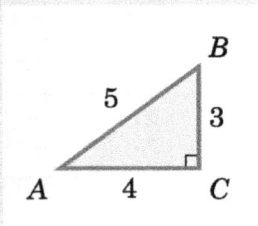

Figure 2: A Given Right Triangle

Answer

The hypotenuse of △ ABC has length 5. For angle A, the opposite side BC has length 3 and the adjacent side AC has length 4. Thus:

$$\sin(A) = \frac{Opposite}{Hypotenuse} = \frac{3}{5} \qquad \cos(A) = \frac{Adjacent}{Hypotenuse} = \frac{4}{5} \qquad \tan(A) = \frac{Opposite}{Adjacent} = \frac{3}{4}$$

$$\csc(A) = \frac{Hypotenuse}{Opposite} = \frac{5}{3} \qquad \sec(A) = \frac{Hypotenuse}{Adjacent} = \frac{5}{4} \qquad \cot(A) = \frac{Adjacent}{Opposite} = \frac{4}{3}$$

For angle B, the opposite side AC has length 4 and the adjacent side BC has length 3. Thus:

$$\sin(B) = \frac{Opposite}{Hypotenuse} = \frac{4}{5} \qquad \cos(B) = \frac{Adjacent}{Hypotenuse} = \frac{3}{5} \qquad \tan(B) = \frac{Opposite}{Adjacent} = \frac{4}{3}$$

$$\csc(B) = \frac{Hypotenuse}{Opposite} = \frac{5}{4} \qquad \sec(B) = \frac{Hypotenuse}{Adjacent} = \frac{5}{3} \qquad \cot(B) = \frac{Adjacent}{Opposite} = \frac{3}{4}$$

Notice in Example 1 that we did not specify the units for the lengths. This raises the possibility that our answers depended on a triangle of a specific physical size. For example, suppose that two different students are reading this textbook: one in the United States and one in Germany. The American student thinks that the lengths 3, 4, and 5 in Example 1 are measured in inches, while the German student thinks that they are measured in centimeters. Since 1 in ≈ 2.54 cm, the students are using triangles of different physical sizes (see Figure 3 below, not drawn to scale).

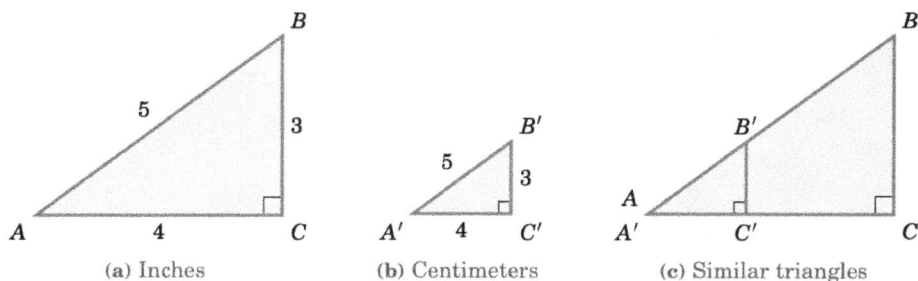

(a) Inches   (b) Centimeters   (c) Similar triangles

Figure 3. △ ABC ~ △ A'B'C' This figure shows two 3,4,5 triangles. The first triangle is a 3,4,5 triangle in inches, the second is a 3,4,5 triangle in centimeters. The third visual shows that these sides correspond to each other.

If the American triangle is △ ABC and the German triangle is △ A 'B 'C ' , then we see from Figure 1 that △ ABC is similar to △ A 'B 'C ', and hence the corresponding angles are equal and the ratios of the corresponding sides are equal. In fact, we know that common ratio: the sides of △ ABC are approximately 2.54 times longer than the corresponding sides of △ A 'B 'C '. So when the American student calculates sin A and the German student calculates sin A ', they get the same answer:

$$\Delta ABC \sim \Delta A'B'C' \Rightarrow \frac{BC}{B'C'} = \frac{AB}{A'B'} \Rightarrow \frac{BC}{AB} = \frac{B'C'}{A'B'} \Rightarrow \sin(A) = \sin(A')$$

Likewise, the other values of the trigonometric functions of A and A ' are the same. In fact, our argument was general enough to work with any similar right triangles. This leads us to the following conclusion:

> When calculating the trigonometric functions of an acute angle A, you may use any right triangle which has A as one of the angles.

Since we defined the trigonometric functions in terms of ratios of sides, you can think of the units of measurement for those sides as canceling out in those ratios. This means that *the values of the trigonometric functions are unitless numbers*. So when the American student calculated 3/5 as the value of sin(A) in Example 1, that is the same as the 3/5 that the German student calculated, despite the different units for the lengths of the sides.

EXAMPLE 2: FINDING TRIGONOMETRIC VALUES FOR A 45 DEGREE ANGLE

Find the values of all six trigonometric functions of 45°.

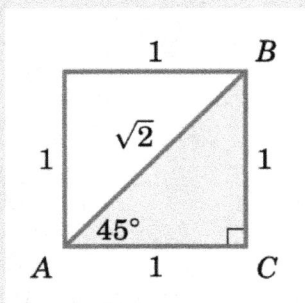

Figure 4. A 45°-45°-90° Right Triangle

Answer

Since we may use any right triangle which has 45° as one of the angles, use the simplest one: take a square whose sides are all 1 unit long and divide it in half diagonally, as in the figure on the right. Since the two legs of the triangle △ ABC have the same length, △ ABC is an isosceles triangle, which means that the angles A and B are equal. So since A + B = 90° , this means that we must have A = B = 45° . By the Pythagorean Theorem, the length c of the hypotenuse is given by

$$c^2 = 1^2 + 1^2 = 2 \Rightarrow c = \sqrt{2}$$

Thus, using the angle *A* we get:

$$\sin\left(45°\right) = \frac{Opposite}{Hypotenuse} = \frac{1}{\sqrt{2}} \quad \cos\left(45°\right) = \frac{Adjacent}{Hypotenuse} = \frac{1}{\sqrt{2}} \quad \tan\left(45°\right) = \frac{Opposite}{Adjacent} = \frac{1}{1} = 1$$

$$\csc\left(45°\right) = \frac{Hypotenuse}{Opposite} = \sqrt{2} \quad \sec\left(45°\right) = \frac{Hypotenuse}{Adjacent} = \sqrt{2} \quad \cot\left(45°\right) = \frac{Adjacent}{Opposite} = \frac{1}{1} = 1$$

Note that we would have obtained the same answers if we had used any right triangle similar to △ ABC. For example, if we multiply each side of △ ABC by $\sqrt{2}$, then we would have a similar triangle with legs of length $\sqrt{2}$ and hypotenuse of length 2. This would give us $\sin\left(45°\right) = \frac{\sqrt{2}}{2}$, which equals $\frac{\sqrt{2}}{\sqrt{2}\sqrt{2}} = \frac{1}{\sqrt{2}}$ as before. The same goes for the other functions.

EXAMPLE 3: FINDING TRIGONOMETRIC VALUES FOR 30 AND 60 DEGREE ANGLES

Find the values of all six trigonometric functions of 60°.

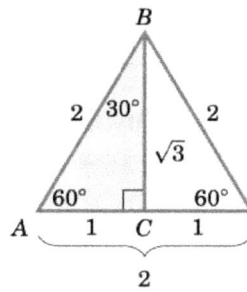

Figure 5. A 30°-60°-90° Right Triangle

Answer

Since we may use any right triangle which has 60° as one of the angles, we will use a simple one: take a triangle whose sides are all 2 units long and divide it in half by drawing the bisector from one vertex to the opposite side, as in the figure above. Since the original triangle was an equilateral triangle (i.e. all three sides had the same length), its three angles were all the same, namely 60° . Recall from elementary geometry that the bisector from the vertex angle of an equilateral triangle to its opposite side bisects both the vertex angle and the opposite side. So as in Figure 5 shown, the triangle △ ABC has angle A = 60° and angle B = 30° , which forces the angle C to be 90° . Thus, △ ABC is a right triangle. We see that the hypotenuse has length c = AB = 2 and the leg AC has length b = AC = 1. By the Pythagorean Theorem, the length a of the leg BC is given by

$$a^2 + b^2 = c^2 \Rightarrow a^2 = 2^2 - 1^2 = 3 \Rightarrow a = \sqrt{3}$$

Thus, using the angle A we get:

$$\sin\left(60°\right) = \frac{Opposite}{Hypotenuse} = \frac{\sqrt{3}}{2} \qquad \cos\left(60°\right) = \frac{Adjacent}{Hypotenuse} = \frac{1}{2} \qquad \tan\left(60°\right) = \frac{Opposite}{Adjacent} = \frac{\sqrt{3}}{1} = \sqrt{3}$$

$$\csc\left(60°\right) = \frac{Hypotenuse}{Opposite} = \frac{2}{\sqrt{3}} \qquad \sec\left(60°\right) = \frac{Hypotenuse}{Adjacent} = 2 \qquad \cot\left(60°\right) = \frac{Adjacent}{Opposite} = \frac{1}{\sqrt{3}}$$

Notice that, as a bonus, we get the values of all six trigonometric functions of 30° , by using angle B = 30° in the same triangle △ ABC above:

$$\sin\left(30°\right) = \frac{Opposite}{Hypotenuse} = \frac{1}{2} \qquad \cos\left(30°\right) = \frac{Adjacent}{Hypotenuse} = \frac{\sqrt{3}}{2} \qquad \tan\left(30°\right) = \frac{Opposite}{Adjacent} = \frac{1}{\sqrt{3}}$$

$$\csc\left(30°\right) = \frac{Hypotenuse}{Opposite} = 2 \qquad \sec\left(30°\right) = \frac{Hypotenuse}{Adjacent} = \frac{2}{\sqrt{3}} \qquad \cot\left(30°\right) = \frac{Adjacent}{opposite} = \frac{\sqrt{3}}{1} = \sqrt{3}$$

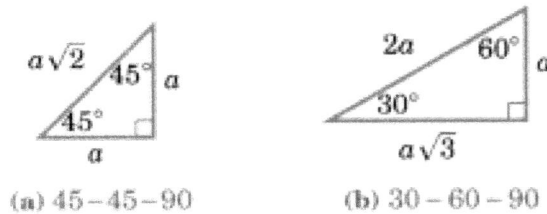

(a) 45 − 45 − 90

(b) 30 − 60 − 90

Figure 6: Two general right triangles (any a > 0)

The angles 30°, 45°, and 60° arise often in applications. We can see what any 45°–45°–90° and 30°–60°–90° right triangles look like, as in Figure 6 above. Notice that the sides can now be any length that fits these ratios, and not just the sides of $1 - 2 - \sqrt{3}$ or $1 - 1 - \sqrt{2}$. For example, the sides of a 30°–60°–90° right triangle might be $4 - 8 - 4\sqrt{3}$.

EXAMPLE 4: GIVEN ONE TRIGONOMETRIC VALUE, FIND REMAINING VALUES

A is an acute angle such that $\sin(A) = \frac{2}{3}$. Find the values of the other trigonometric functions of A.

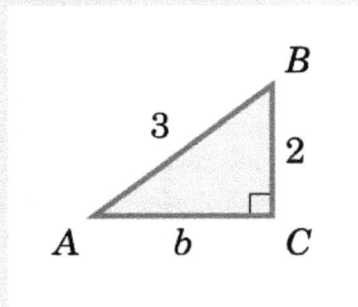

Figure 6. A right triangle where $\sin(A) = \frac{2}{3}$

Answer In general it helps to draw a right triangle to solve problems of this type. The reason is that the trigonometric functions were defined in terms of ratios of sides of a right triangle, and you are given one such function (the sine, in this case) already in terms of a ratio: $\sin(A) = \frac{2}{3}$. Since sin(A) is defined as $\frac{opposite}{hypotenuse}$, use 2 as the length of the side opposite A and use 3 as the length of the hypotenuse in a right triangle $\triangle$ ABC as shown in Figure 6, so that $\sin(A) = \frac{2}{3}$. The adjacent side to A has unknown length b, but we can use the Pythagorean Theorem to find it:

$$2^2 + b^2 = 3^2 \Rightarrow b^2 = 9 - 4 = 5 \Rightarrow b = \sqrt{5}$$

We now know the lengths of all sides of the triangle $\triangle$ ABC, so we have:

$$\cos(A) = \frac{Adjacent}{Hypotenuse} = \frac{\sqrt{5}}{3} \qquad \tan(A) = \frac{Opposite}{Adjacent} = \frac{2}{\sqrt{5}}$$

$$\csc(A) = \frac{Hypotenuse}{Opposite} = \frac{3}{2} \quad \sec(A) = \frac{Hypotenuse}{Adjacent} = \frac{3}{\sqrt{5}} \quad \cot(A) = \frac{Adjacent}{Opposite} = \frac{\sqrt{5}}{2}$$

You may have noticed the connections between the sine and cosine, secant and cosecant, and tangent and cotangent of the complementary angles in Examples 3 and 4. Generalizing those examples gives us the following theorem:

THEOREM 1: COFUNCTION THEOREM

If A and B are the complementary acute angles in a right triangle $\triangle$ ABC, then we know that B = 90° – A, and likewise that A = 90° – B We can then create the following relations:

$$\sin(A) = \cos(90° - A) \qquad \sec(A) = \csc(90° - A) \qquad \tan(A) = \cot(90° - A)$$
$$\sin(B) = \cos(90° - B) \qquad \sec(B) = \csc(90° - B) \qquad \tan(B) = \cot(90° - B)$$

We say that the pairs of functions { sine, cosine }, { sececant, cosecant }, and {tangent, cotangent} are **cofunctions**.

So sine and cosine are cofunctions, secant and cosecant are cofunctions, and tangent and cotangent are cofunctions. That is how the functions cosine, cosecant, and cotangent got the "co" in their names. The Cofunction Theorem says that any trigonometric function of an acute angle is equal to its co-function of the complementary angle.

EXAMPLE 5: USING COFUNCTIONS

Write each of the following numbers as trigonometric functions of an angle less than 45° : **(a)** sin (65°); **(b)** cos (78°) ; **(c)** tan (59º).

Answer

(a) The complement of 65° is 90° – 65° = 25° and the cofunction of sine is cosine, so by the Cofunction Theorem we know that $\sin(65°) = \cos(25°)$.

(b) The complement of 78° is 90° –78° = 12° and the cofunction of cosine is sine, so $\cos(78°) = \sin(12°)$.

(c) The complement of 59° is 90° –59° = 31° and the cofunction of tangent is cotangent, so $\tan(59°) = \cot(31°)$.

KEY CONCEPTS

- The six trigonometric functions are defined in terms of ratios of sides of a right triangle.
- We can use the relationship of the sides of special right triangles with angles of 45°– 45°–90° and 30°–60°–90° to easily generate the values of the 6 trigonometric functions for 30°, 45° and 60°.
- If we are given two sides of a right triangle, we can use the Pythagorean Theorem to find the 3rd side. We can then create the six trigonometric values for either of the acute angles.
- The functions sine and cosine are cofunctions of each other.

3.2 Angles

A golfer swings to hit a ball over a sand trap and onto the green. An airline pilot maneuvers a plane toward a narrow runway. A dress designer creates the latest fashion. What do they all have in common? They all work with angles, and so do all of us at one time or another. Sometimes we need to measure angles exactly with instruments. Other times we estimate them or judge them by eye. Either way, the proper angle can make the difference between success and failure in many undertakings. In this section, we will examine properties of angles.

Drawing Angles in Standard Position

Properly defining an angle first requires that we define a ray. A **ray** consists of one point on a line and all points extending in one direction from that point. The first point is called the endpoint of the ray. We can refer to a specific ray by stating its endpoint and any other point on it. The ray in Figure 1 can be named as ray EF, or in symbol form $\overrightarrow{EF}$.

Ray *EF*

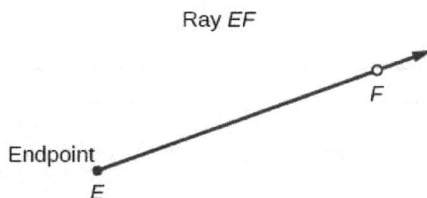

Figure 1.

An **angle** is the union of two rays having a common endpoint. The endpoint is called the **vertex** of the angle, and the two rays are the sides of the angle. The angle in Figure 2 is formed from $\overrightarrow{ED}$ and $\overrightarrow{EF}$. Angles can be named using a point on each ray and the vertex, such as angle *DEF*, or in symbol form $\angle DEF$.

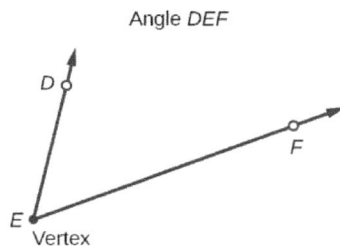

Angle *DEF*

Vertex

Figure 2.

Greek letters are often used as variables for the measure of an angle. Table 1 is a list of Greek letters commonly used to represent angles, and a sample angle is shown in Figure 3.

Table 1

θ	ϕ or φ	α	β	γ
theta	phi	alpha	beta	gamma

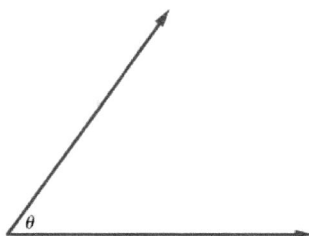

Figure 3. Angle theta, shown as $\angle\theta$

Angle creation is a dynamic process. We start with two rays lying on top of one another. We leave one fixed in place, and rotate the other. The fixed ray is the **initial side,** and the rotated ray is the **terminal side.** In order to identify the different sides, we indicate the rotation with a small arc and arrow close to the vertex as in Figure 4.

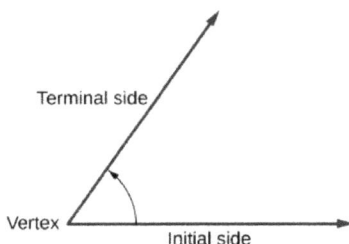

Figure 4.

As we discussed at the beginning of the section, there are many applications for angles, but in order to use them correctly, we must be able to measure them. The **measure of an angle** is the amount of rotation from the initial side to the terminal side. Probably the most familiar unit of angle measurement is the degree. One **degree** is $\frac{1}{360}$ of a circular rotation, so a complete circular rotation contains 360 degrees. An angle measured in degrees should always include the unit "degrees" after the number, or include the degree symbol °. For example, 90 degrees = 90°.

To formalize our work, we will begin by drawing angles on an *x–y* coordinate plane. Angles can occur in any position on the coordinate plane, but for the purpose of comparison, the convention is to illustrate them in the same position whenever possible. An angle is in **standard position** if its vertex is located at the origin, and its initial side extends along the positive *x*-axis. See Figure 5.

Standard Position

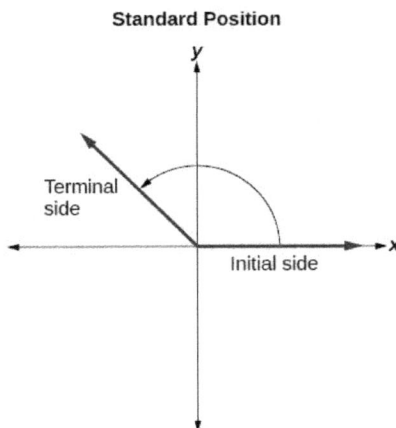

Figure 5.

If the angle is measured in a counterclockwise direction from the initial side to the terminal side, the angle is said to be a **positive angle.** If the angle is measured in a clockwise direction, the angle is said to be a **negative angle**.

Drawing an angle in standard position always starts the same way—draw the initial side along the positive x-axis. To place the terminal side of the angle, we must calculate the fraction of a full rotation the angle represents. We do that by dividing the angle measure in degrees by 360°. For example, to draw a 90° angle, we calculate that $\frac{90}{360} = \frac{1}{4}$. So, the terminal side will be one-fourth of the way around the circle, moving counterclockwise from the positive x-axis. To draw a 360° angle, we calculate that $\frac{360}{360} = 1$. So the terminal side will be 1 complete rotation around the circle, moving counterclockwise from the positive x-axis. In this case, the initial side and the terminal side overlap. See Figure 6.

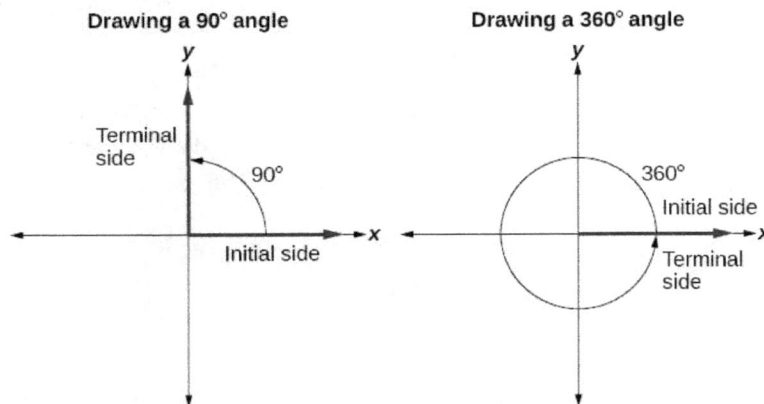

Figure 6.

Since we define an angle in standard position by its initial side, we have a special type of angle whose terminal side lies on an axis, a **quadrantal angle**. This type of angle can have a measure of 0°, 90°, 180°, 270° or 360°. See Figure 7.

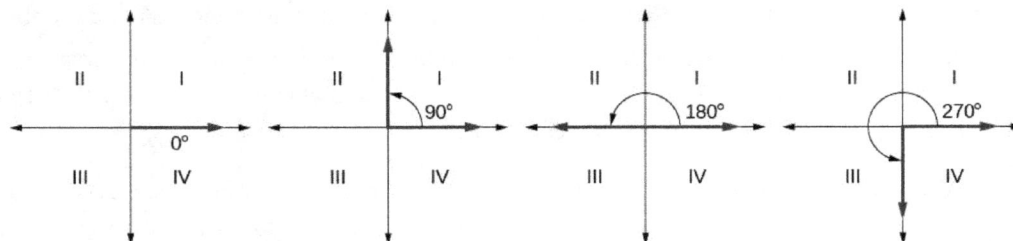

Figure 7. Quadrantal angles are angles in standard position whose terminal side lies along an axis. Examples are shown.

Quadrantal angles are angels in standard position whose terminal side lies on an axis, including 0°, 90°, 180°, 270°, or 360°.

Given an angle measure in degrees, draw the angle in standard position.

1. Express the angle measure as a fraction of 360°.
2. Reduce the fraction to simplest form.
3. Draw an angle that contains that same fraction of the circle, beginning on the positive *x*-axis and moving counterclockwise for positive angles and clockwise for negative angles.

EXAMPLE 1: DRAWING AN ANGLE IN STANDARD POSITION MEASURED IN DEGREES

a. Sketch an angle of 30° in standard position.
b. Sketch an angle of −135° in standard position.

Answer

a. Divide the angle measure by 360°.

$$\frac{30}{360} = \frac{1}{12}$$

To rewrite the fraction in a more familiar fraction, we can recognize that

$$\frac{1}{12} = \frac{1}{3}\left(\frac{1}{4}\right)$$

One-twelfth equals one-third of a quarter, so by dividing a quarter rotation into thirds, we can sketch a line at 30° as in Figure 8.

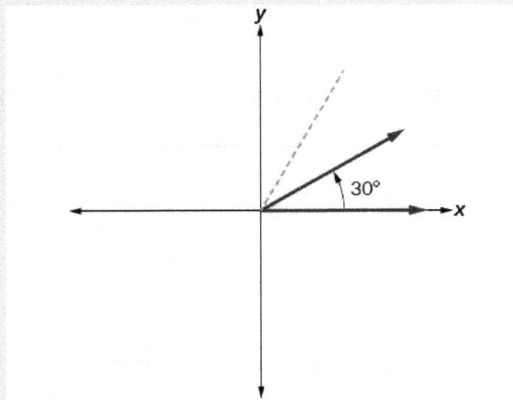

Figure 8.

b. Divide the angle measure by 360°.

$$\frac{-135}{360} = -\frac{3}{8}$$

In this case, we can recognize that

$$-\frac{3}{8} = -\frac{3}{2}\left(\frac{1}{4}\right)$$

Negative three-eighths is one and one-half times a quarter, so we place a line by moving clockwise one full quarter and one-half of another quarter, as in **Figure 9.**

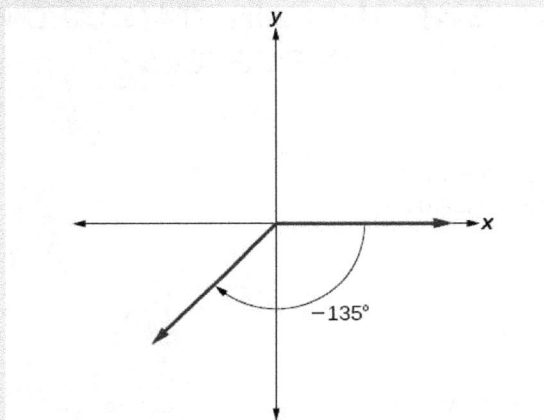

Figure 9.

TRY IT #1

Show an angle of 240° on a circle in standard position.

Answer

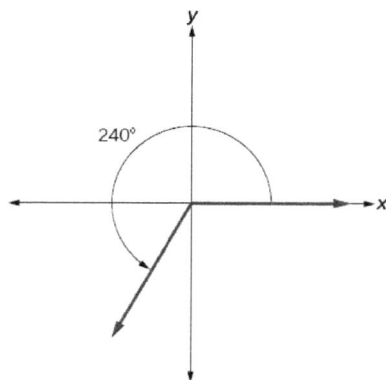

Finding Coterminal Angles

Negative angles and angles greater than a full revolution are more awkward to work with than those in the range of 0° to 360°. It would be convenient to replace those out-of-range angles with a corresponding angle within the range of a single revolution.

It is possible for more than one angle to have the same terminal side. Look at Figure 10. The angle of 140° is a positive angle, measured counterclockwise. The angle of −220° is a negative angle, measured clockwise. But both angles have the same terminal side. If two angles in standard position have the same terminal side, they are **coterminal** angles. Every angle greater than 360° or less than 0° is coterminal with an angle between 0° and 360°, and it is often more convenient to find the coterminal angle within the range of 0° to 360° than to work with an angle that is outside that range.

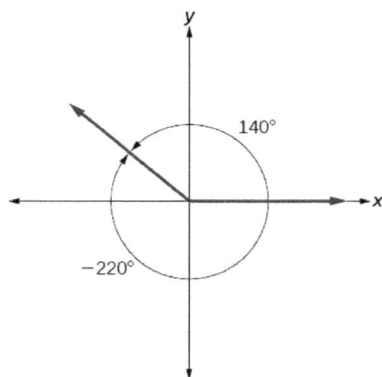

Figure 10. An angle of 140° and an angle of −220° are coterminal angles.

HOW TO

Given an angle with measure greater than 360°, find a coterminal angle having a measure between 0° and 360°.

1. Subtract 360° From the given angle.
2. If the result is still greater than 360°, subtract 360° again until the result is between 0° and 360°.
3. The resulting angle is coterminal with the original angle.

Given an angle with measure less than 0°, find a coterminal angle having a measure between 0° and 360°.

---

1. Add 360° to the given angle.
2. If the result is still less than 0°, add 360° again until the result is between 0° and 360°.
3. The resulting angle is coterminal with the original angle.

DEFINITION

Coterminal angles are two angles in standard position that have the same terminal side.

Any angle has infinitely many coterminal angles because each time we add 360° to that angle—or subtract 360° from it—the resulting value has a terminal side in the same location. For example, 100° and 460° are coterminal for this reason, as is −260°. Recognizing that any angle has infinitely many coterminal angles explains the repetitive shape in the graphs of trigonometric functions, which we will study in great detail in the next sections.

EXAMPLE 2: FINDING AN ANGLE COTERMINAL WITH AN ANGLE OF MEASURE GREATER THAN 360°

Find the least positive angle θ that is coterminal with an angle measuring 800°, where $0 \leq \theta < 360$.

Answer

An angle with measure 800° is coterminal with an angle with measure 800 − 360 = 440°, but 440° is still greater than 360°, so we subtract 360° again to find another coterminal angle: 440 − 360 = 80°.

The angle $\theta = 80$ is coterminal with 800°. To put it another way, 800° equals 80° plus two full rotations, as shown in Figure 11.

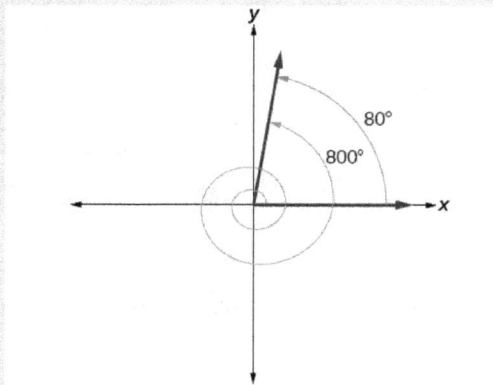

Figure 11.

TRY IT #2

Find an angle α that is coterminal with an angle measuring 870°, where $0 \leq \alpha < 360.$

Answer

$\alpha = 150$

EXAMPLE 3: FINDING AN ANGLE COTERMINAL WITH AN ANGLE MEASURING LESS THAN 0°

Show the angle with measure −45° on a circle and find a positive coterminal angle α such that 0° ≤ α < 360°.

Answer

Since 45° is half of 90°, we can start at the positive horizontal axis and measure clockwise half of a 90° angle.

Because we can find coterminal angles by adding or subtracting a full rotation of 360°, we can find a positive coterminal angle here by adding 360°:

$-45 + 360 = 315$

We can then show the angle on a circle, as in Figure 12.

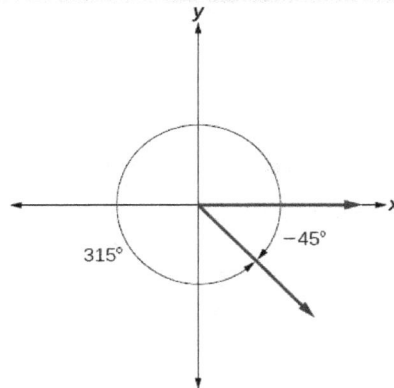

Figure 12.

TRY IT #3

Find an angle β that is coterminal with an angle measuring −300° such that $0 \leq \beta < 360.$

Answer

$\beta = 60$

Converting Between Degrees and Radians

Dividing a circle into 360 parts is an arbitrary choice, although it creates the familiar degree measurement. We may choose other ways to divide a circle. To find another unit, think of the process of drawing a circle. Imagine that you stop before the circle is completed. The portion that you drew is referred to as an **arc**. An arc may be a portion of a full circle, a full circle, or more than a full circle, represented by more than one full rotation. The length of the arc around an entire circle is called the **circumference** of that circle.

The circumference of a circle is $C = 2\pi r$. If we divide both sides of this equation by r, we create the ratio of the circumference to the radius, which is always 2π regardless of the length of the radius. So the circumference of any circle is $2\pi \approx 6.28$ times the length of the radius. That means that if we took a string as long as the radius and used it to measure consecutive lengths around the circumference, there would be room for six full string-lengths and a little more than a quarter of a seventh, as shown in Figure 13.

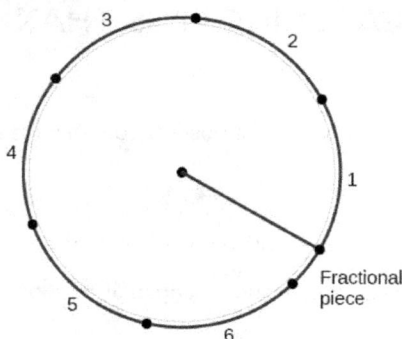

Figure 13.

This brings us to our new angle measure. One radian is the measure of a central angle of a circle that intercepts an arc equal in length to the radius of that circle. A central angle is an angle formed at the center of a circle by two radii. Because the total circumference equals 2π times the radius, a full circular rotation is 2π radians. So

$$2\pi \text{ radians} = 360°$$
$$\pi \text{ radians} = \frac{360°}{2} = 180°$$
$$1 \text{ radian} = \frac{180°}{\pi} \approx 57.3°$$

See Figure 14. Note that when an angle is described without a specific unit, it refers to radian measure. For example, an angle measure of 3 indicates 3 radians. In fact, radian measure is dimensionless, since it is the quotient of a length (circumference) divided by a length (radius) and the length units cancel out.

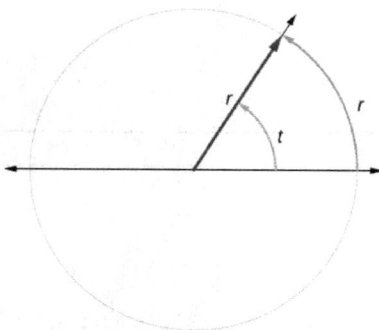

Figure 14. The angle t sweeps out a measure of one radian. Note that the length of the intercepted arc is the same as the length of the radius of the circle.

Relating Arc Lengths to Radius

An **arc length** s is the length of the curve along the arc. Just as the full circumference of a circle always has a constant ratio to the radius, the arc length produced by any given angle also has a constant relation to the radius, regardless of the length of the radius.

This ratio, called the radian measure, is the same regardless of the radius of the circle—it depends only on the angle. This property allows us to define a measure of any angle as the ratio of the arc length s to the radius r. See Figure 15.

$$s = r\theta$$
$$\theta = \frac{s}{r}$$

If $s = r$, then $\theta = \frac{r}{r} = 1$ radian.

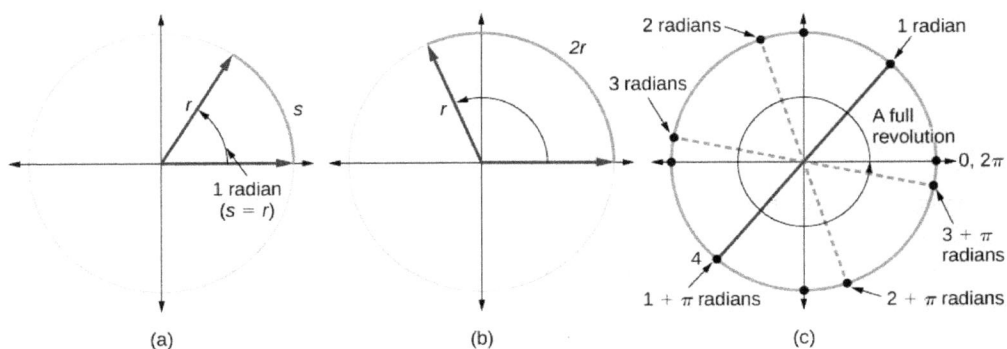

(a)   (b)   (c)

Figure 15. (a) In an angle of 1 radian, the arc length s equals the radius r. (b) An angle of 2 radians has an arc length $s = 2r$. (c) A full revolution is 2π or about 6.28 radians.

To elaborate on this idea, consider two circles, one with radius 2 and the other with radius 3. Recall the circumference of a circle is $C = 2\pi r$, where r is the radius. The smaller circle then has circumference $2\pi\left(2\right) = 4\pi$ and the larger has circumference $2\pi\left(3\right) = 6\pi$. Now we draw a 45° angle on the two circles, as in Figure 16.

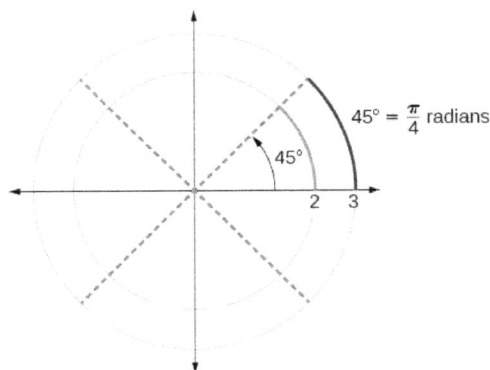

Figure 16. A 45° angle contains one-eighth of the circumference of a circle, regardless of the radius.

Notice what happens if we find the ratio of the arc length divided by the radius of the circle.

$$\text{Smaller circle: } \frac{\frac{1}{2}\pi}{2} = \frac{1}{4}\pi$$
$$\text{Larger circle: } \frac{\frac{3}{4}\pi}{3} = \frac{1}{4}\pi$$

Since both ratios are $\frac{1}{4}\pi$, the angle measures of both circles are the same, even though the arc length and radius differ.

One **radian** is the measure of the central angle of a circle such that the length of the arc between the initial side and the terminal side is equal to the radius of the circle. A full revolution (360°) equals 2π radians. A half revolution (180°) is equivalent to π radians.

The **radian measure** of an angle is the ratio of the length of the arc subtended by the angle to the radius of the circle. In other words, if s is the length of an arc of a circle, and r is the radius of the circle, then the central angle containing that arc measures $\frac{s}{r}$ radians. In a circle of radius 1, the radian measure corresponds to the length of the arc.

A measure of 1 radian looks to be about 60°. Is that correct?

Yes. It is approximately 57.3°. Because 2π *radians equals 360°,* 1 *radian equals* $\frac{360}{2\pi} \approx 57.3$.

Using Radians

Because radian measure is the ratio of two lengths, it is a unit-less measure. For example, suppose the radius were 2 inches and the distance along the arc were also 2 inches. When we calculate the radian measure of the angle, the "inches" cancel, and we have a result without units. Therefore, it is not necessary to write the label "radians" after a radian measure, and if we see an angle that is not labeled with "degrees" or the degree symbol, we can assume that it is a radian measure.

Considering the most basic case, the unit circle (a circle with radius 1), we know that 1 rotation equals 360 degrees, 360°. We can also track one rotation around a circle by finding the circumference, $C = 2\pi r$, and for the unit circle $C = 2\pi$. These two different ways to rotate around a circle give us a way to convert from degrees to radians.

$$\begin{aligned} 1 \text{ rotation} &= 360 &= 2\pi & \quad \text{radians} \\ \tfrac{1}{2} \text{ rotation} &= 180 &= \pi & \quad \text{radians} \\ \tfrac{1}{4} \text{ rotation} &= 90 &= \tfrac{\pi}{2} & \quad \text{radians} \end{aligned}$$

Identifying Special Angles Measured in Radians

In addition to knowing the measurements in degrees and radians of a quarter revolution, a half revolution, and a full revolution, there are other frequently encountered angles in one revolution of a circle with which we should be familiar. It is common to encounter multiples of 30, 45, 60, and 90 degrees. These values are shown in Figure 17. You should become very comfortable with these angles as they will be very useful as we study the properties associated with angles.

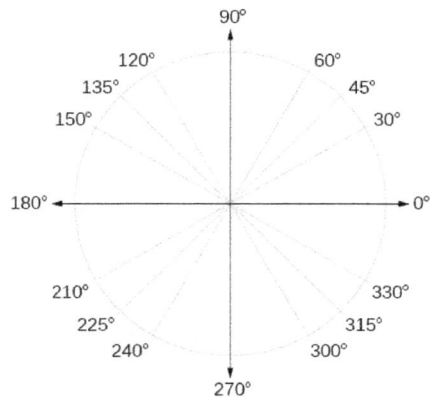

Figure 17. Commonly encountered angles measured in degrees

Now, we can list the corresponding radian values for the common measures of a circle corresponding to those listed in Figure 17, which are shown in Figure 18. Be sure you can verify each of these measures.

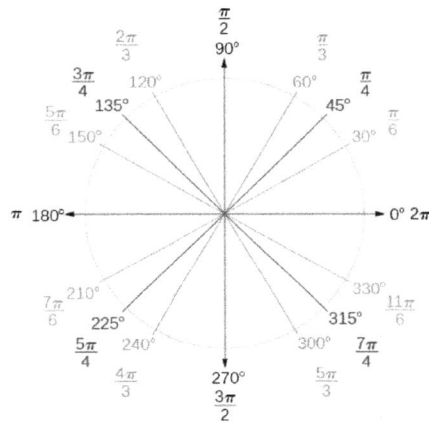

Figure 18. Commonly encountered angles measured in radians

EXAMPLE 4: FINDING A RADIAN MEASURE

Find the radian measure of one-third of a full rotation.

Answer

For any circle, the arc length along such a rotation would be one-third of the circumference. We know that

$$1 \text{ rotation} = 2\pi r$$

So,

$$s = \tfrac{1}{3}\left(2\pi r\right)$$
$$= \tfrac{2\pi r}{3}$$

The radian measure would be the arc length divided by the radius.

$$\text{radian measure} = \frac{\frac{2\pi r}{3}}{r}$$
$$= \frac{2\pi r}{3r}$$
$$= \frac{2\pi}{3}$$

TRY IT #4

Find the radian measure of three-fourths of a full rotation.

Answer

$\frac{3\pi}{2}$

Converting between Radians and Degrees

Because degrees and radians both measure angles, we need to be able to convert between them. We can easily do so using a proportion.

$$\frac{\theta}{180} = \frac{\theta^R}{\pi}$$

This proportion shows that the measure of angle θ in degrees divided by 180° equals the measure of angle θ in radians divided by π. Or, phrased another way, degrees is to 180° as radians is to π.

$$\frac{\text{Degrees}}{180} = \frac{\text{Radians}}{\pi}$$

HOW TO

To convert between degrees and radians, use the proportion

$$\frac{\theta}{180} = \frac{\theta^R}{\pi}$$

EXAMPLE 5: CONVERTING RADIANS TO DEGREES

Convert each radian measure to degrees.

a. $\frac{\pi}{6}$

b. 3

Answer

Because we are given radians and we want degrees, we should set up a proportion and solve for degrees.

a. We use the proportion $\frac{\theta°}{180°} = \frac{\theta^R}{\pi}$.

$$\frac{\theta}{180} = \frac{\frac{\pi}{6}}{\pi} \qquad \text{Substitute the given information.}$$

$$\theta = \frac{180}{6} \qquad \text{Multiply both sides by 180.}$$

$$\theta = 30° \qquad \text{Simplify.}$$

b. We use the proportion $\frac{\theta}{180} = \frac{\theta^R}{\pi}$.

$$\frac{\theta}{180} = \frac{3}{\pi} \qquad \text{Substitue the given information.}$$

$$\theta = \frac{3\,(180)}{\pi} \qquad \text{Multiply both sides by 180.}$$

$$\theta \approx 172° \qquad \text{Simplify.}$$

TRY IT #5

Convert $-\dfrac{3\pi}{4}$ radians to degrees.

Answer

−135°

EXAMPLE 6: CONVERTING DEGREES TO RADIANS

Convert 15 degrees to radians.

Answer

In this example, we start with degrees and want radians, so we again set up a proportion and solve it, but we substitute the given information into a different part of the proportion.

$$\frac{\theta}{180} = \frac{\theta^R}{\pi}$$

$$\frac{15}{180} = \frac{\theta^R}{\pi}$$

$$\frac{15\pi}{180} = \theta^R \qquad \text{Notice that the degrees cancel.}$$

$$\frac{\pi}{12} = \theta^R$$

Analysis

Another way to think about this problem is by remembering that $30° = \frac{\pi}{6}$. Because $15° = \frac{1}{2}(30°)$, we can find that $\frac{1}{2}\left(\frac{\pi}{6}\right)$ is $\frac{\pi}{12}$.

TRY IT #6

Convert 126° to radians.

Answer

$\frac{7\pi}{10}$

Finding Coterminal Angles Measured in Radians

We can find coterminal angles measured in radians in much the same way as we have found them using degrees. In both cases, we find coterminal angles by adding or subtracting one or more full rotations.

HOW TO:

Given an angle greater than 2π, find a coterminal angle between 0 and 2π.

1. Subtract 2π from the given angle.
2. If the result is still greater than 2π, subtract 2π again until the result is between 0 and 2π.
3. The resulting angle is coterminal with the original angle.

Given an angle less than 0, find a coterminous angle between 0 and 2π.

1. Add 2π to the given angle.
2. If the result is still less than 0, add 2π again until the result is a positive number.
3. The resulting angle is coterminal with the original angle.

EXAMPLE 7: FINDING COTERMINAL ANGLES USING RADIANS

Find an angle β that is coterminal with $\frac{19\pi}{4}$, where $0 \leq \beta < 2\pi$.

Answer

When working in degrees, we found coterminal angles by adding or subtracting 360 degrees, a full rotation. Likewise, in radians, we can find coterminal angles by adding or subtracting full rotations of 2π radians:

$$\frac{19\pi}{4} - 2\pi = \frac{19\pi}{4} - \frac{8\pi}{4}$$
$$= \frac{11\pi}{4}$$

The angle $\frac{11\pi}{4}$ is coterminal, but not less than 2π, so we subtract another rotation:

$$\frac{11\pi}{4} - 2\pi = \frac{11\pi}{4} - \frac{8\pi}{4}$$
$$= \frac{3\pi}{4}$$

The angle $\frac{3\pi}{4}$ is coterminal with $\frac{19\pi}{4}$, as shown in Figure 19.

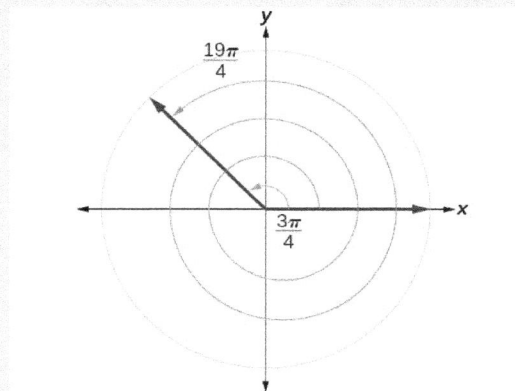

Figure 19.

TRY IT #7

Find an angle of measure θ that is coterminal with an angle of measure $-\frac{17\pi}{6}$ where $0 \leq \theta < 2\pi$.

Answer

$\frac{7\pi}{6}$

Determining the Length of an Arc

Recall that the radian measure θ of an angle was defined as the ratio of the arc length s of a circular arc to the radius r of the circle, $\theta = \frac{s}{r}$. From this relationship, we can find arc length along a circle, given an angle.

ARC LENGTH ON A CIRCLE

In a circle of radius r, the length of an arc s subtended by an angle with measure θ in radians, shown in Figure 20, is

$$s = r\theta$$

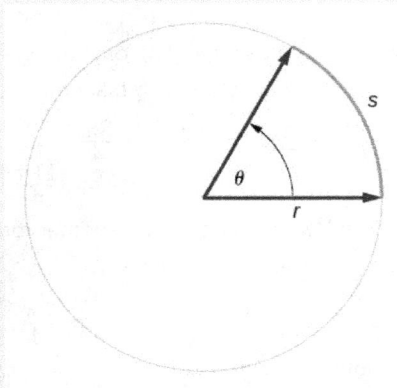

Figure 20.

HOW TO:

Given a circle of radius r, calculate the length s of the arc subtended by a given angle of measure θ.

1. If necessary, convert θ to radians.
2. Multiply the radius r by the radian measure of θ : $s = r\theta$.

EXAMPLE 8: FINDING THE LENGTH OF AN ARC

Assume the orbit of Mercury around the sun is a perfect circle. Mercury is approximately 36 million miles from the sun.

a. In one Earth day, Mercury completes 0.0114 of its total revolution. How many miles does it travel in one day?
b. Use your answer from part (a) to determine the radian measure for Mercury's movement in one Earth day.

Answer

a. Let's begin by finding the circumference of Mercury's orbit.

$$C = 2\pi r$$
$$= 2\pi \left(36 \text{ million miles}\right)$$
$$\approx 226 \text{ million miles}$$

Since Mercury completes 0.0114 of its total revolution in one Earth day, we can now find the distance traveled:

$$(0.0114)\,226 \text{ million miles} = 2.58 \text{ million miles}$$

b. Now, we convert to radians:

$$\text{radian} = \frac{\text{arclength}}{\text{radius}}$$
$$= \frac{2.58 \text{ million miles}}{36 \text{ million miles}}$$
$$= 0.0717$$

TRY IT #8

Find the arc length along a circle of radius 10 units subtended by an angle of 215°.

Answer

$$\frac{215\pi}{18} = 37.525 \text{ units}$$

Finding the Area of a Sector of a Circle (Optional)

In addition to arc length, we can also use angles to find the area of a sector of a circle. A sector is a region of a circle bounded by two radii and the intercepted arc, like a slice of pizza or pie. Recall that the area of a circle with radius r can be found using the formula $A = \pi r^2$. If the two radii form an angle of θ, measured in radians, then $\frac{\theta}{2\pi}$ is the ratio of the angle measure to the measure of a full rotation and is also, therefore, the ratio of the area of the sector to the area of the circle. Thus, the **area of a sector** is the fraction $\frac{\theta}{2\pi}$ multiplied by the entire area. (Always remember that this formula only applies if θ is in radians.)

$$\text{Area of sector} = \left(\frac{\theta}{2\pi}\right)\pi r^2$$
$$= \frac{\theta\pi r^2}{2\pi}$$
$$= \frac{1}{2}\theta r^2$$

AREA OF A SECTOR (OPTIONAL)

The area of a sector of a circle with radius r subtended by an angle θ, measured in radians, is

$$A = \tfrac{1}{2}\theta r^2$$

See Figure 21.

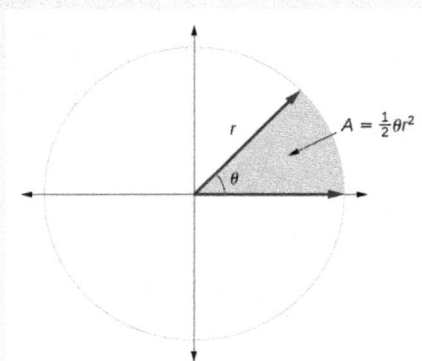

Figure 21. The area of the sector equals half the square of the radius times the central angle measured in radians.

HOW TO

Given a circle of radius r, find the area of a sector defined by a given angle θ.

1. If necessary, convert θ to radians.
2. Multiply half the radian measure of θ by the square of the radius r : $A = \tfrac{1}{2}\theta r^2$.

EXAMPLE 9: FINDING THE AREA OF A SECTOR

An automatic lawn sprinkler sprays a distance of 20 feet while rotating 30 degrees, as shown in Figure 22. What is the area of the sector of grass the sprinkler waters?

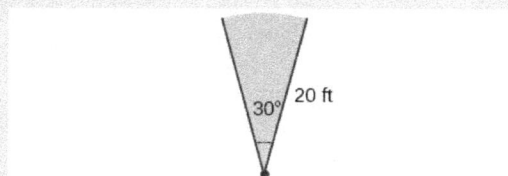

Figure 22. The sprinkler sprays 20 ft within an arc of 30°.

Answer

First, we need to convert the angle measure into radians. Because 30 degrees is one of our special angles, we already know the equivalent radian measure, but we can also convert:

$$30 = 30 \cdot \frac{\pi}{180}$$
$$= \frac{\pi}{6} \text{ radians.}$$

The area of the sector is then

$$\text{Area} = \frac{1}{2}\left(\frac{\pi}{6}\right)(20)^2$$
$$\approx 104.72$$

So the area is about 104.72 ft^2.

TRY IT #9

In central pivot irrigation, a large irrigation pipe on wheels rotates around a center point. A farmer has a central pivot system with a radius of 400 meters. If water restrictions only allow her to water 150 thousand square meters a day, what angle should she set the system to cover? Write the answer in radian measure to two decimal places.

Answer

1.88

Use Linear and Angular Speed to Describe Motion on a Circular Path (Optional)

In addition to finding the area of a sector, we can use angles to describe the speed of a moving object. An object traveling in a circular path has two types of speed. **Linear speed** is speed along a straight path and can be determined by the distance it moves along (its **displacement**) in a given time interval. For instance, if a wheel with radius 5 inches rotates once a second, a point on the edge of the wheel moves a distance equal to the circumference, or 10π inches, every second. So the linear speed of the point is 10π in./s. The equation for linear speed is as follows where v is linear speed, s is displacement, and t is time.

$$v = \frac{s}{t}$$

Angular speed results from circular motion and can be determined by the angle through which a point rotates in a given time interval. In other words, angular speed is angular rotation per unit time. So, for instance, if a gear makes a full rotation every 4 seconds, we can calculate its angular speed as $\frac{360 \text{ degrees}}{4 \text{ seconds}} = 90$ degrees per second. Angular speed can be given in radians per second, rotations per minute, or degrees per hour for example. The equation for angular speed is as follows, where ω (read as omega) is angular speed, θ is the angle traversed, and t is time.

$$\omega = \frac{\theta}{t}$$

Combining the definition of angular speed with the arc length equation, $s = r\theta$, we can find a relationship between angular and linear speeds. The angular speed equation can be solved for θ, giving $\theta = \omega t$. Substituting this into the arc length equation gives:

$$s = r\theta$$
$$= r\omega t$$

Substituting this into the linear speed equation gives:

$$v = \frac{s}{t}$$
$$= \frac{r\omega t}{t}$$
$$= r\omega$$

ANGULAR AND LINEAR SPEED (OPTIONAL)

As a point moves along a circle of radius r, its **angular speed**, ω, is the angular rotation θ per unit time, t.

$$\omega = \frac{\theta}{t}$$

The **linear speed**, v, of the point can be found as the distance traveled, arc length s, per unit time, t.

$$v = \frac{s}{t}$$

When the angular speed is measured in radians per unit time, linear speed and angular speed are related by the equation

$$v = r\omega$$

This equation states that the angular speed in radians, ω, representing the amount of rotation occurring in a unit of time, can be multiplied by the radius r to calculate the total arc length traveled in a unit of time, which is the definition of linear speed.

HOW TO:

Given the amount of angle rotation and the time elapsed, calculate the angular speed.

1. If necessary, convert the angle measure to radians.
2. Divide the angle in radians by the number of time units elapsed: $\omega = \frac{\theta}{t}$.
3. The resulting speed will be in radians per time unit.

EXAMPLE 10: FINDING ANGULAR SPEED

A water wheel, shown in Figure 23, completes 1 rotation every 5 seconds. Find the angular speed in radians per second.

Figure 23.

Answer

The wheel completes 1 rotation, or passes through an angle of 2π radians in 5 seconds, so the angular speed would be $\omega = \frac{2\pi}{5} \approx 1.257$ radians per second.

TRY IT #10

An old vinyl record is played on a turntable rotating clockwise at a rate of 45 rotations per minute. Find the angular speed in radians per second.

Answer

$\frac{-3\pi}{2}$ rad/s

HOW TO:

Given the radius of a circle, an angle of rotation, and a length of elapsed time, determine the linear speed.

1. Convert the total rotation to radians if necessary.
2. Divide the total rotation in radians by the elapsed time to find the angular speed: apply $\omega = \frac{\theta}{t}$.
3. Multiply the angular speed by the length of the radius to find the linear speed, expressed in terms of the length unit used for the radius and the time unit used for the elapsed time: apply $v = r\omega$.

EXAMPLE 11: FINDING A LINEAR SPEED

A bicycle has wheels 28 inches in diameter. A tachometer determines the wheels are rotating at 180 RPM (revolutions per minute). Find the speed the bicycle is traveling down the road.

Answer

Here, we have an angular speed and need to find the corresponding linear speed, since the linear speed of the outside of the tires is the speed at which the bicycle travels down the road.

We begin by converting from rotations per minute to radians per minute. It can be helpful to utilize the units to make this conversion:

$$180\frac{\text{rotations}}{\text{minute}} \cdot \frac{2\pi \text{ radians}}{\text{rotation}} = 360\pi\frac{\text{radians}}{\text{minute}}$$

Using the formula from above along with the radius of the wheels, we can find the linear speed:

$$v = (14 \text{ inches})\left(360\pi\frac{\text{radians}}{\text{minute}}\right)$$
$$= 5040\pi\frac{\text{inches}}{\text{minute}}$$

Remember that radians are a unitless measure, so it is not necessary to include them.

Finally, we may wish to convert this linear speed into a more familiar measurement, like miles per hour.

$$5040\pi\frac{\text{inches}}{\text{minute}} \cdot \frac{1 \text{ feet}}{12 \text{ inches}} \cdot \frac{1 \text{ mile}}{5280 \text{ feet}} \cdot \frac{60 \text{ minutes}}{1 \text{ hour}}$$
$$\approx 14.99 \text{ miles per hour (mph)}$$

TRY IT #11

A satellite is rotating around Earth at 0.25 radians per hour at an altitude of 242 km above Earth. If the radius of Earth is 6378 kilometers, find the linear speed of the satellite in kilometers per hour.

Answer

1655 kilometers per hour

Access these online resources for additional instruction and practice with angles, arc length, and areas of sectors.

- Angles in Standard Position
- Angle of Rotation
- Coterminal Angles
- Determining Coterminal Angles
- Positive and Negative Coterminal Angles
- Radian Measure
- Coterminal Angles in Radians
- Arc Length and Area of a Sector

Key Equations

arc length	$s = r\theta$
area of a sector (Optional)	$A = \frac{1}{2}\theta r^2$
angular speed (Optional)	$\omega = \frac{\theta}{t}$
linear speed (Optional)	$v = \frac{s}{t}$
linear speed related to angular speed (Optional)	$v = r\omega$

KEY CONCEPTS

- An angle is formed from the union of two rays, by keeping the initial side fixed and rotating the terminal side. The amount of rotation determines the measure of the angle.
- An angle is in standard position if its vertex is at the origin and its initial side lies along the positive *x*-axis. A positive angle is measured counterclockwise from the initial side and a negative angle is measured clockwise.
- To draw an angle in standard position, draw the initial side along the positive *x*-axis and then place the terminal side according to the fraction of a full rotation the angle represents.
- Two angles that have the same terminal side are called coterminal angles.
- We can find coterminal angles by adding or subtracting 360°.
- In addition to degrees, the measure of an angle can be described in radians.
- To convert between degrees and radians, use the proportion $\frac{\theta}{180} = \frac{\theta^R}{\pi}$.
- Coterminal angles can be found using radians just as they are for degrees.
- The length of a circular arc is a fraction of the circumference of the entire circle.

Optional:

- The area of sector is a fraction of the area of the entire circle.
- An object moving in a circular path has both linear and angular speed.
- The angular speed of an object traveling in a circular path is the measure of the angle through which it turns in a unit of time.
- The linear speed of an object traveling along a circular path is the distance it travels in a unit of time.

GLOSSARY

angle
 the union of two rays having a common endpoint

angular speed
 the angle through which a rotating object travels in a unit of time

arc length
 the length of the curve formed by an arc

area of a sector

area of a portion of a circle bordered by two radii and the intercepted arc; the fraction $\frac{\theta}{2\pi}$ multiplied by the area of the entire circle

coterminal angles

description of positive and negative angles in standard position sharing the same terminal side

degree

a unit of measure describing the size of an angle as one-360th of a full revolution of a circle

initial side

the side of an angle from which rotation begins

linear speed

the distance along a straight path a rotating object travels in a unit of time; determined by the arc length

measure of an angle

the amount of rotation from the initial side to the terminal side

negative angle

description of an angle measured clockwise from the positive x-axis

positive angle

description of an angle measured counterclockwise from the positive x-axis

quadrantal angle

an angle whose terminal side lies on an axis

radian measure

the ratio of the arc length formed by an angle divided by the radius of the circle

radian

the measure of a central angle of a circle that intercepts an arc equal in length to the radius of that circle

ray

one point on a line and all points extending in one direction from that point; one side of an angle

reference angle

the measure of the acute angle formed by the terminal side of the angle and the horizontal axis

standard position

the position of an angle having the vertex at the origin and the initial side along the positive x-axis

terminal side

the side of an angle at which rotation ends

vertex

the common endpoint of two rays that form an angle

3.3 Unit Circle: Sine and Cosine

Figure 1: The Singapore Flyer is the world's tallest Ferris wheel. (credit: "Vibin JK"/Flickr)

Looking for a thrill? Then consider a ride on the Singapore Flyer, the world's tallest Ferris wheel. Located in Singapore, the Ferris wheel soars to a height of 541 feet—a little more than a tenth of a mile! Described as an observation wheel, riders enjoy spectacular views as they travel from the ground to the peak and down again in a repeating pattern. In this section, we will examine this type of revolving motion around a circle. To do so, we need to define the type of circle first, and then place that circle on a coordinate system. Then we can discuss circular motion in terms of the coordinate pairs.

Finding Function Values for the Sine and Cosine

To define our trigonometric functions, we begin by drawing a unit circle, a circle centered at the origin with radius 1, as shown in Figure 2. The angle (in radians) that t intercepts forms an arc of length s. Using the formula $s = rt$, and knowing that $r = 1$, we see that for a unit circle, $s = t$.

Recall that the x- and y-axes divide the coordinate plane into four quarters called quadrants. We label these quadrants to mimic the direction a positive angle would sweep. The four quadrants are labeled I, II, III, and IV.

For any angle t, we can label the intersection of the terminal side and the unit circle by its coordinates, (x, y). Consider an angle that is in the first quadrant. We can drop a perpendicular line to the x-axis to create a right triangle. The sides of the right triangle will be x and y. If we use our right trigonometric definitions from Section 3.1, we can see that $\cos(t) = \frac{x}{1}$ and $\sin(t) = \frac{y}{1}$. This means $(x, y) = (\cos(t), \sin(t))$.

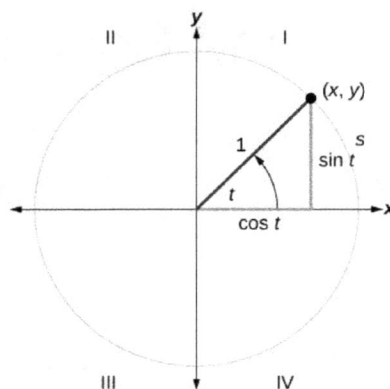

Figure 2: Unit circle where the central angle is t radians

DEFINITION

A **unit circle** has a center at $(0, 0)$ and radius 1 . In a unit circle, the length of the intercepted arc is equal to the radian measure of the central angle t.

Let (x, y) be the endpoint on the unit circle of an arc of arc length t. The (x, y) coordinates of this point can be described as functions of the angle t where:

$$\cos{(t)} = x \text{ and}$$
$$\sin{(t)} = y.$$

Note that this definition allows us to use angles which are not acute or in other words, angles in standard position whose terminal side is not in the first quadrant.

Now that we have our unit circle labeled, we can learn how the (x, y) coordinates relate to the arc length and angle. The sine function relates a real number t to the y-coordinate of the point where the corresponding angle intercepts the unit circle. More precisely, the sine of an angle t equals the y-value of the endpoint on the unit circle of an arc of length t. In Figure 3, the sine is equal to y. Like all functions, the sine function has an input and an output. Its input is the measure of the angle; its output is the y-coordinate of the corresponding point on the unit circle.

The cosine function of an angle t equals the x-value of the endpoint on the unit circle of an arc of length t. In Figure 3, the cosine is equal to x.

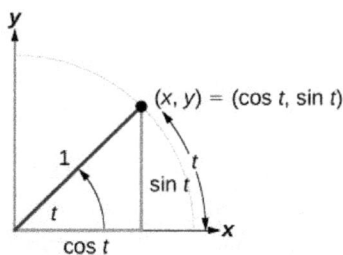

Figure 3: Illustration of an angle t, with terminal side length equal to 1, and an arc created by angle with length t.

Important Note: Because it is understood that sine and cosine are functions, we do not always need to write them with parentheses: $\sin t$ is the same as $\sin(t)$ and $\cos t$ is the same as $\cos(t)$. Likewise, $\cos^2(t)$ is a commonly used shorthand notation for $(\cos(t))^2$. Be aware that many calculators and computers do not recognize the shorthand notation. When in doubt, use the extra parentheses when entering calculations into a calculator or computer.

HOW TO

Given a point $P(x, y)$ on the unit circle corresponding to an angle of t, find the sine and cosine.

1. The sine of t is equal to the y-coordinate of point P : $\sin(t) = y$.
2. The cosine of t is equal to the x-coordinate of point P : $\cos(t) = x$.

EXAMPLE 1: FINDING FUNCTION VALUES FOR SINE AND COSINE

Point P is a point on the unit circle corresponding to an angle of t, as shown in Figure 4. Find $\cos(t)$ and $\sin(t)$.

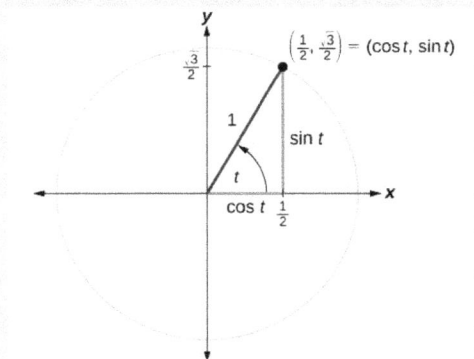

Figure 4: Graph of a circle with angle t, radius of 1, and a terminal side that intersects the circle at the given point

Answer

We know that $\cos(t)$ is the x-coordinate of the corresponding point on the unit circle and $\sin(t)$ is the y-coordinate of the corresponding point on the unit circle. So:

$$x = \cos(t) = \frac{1}{2}$$

$$y = \sin(t) = \frac{\sqrt{3}}{2}$$

A certain angle t corresponds to a point on the unit circle at $\left(-\frac{\sqrt{2}}{2}, \frac{\sqrt{2}}{2}\right)$ as shown in Figure 5. Find $\cos(t)$ and $\sin(t)$.

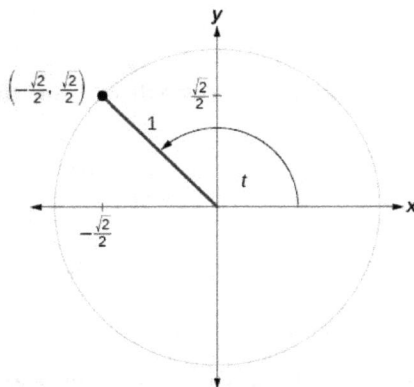

Figure 5: Graph of a circle with angle t, radius of 1, and a terminal side that intersects the circle at the given point.

Answer

$$\cos(t) = -\frac{\sqrt{2}}{2}, \quad \sin(t) = \frac{\sqrt{2}}{2}$$

Finding Sines and Cosines of Angles on an Axis

For **quadrantral angles**, the corresponding point on the unit circle falls on the *x*- or *y*-axis. In that case, we can easily calculate cosine and sine from the values of x and y.

EXAMPLE 2: CALCULATING SINES AND COSINES ALONG AN AXIS

Find $\cos(90°)$ and $\sin(90°)$.

Answer

Moving $90°$ counterclockwise around the unit circle from the positive *x*-axis brings us to the top of the circle, where the (x, y) coordinates are (0, 1), as shown in Figure 6.

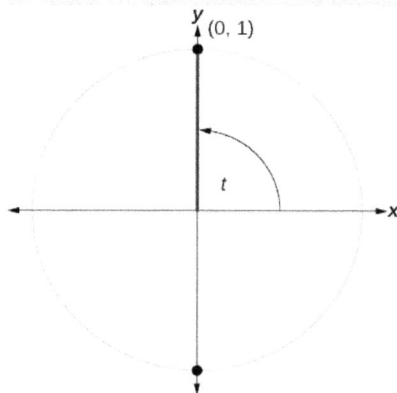

Figure 6: Graph of a circle with angle t, radius of 1, and a terminal side that intersects the circle at the point (0, 1).

Using our definitions of cosine and sine,

$$x = \cos\left(t\right) = \cos\left(90°\right) = 0$$
$$y = \sin\left(t\right) = \sin\left(90°\right) = 1$$

The cosine of 90^{\circ} is 0; the sine of 90^{\circ} is 1.

TRY IT #2

Find cosine and sine of the angle π.

Answer

$$\cos\left(\pi\right) = -1, \;\; \sin\left(\pi\right) = 0$$

The Pythagorean Identity

Now that we can define sine and cosine, we will learn how they relate to each other and the unit circle. Recall that the equation for the unit circle is $x^2 + y^2 = 1$. Because $x = \cos\left(t\right)$ and $y = \sin\left(t\right)$, we can substitute for x and y to get $\cos^2\left(t\right) + \sin^2\left(t\right) = 1$. This equation, $\cos^2\left(t\right) + \sin^2\left(t\right) = 1$, is known as the **Pythagorean Identity**. See Figure 7.

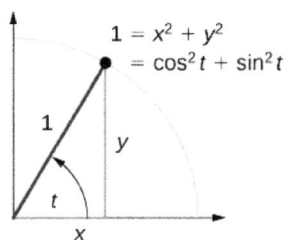

Figure 7

We can use the Pythagorean Identity to find the cosine of an angle if we know the sine, or vice versa. However, because the equation yields two solutions, we need additional knowledge of the angle to choose the solution with the correct sign. If we know the quadrant where the angle is, we can easily choose the correct solution.

DEFINITION

The **Pythagorean Identity** states that, for any real number t,

$$\cos^2(t) + \sin^2(t) = 1.$$

HOW TO

Given the sine of some angle t and its quadrant location, find the cosine of t.

1. Substitute the known value of $\sin(t)$ into the Pythagorean Identity.
2. Solve for $\cos(t)$.
3. Choose the solution with the appropriate sign for the *x*-values in the quadrant where t is located.

EXAMPLE 3: FINDING A COSINE FROM A SINE OR A SINE FROM A COSINE

If $\sin(t) = \frac{3}{7}$ and t is in the second quadrant, find $\cos(t)$.

Answer

If we drop a vertical line from the point on the unit circle corresponding to t, we create a right triangle, from which we can see that the Pythagorean Identity is simply one case of the Pythagorean Theorem. See Figure 8.

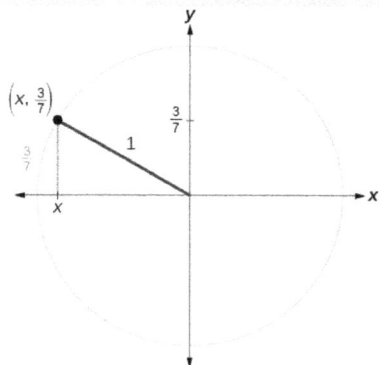

Figure 8: Graph of a unit circle with an angle that intersects the circle at a point with the y-coordinate equal to 3/7.

Substituting the known value for sine into the Pythagorean Identity,

$$\cos^2(t) + \sin^2(t) = 1$$

$$\cos^2(t) + \frac{9}{49} = 1$$

$$\cos^2(t) = \frac{40}{49}$$

$$\cos(t) = \sqrt{\frac{40}{49}} = \frac{\sqrt{40}}{7} = \frac{2\sqrt{10}}{7}$$

Because the angle is in the second quadrant, we know the x-value is a negative real number, so the cosine is also negative. So $\cos(t) = -\frac{2\sqrt{10}}{7}$.

TRY IT #3

If $\cos(t) = \frac{24}{25}$ and t is in the fourth quadrant, find $\sin(t)$.

Answer

$\sin(t) = -\frac{7}{25}$

Finding Sines and Cosines of Special Angles

We have already learned some properties of the special angles, such as the conversion from radians to degrees. In section 3.1, we also calculated sines and cosines of the special angles using the Pythagorean Identity and our knowledge of triangles.

Finding Sines and Cosines of 45° Angles and 30° and 60° Angles

We have already found the cosine and sine values for all of the most commonly encountered angles in the first quadrant of the unit circle. Figure 9 summarizes these values.

Angle	0	$\frac{\pi}{6}$, or 30°	$\frac{\pi}{4}$, or 45°	$\frac{\pi}{3}$, or 60°	$\frac{\pi}{2}$, or 90°
Cosine	1	$\frac{\sqrt{3}}{2}$	$\frac{\sqrt{2}}{2}$	$\frac{1}{2}$	0
Sine	0	$\frac{1}{2}$	$\frac{\sqrt{2}}{2}$	$\frac{\sqrt{3}}{2}$	1

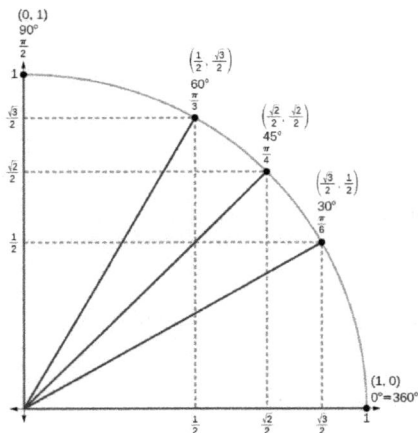

Figure 9: Graph of a quarter circle with angles of 0, 30, 45, 60, and 90 degrees inscribed. Equivalence of angles in radians shown.

Using a Calculator to Find Sine and Cosine

To find the cosine and sine of angles other than the special angles, we turn to a computer or calculator. **Be aware**: Most calculators can be set into "degree" or "radian" mode, which tells the calculator the units for the input value. When we evaluate $\cos(30)$ on our calculator, it will evaluate it as the cosine of 30 degrees if the calculator is in degree mode, or the cosine of 30 radians if the calculator is in radian mode.

HOW TO

Given an angle in radians, use a graphing calculator to find the cosine.

1. If the calculator has degree mode and radian mode, set it to radian mode.
2. Press the COS key.
3. Enter the radian value of the angle and press the close-parentheses key ")".
4. Press ENTER.

EXAMPLE 4: USING A GRAPHING CALCULATOR TO FIND SINE AND COSINE

Evaluate $\cos\left(5.1\right)$ using a graphing calculator or computer.

Answer

Make sure your calculator is in radian mode. Enter the following keystrokes:

$$\text{COS} \ (5.1) \ \text{ENTER}$$
$$\cos\left(5.1\right) \approx 0.37798$$

TRY IT #4

Evaluate $\sin\left(2.3\right)$.

Answer

approximately 0.74571

Reference Angles

For any given angle in the first quadrant, there is an angle in the second quadrant with the same *y*-value and therefore the same sine value. Because the sine value is the *y*-coordinate on the unit circle, the other angle with the same sine will share the same *y*-value, but have the opposite *x*-value. Therefore, its cosine value will be the opposite of the first angle's cosine value.

Likewise, there will be an angle in the fourth quadrant with the same *x*-value and therefore the same cosine as the original angle in the first quadrant. The angle with the same cosine will share the same *x*-value but will have the opposite *y*-value. Therefore, its sine value will be the opposite of the original angle's sine value.

As shown in Figure 11, angle α has the same sine value as angle t; the cosine values are opposites. Angle β has the same cosine value as angle t; the sine values are opposites.

$$\sin\left(t\right) = \ \sin\left(\alpha\right) \quad \text{and} \quad \cos\left(t\right) = -\cos\left(\alpha\right)$$
$$\sin\left(t\right) = -\sin\left(\beta\right) \quad \text{and} \quad \cos\left(t\right) = \ \cos\left(\beta\right)$$

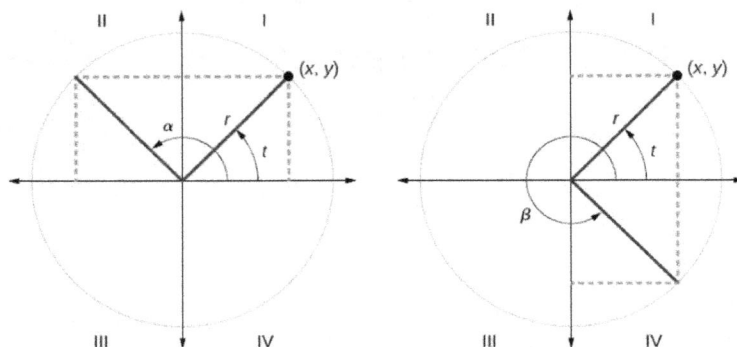

Figure 11

Finding Reference Angles

An angle's **reference angle** is the measure of the smallest, positive, acute angle t formed by the terminal side of the angle t and the horizontal axis. Thus positive reference angles have terminal sides that lie in the first quadrant and can be used as models for angles in other quadrants.

DEFINITION

An angle's **reference angle** is the size of the smallest acute angle, t', formed by the terminal side of the angle t and the horizontal axis.

We can see that reference angle is always an angle between $0°$ and $90°$, or 0 and $\frac{\pi}{2}$ radians. As we can see from Figure 12, for any angle in quadrants II, III, or IV, there is a reference angle in quadrant I.

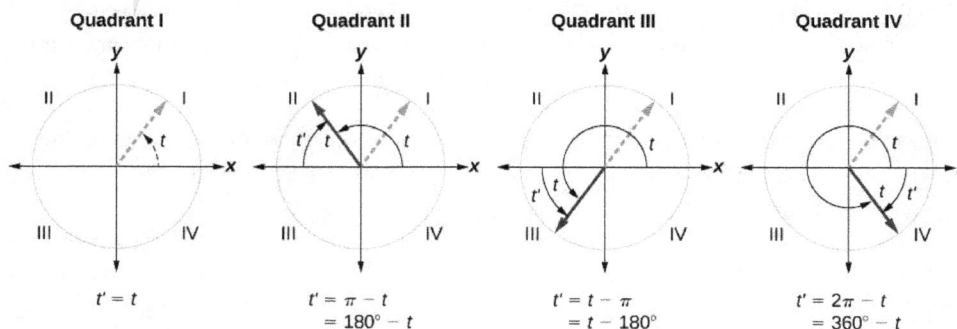

Figure 12: Reference Angles

HOW TO

Given an angle between 0 and 2π, or between $0°$ and $360°$, find its reference angle.

1. An angle in the first quadrant is its own reference angle.
2. For an angle in the second quadrant, the reference angle is $\pi - t$ or $180° - t.$

---

3. For an angle in the third quadrant, the reference angle is $t - \pi$ or $t - 180°$.
4. For an angle in the fourth quadrant, the reference angle is $2\pi - t$ or $360° - t$.

Given an angle less than 0 or greater than 2π, or less than $0°$ or greater than $360°$, find its reference angle.

1. If an angle is less than 0 or greater than 2π, add or subtract 2π as many times as needed to find a coterminal angle between 0 and 2π. Similarly, if working in degrees add or subtract $360°$ as many times as needed to find a coterminal angle between $0°$ and $360°$.
2. Once you have an angle between 0 and 2π, or between $0°$ and $360°$, follow steps 1 – 4 above.

EXAMPLE 5: FINDING A REFERENCE ANGLE

Find the reference angle of $225°$ as shown in Figure 13.

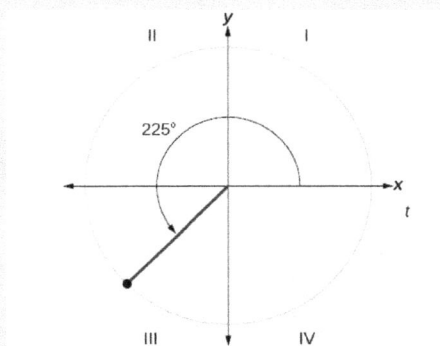

Figure 13: Graph of circle with 225 degree angle inscribed.

Answer

Because $225°$ is in the third quadrant, the reference angle is $225° - 180° = 45°$.

TRY IT # 5

Find the reference angle of $\frac{5\pi}{3}$.

Answer

$\frac{\pi}{3}$

Using Reference Angles to Find Exact Values for Cosine and Sine

We can find the cosine and sine of any angle in any quadrant if we know the cosine or sine of its reference angle. The absolute values of the cosine and sine of an angle are the same as those of the reference angle. The sign depends on the quadrant of the original angle. The cosine will be positive or negative depending on the sign of the *x*-values in that quadrant. The sine will be positive or negative depending on the sign of the *y*-values in that quadrant.

For an angle which has a special angle as the reference angle ($\frac{\pi}{6}$, $\frac{\pi}{4}$, or $\frac{\pi}{3}$,) we can produce exact value outputs for sine and cosine.

HOW TO

Given an angle in standard position, find the reference angle, and the cosine and sine of the original angle.

1. Measure the angle between the terminal side of the given angle and the horizontal axis. That is the reference angle.
2. Determine the values of the cosine and sine of the reference angle.
3. Give the cosine the same sign as the *x*-values in the quadrant of the original angle.
4. Give the sine the same sign as the *y*-values in the quadrant of the original angle.

EXAMPLE 6: USING REFERENCE ANGLES TO FIND SINE AND COSINE

1. Using a reference angle, find the exact values of $\cos\left(150°\right)$ and $\sin\left(150°\right)$.
2. Find angles between $0°$ and $360°$ which have the same exact values as $\cos\left(150°\right)$ and $\sin\left(150°\right)$.

Answer

1. 150° is located in the second quadrant. The angle it makes with the x-axis is 180° − 150° = 30°, so the reference angle is 30°.

   This tells us that 150° has the same sine and cosine values as 30°, except for the sign. We know that $\cos\left(30°\right) = \frac{\sqrt{3}}{2}$, and $\sin\left(30°\right) = \frac{1}{2}$. Since 150° is in the second quadrant, the x-coordinate of the point on the circle is negative, so the cosine value is negative. The y-coordinate is positive, so the sine value is positive.

$$\cos\left(150°\right) = -\frac{\sqrt{3}}{2}$$
$$\sin\left(150°\right) = \frac{1}{2}$$

2. From part 1, we know that 150° is in quadrant 2 and has a reference angle of 30°. We also discussed that the cosine value in quadrant 2 is negative. The other quadrant where cosine is negative is quadrant 3, since the x coordinate of any point in quadrant 3 is negative. Therefore, we need an angle in quadrant 3 that has a reference angle of 30°. Since we know that the reference angle in quadrant 3 is $t - 180°$ where t is an angle in standard position, the angle we want can be found by solving the equation

$$30° = t - 180°$$
$$t = 180° + 30°$$
$$t = 210°.$$

Therefore $\cos\left(150°\right) = \cos\left(210°\right)$.

We discussed in part 1 that the sine value in quadrant 2 is positive. The other quadrant where sine is positive is quadrant 1, since the y coordinate is positive in quadrant 1.

Therefore, $\sin\left(150°\right) = \sin\left(30°\right)$.

TRY IT #6

1. Use the reference angle of $315°$ to find the exact values for $\cos\left(315°\right)$ and $\sin\left(315°\right)$.
2. Find angles between $0°$ and $360°$ which have the same exact values as $\cos\left(315°\right)$ and $\sin\left(315°\right)$.

Answer

a. The reference angle is 45° and the angle is in the fourth quadrant. $\cos\left(315°\right) = \frac{\sqrt{2}}{2}$, $\sin\left(315°\right) = \frac{-\sqrt{2}}{2}$
$$\cos\left(315°\right) = \cos\left(45°\right)$$
b. $\sin\left(315°\right) = \sin\left(225°\right)$

EXAMPLE 7: USING REFERENCE ANGLES TO FIND SINE AND COSINE

1. Using a reference angle, find the exact values of $\cos\left(\frac{5\pi}{4}\right)$ and $\sin\left(\frac{5\pi}{4}\right)$.
2. Find angles between 0 and 2π which have the same exact values as $\cos\left(\frac{5\pi}{4}\right)$ and $\sin\left(\frac{5\pi}{4}\right)$.

Answer

1. $\frac{5\pi}{4}$ is in the third quadrant. Its reference angle is $\frac{5\pi}{4} - \pi = \frac{\pi}{4}$. The cosine and sine of $\frac{\pi}{4}$ are both $\frac{\sqrt{2}}{2}$. In the third quadrant, both x and y are negative, so:
$$\cos\left(\frac{5\pi}{4}\right) = -\frac{\sqrt{2}}{2}$$
$$\sin\left(\frac{5\pi}{4}\right) = -\frac{\sqrt{2}}{2}$$

2. We know from part 3 that $\frac{5\pi}{4}$ is in the third quadrant and its reference angle is $\frac{5\pi}{4} - \pi = \frac{\pi}{4}$. We also know that both cosine and sine are negative in quadrant 3. Cosine is also negative in quadrant 2 since the x-coordinate is negative in quadrant 2, so we can use the fact that in quadrant 2, the reference angle is:

$$\frac{\pi}{4} = \pi - t \qquad\qquad \text{t is an angle in standard position.}$$

$$t = \pi - \frac{\pi}{4}$$

$$t = \frac{3\pi}{4}$$

We therefore know that $\cos\left(\frac{5\pi}{4}\right) = \cos\left(\frac{3\pi}{4}\right)$. Sine is also negative in quadrant 4 since the y-coordinate is negative in quadrant 4. We know that in quadrant 4 the reference angle is:

$$\frac{\pi}{4} = 2\pi - t \qquad\qquad \text{Now solve for t.}$$

$$t = 2\pi - \frac{\pi}{4}$$

$$t = \frac{7\pi}{4}$$

We therefore know that $\sin\left(\frac{5\pi}{4}\right) = \sin\left(\frac{7\pi}{4}\right)$.

TRY IT #7

a. Use the reference angle of $-\frac{\pi}{6}$ to find the exact values of $\cos\left(-\frac{\pi}{6}\right)$ and $\sin\left(-\frac{\pi}{6}\right)$.
b. Find angles between 0 and 2π which have the same exact values as $\cos\left(-\frac{\pi}{6}\right)$ and $\sin\left(-\frac{\pi}{6}\right)$.

Answer

a. The reference angle is $\frac{\pi}{6}$ and the angle is in the fourth quadrant. $\cos\left(-\frac{\pi}{6}\right) = \frac{\sqrt{3}}{2}$, $\sin\left(-\frac{\pi}{6}\right) = -\frac{1}{2}$
b. $\cos\left(-\frac{\pi}{6}\right) = \cos\left(\frac{\pi}{6}\right) = \cos\left(\frac{11\pi}{6}\right)$. $\sin\left(-\frac{\pi}{6}\right) = \sin\left(\frac{7\pi}{6}\right) = \cos\left(\frac{11\pi}{6}\right)$.

Using Reference Angles to Find Coordinates

Now that we have learned how to find the cosine and sine values for angles whose reference angles are acute special angles, the rest of the special angles on the unit circle can be determine. They are shown in Figure 14. Take time to learn the (x, y) coordinates of all of the major angles in the first quadrant.

90°, $\frac{\pi}{2}$, (0, 1)

120°, $\frac{2\pi}{3}$, $\left(-\frac{1}{2}, \frac{\sqrt{3}}{2}\right)$

60°, $\frac{\pi}{3}$, $\left(\frac{1}{2}, \frac{\sqrt{3}}{2}\right)$

135°, $\frac{3\pi}{4}$, $\left(-\frac{\sqrt{2}}{2}, \frac{\sqrt{2}}{2}\right)$

45°, $\frac{\pi}{4}$, $\left(\frac{\sqrt{2}}{2}, \frac{\sqrt{2}}{2}\right)$

150°, $\frac{5\pi}{6}$, $\left(-\frac{\sqrt{3}}{2}, \frac{1}{2}\right)$

30°, $\frac{\pi}{6}$, $\left(\frac{\sqrt{3}}{2}, \frac{1}{2}\right)$

180°, π, $(-1, 0)$

0°, 0, (1, 0)
360°, 2π, (1, 0)

210°, $\frac{7\pi}{6}$, $\left(-\frac{\sqrt{3}}{2}, -\frac{1}{2}\right)$

330°, $\frac{11\pi}{6}$, $\left(\frac{\sqrt{3}}{2}, -\frac{1}{2}\right)$

225°, $\frac{5\pi}{4}$, $\left(-\frac{\sqrt{2}}{2}, -\frac{\sqrt{2}}{2}\right)$

315°, $\frac{7\pi}{4}$, $\left(\frac{\sqrt{2}}{2}, -\frac{\sqrt{2}}{2}\right)$

240°, $\frac{4\pi}{3}$, $\left(-\frac{1}{2}, -\frac{\sqrt{3}}{2}\right)$

300°, $\frac{5\pi}{3}$, $\left(\frac{1}{2}, -\frac{\sqrt{3}}{2}\right)$

270°, $\frac{3\pi}{2}$, (0, −1)

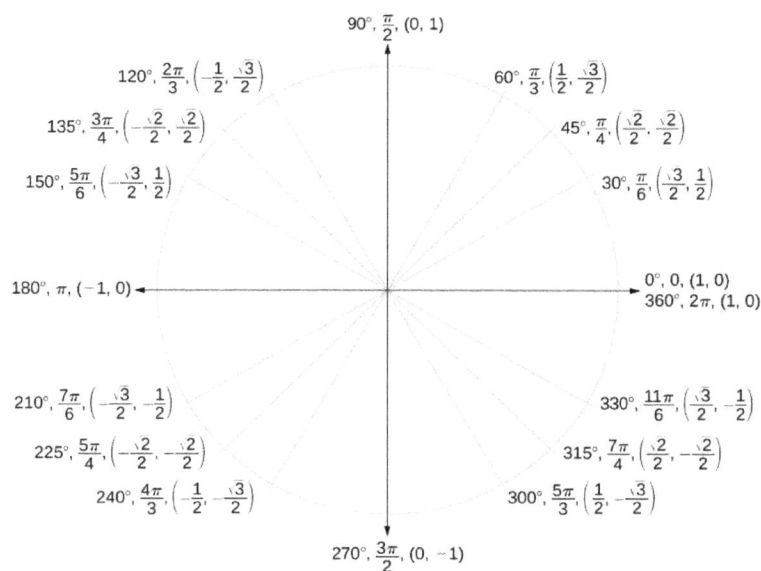

Figure 14: Special angles and coordinates of corresponding points on the unit circle

In addition to learning the values for special angles, we can use reference angles to find (x, y) coordinates of any point on the unit circle, using what we know of reference angles along with the identities

$$x = \cos(t)$$
$$y = \sin(t).$$

First we find the reference angle corresponding to the given angle. Then we take the sine and cosine values of the reference angle, and give them the signs corresponding to the y– and x-values of the quadrant.

HOW TO

Given the angle of a point on a unit circle, find the (x, y) coordinates of the point.

1. Find the reference angle by measuring the smallest angle to the x-axis.
2. Find the cosine and sine of the reference angle.
3. Determine the appropriate signs for x and y in the given quadrant.

EXAMPLE 8: USING THE UNIT CIRCLE TO FIND COORDINATES

Find the coordinates of the point on the unit circle at an angle of $\frac{7\pi}{6}$.

Answer

We know that the angle $\frac{7\pi}{6}$ is in the third quadrant.

First, let's find the reference angle by measuring the angle to the *x*-axis. To find the reference angle of an angle whose terminal side is in quadrant III, we find the difference of the angle and π.

$$\frac{7\pi}{6} - \pi = \frac{\pi}{6}$$

Next, we will find the cosine and sine of the reference angle:

$$\cos\left(\frac{\pi}{6}\right) = \frac{\sqrt{3}}{2} \qquad \sin\left(\frac{\pi}{6}\right) = \frac{1}{2}$$

We must determine the appropriate signs for *x* and *y* in the given quadrant. Because our original angle is in the third quadrant, where both x and y are negative, both cosine and sine are negative.

$$\cos\left(\frac{7\pi}{6}\right) = -\frac{\sqrt{3}}{2}$$

$$\sin\left(\frac{7\pi}{6}\right) = -\frac{1}{2}$$

Now we can calculate the (x, y) coordinates using the identities $x = \cos(\theta)$ and $y = \sin(\theta)$.

The coordinates of the point are $\left(-\frac{\sqrt{3}}{2}, -\frac{1}{2}\right)$ on the unit circle.

TRY IT # 8

Find the coordinates of the point on the unit circle at an angle of $\frac{5\pi}{3}$.

Answer

$$\left(\frac{1}{2}, -\frac{\sqrt{3}}{2}\right)$$

Circles with Radius Different Than 1

Suppose we have a circle centered at the origin other than the unit circle. How can we find the coordinates of a point where the terminal side of the angle in standard position intersects the circle? We know from Section 3.1 that the sine and cosine values of an acute angle do not change regardless of the size of the triangle. This means that $\cos(t)$ will be the same whether t is in a triangle with small or large sides. However, it is clear that the coordinates of the point of intersection of the terminal side with the circle of radius r will not be the same as the coordinates of the point on the unit circle.

We can apply similar reasoning to considering values of sine and cosine values on a circle that is not a unit circle as we did earlier in this section. By drawing an angle in standard position in quadrant 1, we can drop a perpendicular from the point of intersection of the terminal side with the circle whose center is at the origin to the x axis. If the point on the circle is (x, y), then the sides of the right triangle formed by dropping the perpendicular will also be x and y. Now, instead of the radius being 1, we would have a radius of r. Therefore, $\cos(t) = \frac{x}{r}$ and $\sin(t) = \frac{y}{r}$. This means that $x = r\cos(t)$ and $y = r\sin(t)$.

DEFINITION

A point (x, y) on the circle, centered at the origin, with radius r corresponds to

$$x = r\cos(t) \text{ and}$$
$$y = r\sin(t)$$

where t is the angle in standard position.

EXAMPLE 9: FINDING COORDINATES ON A CIRCLE WITH RADIUS r

Find the coordinates of the point on the circle with radius 5 at an angle of $\frac{7\pi}{6}$.

Answer

We already know from Example 7 that:

$$\cos\left(\frac{7\pi}{6}\right) = -\frac{\sqrt{3}}{2}$$

$$\sin\left(\frac{7\pi}{6}\right) = -\frac{1}{2}$$

Now we can calculate the (x, y) coordinates using the identities

$$x = r\cos(\theta) \text{ and } y = r\sin(\theta).$$

The coordinates of the point are $\left(-\frac{5\sqrt{3}}{2}, -\frac{5}{2}\right)$ on the circle with radius 5.

TRY IT # 9

Find the coordinates of the point on the circle with radius 7 at an angle of $\frac{5\pi}{3}$.

Answer

$\left(\frac{7}{2}, -\frac{7\sqrt{3}}{2}\right)$

Even and Odd Functions

To be able to use our sine and cosine functions freely with both positive and negative angle inputs, we should examine how each function treats a negative input. As it turns out, there is an important difference among the functions in this regard.

Recall that:

- An even function is one in which $f(-x) = f(x)$.
- An odd function is one in which $f(-x) = -f(x)$.

We can test whether a trigonometric function is even or odd by drawing a unit circle with a positive and a negative angle, as in Figure 15. The sine of the positive angle is y. The sine of the negative angle is $-y$. The sine function, then, is an odd function. The cosine of the positive angle is x, as is the cosine of the negative angle. Therefore, the cosine function is an even function.

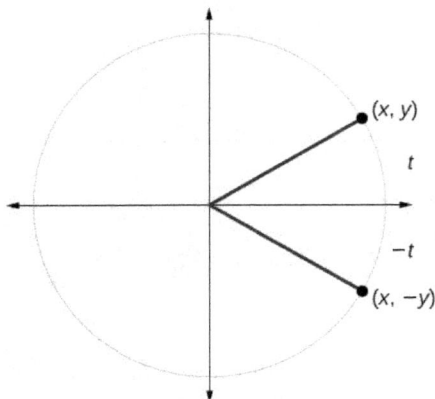

Figure 15

We can summarize this by saying:

$$\sin(-x) = -\sin(x)$$
$$\cos(-x) = \cos(x)$$

Access these online resources for additional instruction and practice with sine and cosine functions.

- Trigonometric Functions Using the Unit Circle
- Sine and Cosine from the Unit Circle
- Sine and Cosine from the Unit Circle and Multiples of Pi Divided by Six
- Sine and Cosine from the Unit Circle and Multiples of Pi Divided by Four
- Trigonometric Functions Using Reference Angles

Key Equations

Cosine	$\cos(t) = x$
Sine	$\sin(t) = y$
Pythagorean Identity	$\cos^2(t) + \sin^2(t) = 1$

KEY CONCEPTS

- Finding the function values for the sine and cosine begins with drawing a unit circle, which is centered at the origin and has a radius of 1 unit.
- Using the unit circle, the sine of an angle t equals the y-value of the endpoint on the unit circle of an arc of length t whereas the cosine of an angle t equals the x-value of the endpoint.
- When the sine or cosine is known, we can use the Pythagorean Identity to find the other. The Pythagorean Identity is also useful for determining the sines and cosines of special angles.
- Calculators and graphing software are helpful for finding sines and cosines if the proper procedure for entering information is known.
- The domain of the sine and cosine functions is all real numbers.
- The range of both the sine and cosine functions is $[-1, 1]$.
- The sine and cosine of an angle have the same absolute value as the sine and cosine of its reference angle.
- The signs of the sine and cosine are determined from the $x-$ and y-values in the quadrant of the original angle.
- An angle's reference angle is the size angle, t, formed by the terminal side of the angle t and the horizontal axis.
- Reference angles can be used to find the sine and cosine of the original angle.
- Reference angles can also be used to find the coordinates of a point on a unit circle.
- When the radius of a circle centered at the origin is not 1, we can find coordinates of a point on the circle by multiplying the sine and cosine of the angle by r.

GLOSSARY

cosine function
   the x-value of the point on a unit circle corresponding to a given angle

Pythagorean Identity
   a corollary of the Pythagorean Theorem stating that the square of the cosine of a given angle plus the square of the sine of that angle equals 1

sine function
   the y-value of the point on a unit circle corresponding to a given angle

unit circle
   a circle with a center at $(0, 0)$ and radius 1.

3.4 Graphs of the Sine and Cosine Functions

Figure 1. The tide rises and falls at regular, predictable intervals. (credit: Andrea Schaffer, Flickr)

Life is dense with phenomena that repeat in regular intervals. Each day, for example, the tides rise and fall in response to the gravitational pull of the moon. Similarly, the progression from day to night occurs as a result of Earth's rotation, and the pattern of the seasons repeats in response to Earth's revolution around the sun. Outside of nature, many stocks that mirror a company's profits are influenced by changes in the economic business cycle.

In mathematics, a function that repeats its values in regular intervals is known as a **periodic function**. The graphs of such functions show a general shape reflective of a pattern that keeps repeating. This means the graph of the function has the same output at exactly the same place in every cycle. And this translates to all the cycles of the function having exactly the same length. So, if we know all the details of one full cycle of a true periodic function, then we know the state of the function's outputs at all times, future and past. In this chapter, we will investigate various examples of periodic functions.

Graphing Sine and Cosine Functions

Recall that the cosine and sine functions relate real number values to the $x-$ and y-coordinates of a point on the unit circle. So what do they look like on a graph on a coordinate plane? Let's start with the sine function. We can create a table of values and use them to sketch a graph. Table 1 lists some of the values for the sine function on a unit circle.

Table 1

t	0	$\frac{\pi}{6}$	$\frac{\pi}{4}$	$\frac{\pi}{3}$	$\frac{\pi}{2}$	$\frac{2\pi}{3}$	$\frac{3\pi}{4}$	$\frac{5\pi}{6}$	π
$\sin(t)$	0	$\frac{1}{2}$	$\frac{\sqrt{2}}{2}$	$\frac{\sqrt{3}}{2}$	1	$\frac{\sqrt{3}}{2}$	$\frac{\sqrt{2}}{2}$	$\frac{1}{2}$	0

Plotting the points from the table and continuing along the horizontal axis gives the shape of the sine function. See Figure 2.

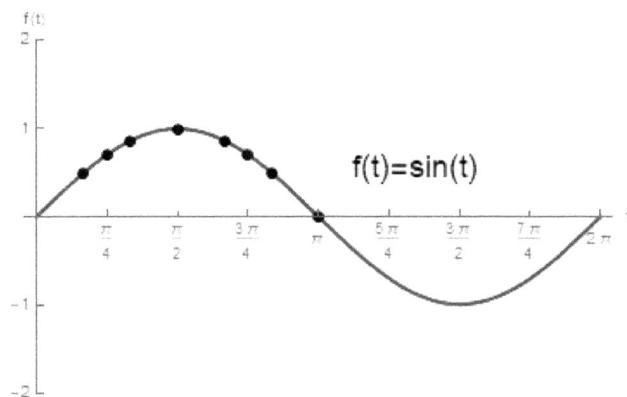

Figure 2: A graph of sin(t).

Notice how the sine values are positive between 0 and π, which correspond to the values of the sine function in quadrants I and II on the unit circle, and the sine values are negative between π and 2π, which correspond to the values of the sine function in quadrants III and IV on the unit circle. See Figure 3.

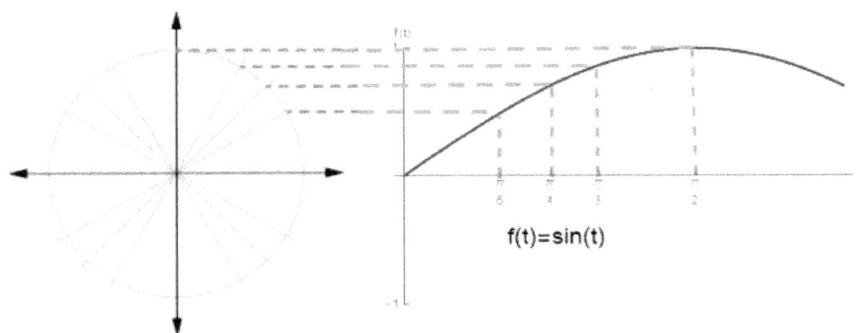

Figure 3: Plotting values of the sine function and relating them to the unit circle.

Now let's take a similar look at the cosine function. Again, we can create a table of values and use them to sketch a graph. Table 2 lists some of the values for the cosine function on a unit circle.

Table 2

t	0	$\frac{\pi}{6}$	$\frac{\pi}{4}$	$\frac{\pi}{3}$	$\frac{\pi}{2}$	$\frac{2\pi}{3}$	$\frac{3\pi}{4}$	$\frac{5\pi}{6}$	π
$\cos(t)$	1	$\frac{\sqrt{3}}{2}$	$\frac{\sqrt{2}}{2}$	$\frac{1}{2}$	0	$-\frac{1}{2}$	$-\frac{\sqrt{2}}{2}$	$-\frac{\sqrt{3}}{2}$	-1

As with the sine function, we can plots points to create a graph of the cosine function as in Figure 4.

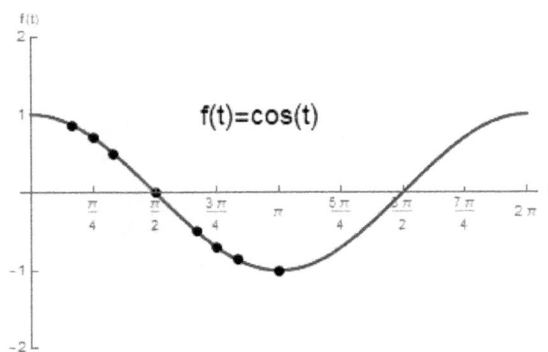

Figure 4: A graph of cos(t).

Domain and Range

We have created the graphs of the sine and cosine functions by following a point from the positive x axis one complete rotation counterclockwise around the unit circle. This rotation generates input values between 0 and 2π. We know that we can continue to travel counterclockwise around the unit circle over and over again, and generate larger and larger positive input values. That means that we can continue to generate sine and cosine values for any positive input value. Likewise, we can start at the positive x axis, and travel clockwise around the circle. One revolution in the clockwise direction generates inputs starting at 0 and decreasing to -2π. Again, since we can continue to travel around the circle over and over again in the clockwise direction, we can generate sine and cosine values for any negative input value. Because we can evaluate the sine and cosine of any real number, both of these functions have a domain of all real numbers.

What are the ranges of the sine and cosine functions? By thinking of the sine and cosine values as coordinates of points on a unit circle, it becomes clear that since the x and y coordinates of points on the circle must each be in the interval $[-1, 1]$, the range of both functions must be the interval $[-1, 1]$.

Period

In both graphs, the shape of the graph repeats after 2π, which means the functions are periodic with a period of 2π. A periodic function is a function for which a specific horizontal shift, P, results in a function equal to the original function: $f\left(t + P\right) = f\left(t\right)$ for all values of t in the domain of f. When this occurs, we call the smallest such horizontal shift with $P > 0$ the **period** of the function. Figure 5a and 5b show several periods of the sine and cosine functions.

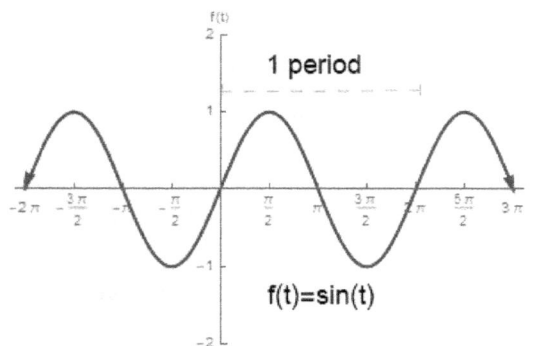

Figure 5a: Sine graph demonstrating a period.

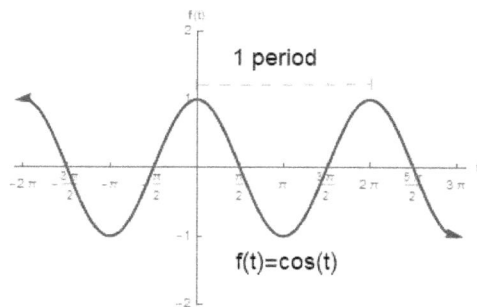

Figure 5b: Cosine graph demonstrating a period.

Symmetries

Looking again at the sine and cosine functions on a domain centered at the vertical axis helps reveal symmetries. As we can see in Figure 6, the sine function is symmetric about the origin. Recall that in Section 3.3, we determined from the unit circle that the sine function is an odd function because $\sin(-t) = -\sin(t)$. Now we can clearly see this property from the graph.

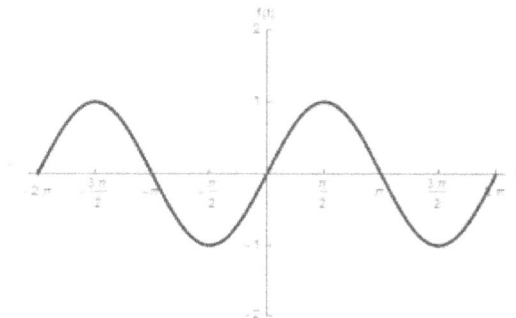

Figure 6: Odd symmetry of the sine function.

Figure 7 shows that the cosine function is symmetric about the *y*-axis. Again, we determined from the unit circle that the cosine function is an even function. Now we can see from the graph that $\cos(-t) = \cos(t)$.

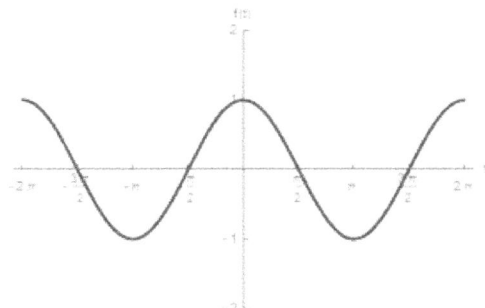

Figure 7: Even symmetry of the cosine function.

CHARACTERISTICS OF SINE AND COSINE FUNCTIONS

The sine and cosine functions have several distinct characteristics:

- They are periodic functions with a period of 2π.
- The domain of each function is $(-\infty, \infty)$ and the range is $[-1, 1]$.

- The graph of $f(t) = \sin(t)$ is symmetric about the origin, because it is an odd function.
- The graph of $f(t) = \cos(t)$ is symmetric about the vertical axis, because it is an even function.

Investigating Sinusoidal Functions

Recall when we first defined sine and cosine, we referenced an acute angle t in standard position as the input and associated each function with either the x or y coordinate of a point on the unit circle. In the material that follows, we switch back to typical function notation in the coordinate plane where $y = f(x)$ and x will represent the angle and not the coordinate.

As we can see, sine and cosine functions have a regular period and range. If we watch ocean waves or ripples on a pond, we will see that they resemble the sine or cosine functions. However, they are not necessarily identical. Some are taller or longer than others. A function that has the same general shape as a sine or cosine function is known as a **sinusoidal function**. The general forms of sinusoidal functions are:

$$y = A\sin(B(x-h)) + k \text{ and } y = A\cos(B(x-h)) + k.$$

Determining the Period of Sinusoidal Functions

Looking at the forms of sinusoidal functions, we can see that they are transformations of the sine and cosine functions. We can use what we know about transformations to determine the period.

In the general formula, B is related to the period by $P = \frac{2\pi}{|B|}$. If $|B| > 1$, then the period is less than 2π and the function undergoes a horizontal compression, whereas if $|B| < 1$, then the period is greater than 2π and the function undergoes a horizontal stretch. For example, $f(x) = \sin(x)$, $B = 1$, so the period is 2π, which we knew. If $f(x) = \sin(2x)$, then $B = 2$, so the period is π since the graph is compressed horizontally by a factor of 1/2. If $f(x) = \sin\left(\frac{x}{2}\right)$, then $B = \frac{1}{2}$, so the period is 4π since the graph is stretched horizontally by a factor of 2. Notice in Figure 8 how the period is indirectly related to $|B|$.

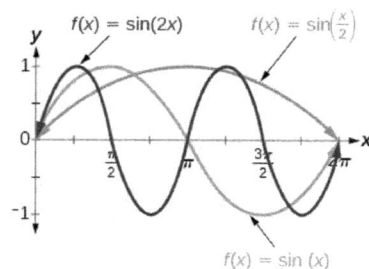

Figure 8: Three sine functions with different periods.

If we let $h = 0$ and $k = 0$ in the general form equations of the sine and cosine functions, we obtain the forms

$$y = A\sin(Bx)$$
$$y = A\cos(Bx)$$

The period is $\frac{2\pi}{|B|}$.

EXAMPLE 1: IDENTIFYING THE PERIOD OF A SINE OR COSINE FUNCTION

Determine the period of the function $f(x) = \sin\left(\frac{\pi}{6}x\right)$.

Answer

Let's begin by comparing the equation to the general form $y = A\sin(Bx)$.

In the given equation, $B = \frac{\pi}{6}$, so the period will be

$$
\begin{aligned}
P &= \frac{2\pi}{|B|} \\
&= \frac{2\pi}{\frac{\pi}{6}} \\
&= 2\pi \cdot \frac{6}{\pi} \\
&= 12.
\end{aligned}
$$

TRY IT #1

Determine the period of the function $g(x) = \cos\left(\frac{x}{3}\right)$.

Answer

6π

Determining Amplitude

Returning to the general formula for a sinusoidal function, we have analyzed how the variable B relates to the period. Now let's turn to the variable A so we can analyze how it is related to the **amplitude**, or greatest distance from rest. A represents the vertical stretch or compression factor, and its absolute value $|A|$ is the amplitude. The local maxima will be a distance $|A|$ above the horizontal **midline** of the graph, which is the line $y = k$; because $k = 0$ in this case, the midline is the x-axis. The local minima will be the same distance below the midline. If $|A| > 1$, the function is stretched vertically by a factor of $|A|$. For example, the amplitude of $f(x) = 4\sin(x)$ is twice the amplitude of $f(x) = 2\sin(x)$. If $|A| < 1$, the function is compressed vertically by a factor of $|A|$. Figure 9 compares several sine functions with different amplitudes.

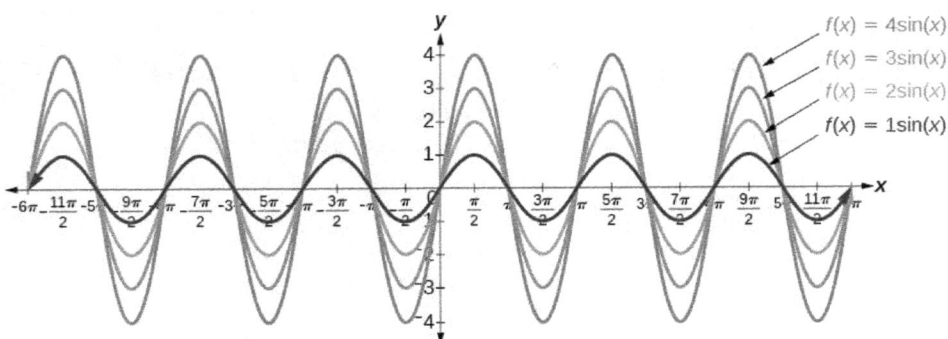

Figure 9: Four sine graphs with different amplitudes.

If we let $h = 0$ and $k = 0$ in the general form equations of the sine and cosine functions, we obtain the forms:

$$y = A\sin(Bx) \text{ and } y = A\cos(Bx).$$

The amplitude is A, and the vertical height from the midline is $|A|$. In addition, notice in the example that:

$$|A| = \text{amplitude} = \tfrac{1}{2}|\text{maximum} - \text{minimum}|.$$

EXAMPLE 2: IDENTIFYING THE AMPLITUDE OF A SINE OR COSINE FUNCTION

What is the amplitude of the sinusoidal function $f(x) = -4\sin(x)$? Is the function stretched or compressed vertically?

Answer

Let's begin by comparing the function to the simplified form $y = A\sin(Bx)$.

In the given function, $A = -4$, so the amplitude is $|A| = |-4| = 4$. The function is stretched vertically by a factor of 4.

The negative value of A results in a reflection across the x-axis of the sine function, as shown in Figure 10.

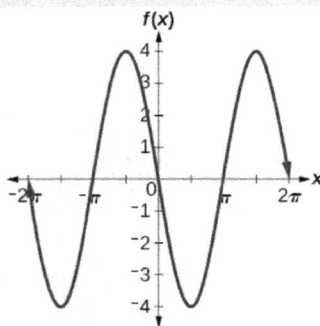

Figure 10: A graph of -4sin(x).

Analyzing Shifts of Sinusoidal Functions

Now that we understand how A and B relate to the general form equation for the sine and cosine functions, we will explore the variables h and k. Recall the general form:

$$y = A\sin(B(x-h)) + k \text{ and } y = A\cos(B(x-h)) + k$$

The value k is the vertical shift of the function and for sinusoidal functions k shifts the midline up or down k units. The function $y = A\sin(x) + k$ has a midline of $y = k$. See Figure 11.

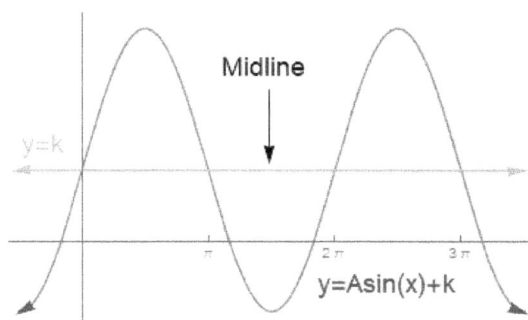

Figure 11: A graph of $y = A\sin(x) + k$.

In the equation $y = A\sin(B(x-h)) + k$, any value of k other than zero shifts the graph of $y = A\sin(B(x-h))$ up or down. Figure 12 compares $f(x) = \sin(x)$ with $f(x) = \sin(x) + 2$, which is shifted 2 units up on a graph.

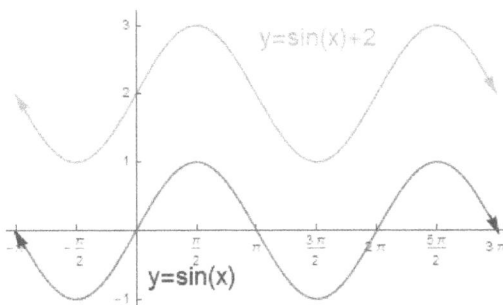

Figure 12: Vertically shifted sine function.

The value h for a sinusoidal function is called the the horizontal displacement of the basic sine or cosine function. If $h > 0$, the graph shifts to the right. If $h < 0$, the graph shifts to the left. The greater the value of $|h|$, the more the graph is shifted. Figure 13 shows that the graph of $g(x) = \sin(x - \pi)$ shifts $f(x) = \sin(x)$ to the right by π units, which is more than we see in the graph of $p(x) = \sin\left(x - \frac{\pi}{4}\right)$, which shifts $f(x) = \sin(x)$ to the right by $\frac{\pi}{4}$ units.

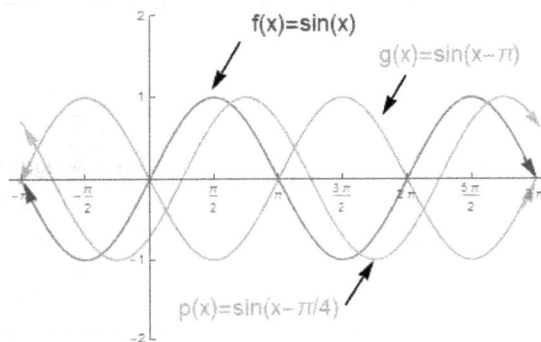

Figure 13: Horizontally shifted sine functions.

Important Note: Some books and videos use the form $y = A\cos(Bx - C) + D$ and then factor out B leaving the form $y = A\cos\left(B\left(x - \frac{C}{B}\right)\right) + D$. In this case, the value of $\frac{C}{B}$ represents what is shown previously as h (the horizontal shift). We also see the D replaces what we have previously referred to as k. It is simply a matter of getting familiar with the form that is being used in a particular problem. For the most part, we will prefer the form that uses h and k. The same pattern can be used with the sine function.

EXAMPLE 3: IDENTIFYING THE VERTICAL SHIFT OF A FUNCTION

Determine the direction and magnitude of the vertical shift for $f(x) = \cos(x) - 3$.

Answer

Let's begin by comparing the equation to the general form $y = A\cos(B(x - h)) + k,$.

In the given equation, $k = -3$ so the shift is 3 units downward.

TRY IT #3

Determine the direction and magnitude of the vertical shift for $f(x) = 3\sin(x) + 2$.

Answer

2 units up

EXAMPLE 4: IDENTIFYING THE HORIZONTAL SHIFT OF A FUNCTION

Determine the direction and magnitude of the horizontal shift for $f(x) = \sin\left(x + \frac{\pi}{6}\right) - 2$.

Answer

Let's begin by comparing the equation to the general form $y = A\sin\left(B\left(x-h\right)\right) + k.$

In the given equation, notice that $B = 1$ and $h = -\frac{\pi}{6}$. So the horizontal shift is $\frac{\pi}{6}$ units to the left.

We must pay attention to the sign in the equation for the general form of a sinusoidal function. The equation shows a minus sign before h. Therefore $f\left(x\right) = \sin\left(x + \frac{\pi}{6}\right) - 2$ can be rewritten as $f\left(x\right) = \sin\left(x - \left(-\frac{\pi}{6}\right)\right) - 2.$ If the value of h is negative, the shift is to the left.

TRY IT #4

Determine the direction and magnitude of the horizontal shift for $f\left(x\right) = 3\cos\left(x - \frac{\pi}{2}\right).$

Answer

$\frac{\pi}{2}$; right

Sine and Cosine are Co-Functions

You should recall that in Section 3.1, we described sine and cosine as co-functions. This was in relationship to the acute angles in a right triangle. We can also see this relationship when we consider the functions as functions of real numbers now that we understand how to horizontally shift these functions. See Figure 14 below.

We know that $\sin\left(x + \frac{\pi}{2}\right)$ shifts the sine function $\frac{\pi}{2}$ units to the left. Notice that this shift makes the graph of the sine curve (dotted curve) overlap the graph of the cosine curve (solid curve).

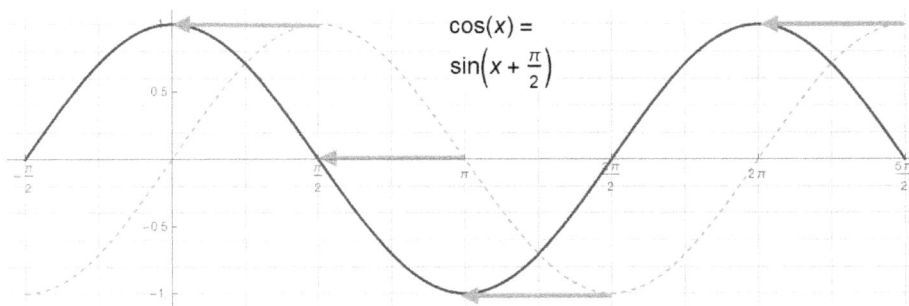

Figure 14: Shifted Sine function yields the Cosine function

Likewise, we know that $\cos\left(x - \frac{\pi}{2}\right)$ shifts the cosine function $\frac{\pi}{2}$ units to the right. Looking at the diagram above, you can see that this shift would make the cosine curve (solid curve) align with the sine curve. (dotted curve)

Putting It All Together

Given a sinusoidal function in the form $f(x) = A\sin(B(x-h)) + k$, we can identify the midline, the amplitude, the period, and the horizontal shift. Recall, the amplitude is $|A|$, the midline is $y = k$, and the horizontal shift is h. The period is related to the horizontal stretch or compression and must be calculated using the formula $P = \frac{2\pi}{|B|}$.

EXAMPLE 5: IDENTIFYING THE VARIATIONS OF A SINUSOIDAL FUNCTION FROM AN EQUATION

Determine the midline, amplitude, period, and horizontal shift of the function $y = 3\sin(2x) + 1$.

Answer

Let's begin by comparing the equation to the general form $y = A\sin(B(x-h)) + k$.

$A = 3$, so the amplitude is $|A| = 3$.

Next, $B = 2$, so the period is $P = \frac{2\pi}{|B|} = \frac{2\pi}{2} = \pi$.

There is no subtracted constant inside the parentheses, so $h = 0$ and the horizontal shift is 0.

Finally, $k = 1$, so the midline is $y = 1$.

Analysis

Inspecting the graph, we can determine that the period is π, the midline is $y = 1$, and the amplitude is 3. See Figure 15.

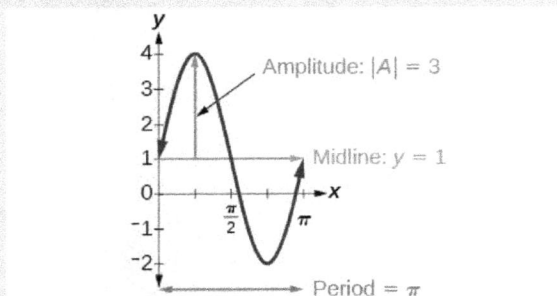

Figure 15: A graph of y=3sin(2x)+1.

TRY IT #5

Determine the midline, amplitude, period, and horizontal shift of the function $y = \frac{1}{2}\cos\left(\frac{x}{3} - \frac{\pi}{3}\right)$.

Answer

First rewrite the equation in the form: $y = \frac{1}{2}\cos\left(\frac{1}{3}(x - \pi)\right)$.

midline: $y = 0$; amplitude: $|A| = \frac{1}{2}$; period: $P = \frac{2\pi}{|B|} = 6\pi$; horizontal shift: π units right

EXAMPLE 6: IDENTIFYING THE EQUATION FOR A SINUSOIDAL FUNCTION FROM A GRAPH

Determine the formula for the cosine function in Figure 16.

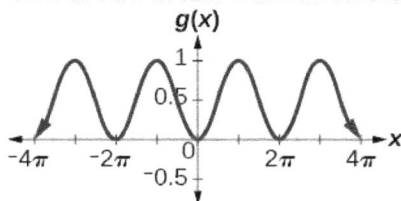

Figure 16: A graph of -0.5cos(x)+0.5.

Answer

To determine the equation, we need to identify each value in the general form of a sinusoidal function.

$$y = A\sin\left(B\left(x - h\right)\right) + k$$
$$y = A\cos\left(B\left(x - h\right)\right) + k$$

The graph could represent either a sine or a cosine function that is shifted and/or reflected. When $x = 0$, the graph has an extreme point, $(0, 0)$. Since the cosine function has an extreme point for $x = 0$, let us write our equation in terms of a cosine function.

Notice that the extreme point for $x = 0$, is a minimum. That would indicate that the cosine function has been reflected. Let's consider the amplitude and then incorporate the reflection as part of our value for A.

We can determine the amplitude by recognizing that the difference between the height of local maxima and minima is 1, so $|A| = \frac{1}{2} = 0.5$. Remember, we determined the graph is reflected about the x-axis so that $A = -0.5$.

Now let's consider the midline. We can see that the graph rises and falls an equal distance above and below $y = 0.5$. This value, which is the midline, is k in the equation, so $k = 0.5$.

The graph is not horizontally stretched or compressed since we can see that we have a minimum, maximum and minimum within 2π units, so $B = 1$; and the graph is not shifted horizontally, so $h = 0$.

Putting this all together,

$$g\left(x\right) = -0.5\cos\left(x\right) + 0.5.$$

TRY IT #6

Determine the formula for the sine function in Figure 17.

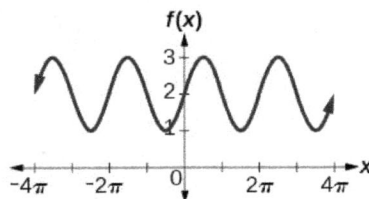

Figure 17: Determine the function for this graph.

Answer

$$f(x) = \sin(x) + 2$$

EXAMPLE 7: IDENTIFYING THE EQUATION FOR A SINUSOIDAL FUNCTION FROM A GRAPH

Determine the equation for the sinusoidal function in Figure 18.

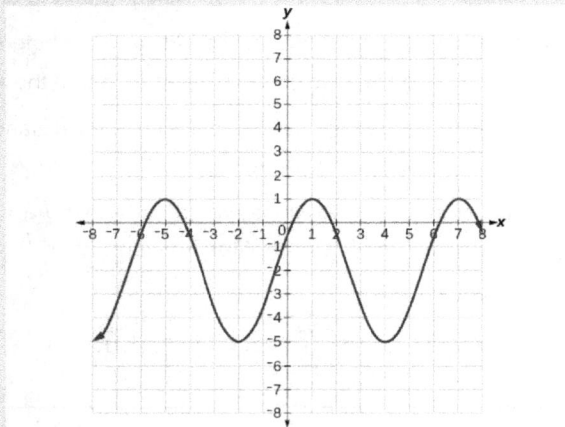

Figure 18: Determine the function for this graph.

Answer

With the highest value at 1 and the lowest value at -5, the amplitude will be $|A| = 3$, and the midline will be halfway between at -2. So $k = -2$.

The period of the graph is 6, which can be measured from the peak at $x = 1$ to the next peak at $x = 7$, or from the distance between the lowest points. Therefore, $P = \frac{2\pi}{|B|} = 6$. Using the positive value for B, we find that

$$B = \frac{2\pi}{P} = \frac{2\pi}{6} = \frac{\pi}{3}.$$

So far, our equation is either $y = 3\sin\left(\frac{\pi}{3}\left(x-h\right)\right) - 2$ or $y = 3\cos\left(\frac{\pi}{3}\left(x-h\right)\right) - 2$. For the shape and shift, we have more than one option. We could write this as any one of the following:

- a cosine shifted to the right
- a negative cosine shifted to the left
- a sine shifted to the left
- a negative sine shifted to the right

While any of these would be correct, the cosine shifts are easier to work with than the sine shifts in this case because they involve integer values. So our function becomes

$$y = 3\cos\left(\frac{\pi}{3}\left(x-1\right)\right) - 2 \text{ or } y = -3\cos\left(\frac{\pi}{3}\left(x+2\right)\right) - 2.$$

Again, these functions are equivalent, so both yield the same graph.

TRY IT #7

Write a formula for the function graphed in Figure 19.

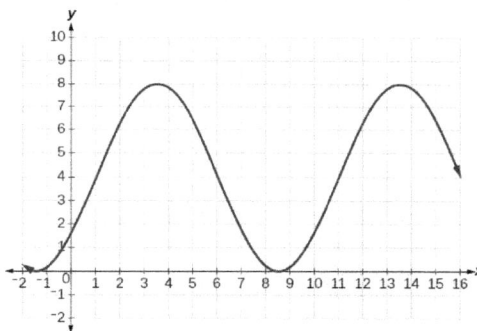

Figure 19: Determine the function for this graph.

Answer

two possibilities: $y = 4\sin\left(\frac{\pi}{5}\left(x-1\right)\right) + 4$ or $y = -4\sin\left(\frac{\pi}{5}\left(x+4\right)\right) + 4$

Throughout this section, we have learned about types of variations of sine and cosine functions and used that information to write equations from graphs. Now we can use the same information to create graphs from equations.

Instead of focusing on the general form equations

$$y = A\sin\left(B\left(x-h\right)\right) + k \text{ and } y = A\cos\left(B\left(x-h\right)\right) + k,$$

we will let $h = 0$ and $k = 0$ and work with a simplified form of the equations in the following examples.

HOW TO

Given the function $y = A\sin(Bx)$ **, sketch its graph. Note that** $h = k = 0$.

1. Identify the amplitude, $|A|$.
2. Identify the period, $P = \frac{2\pi}{|B|}$.
3. Start at the origin, with the function increasing to the right if A is positive or decreasing if A is negative.
4. At $x = \frac{\pi}{2|B|}$ there is a local maximum for $A > 0$ or a minimum for $A < 0$, with $y = A$.
5. The curve returns to the x-axis at $x = \frac{\pi}{|B|}$.
6. There is a local minimum for $A > 0$ (maximum for $A < 0$) at $x = \frac{3\pi}{2|B|}$ with $y = -A$.
7. The curve returns again to the x-axis at $x = \frac{2\pi}{|B|}$.

EXAMPLE 8: GRAPHING A FUNCTION AND IDENTIFYING THE AMPLITUDE AND PERIOD

Sketch a graph of $f(x) = -2\sin\left(\frac{\pi x}{2}\right)$.

Answer

Let's begin by comparing the equation to the form $y = A\sin(Bx)$.

- *Step 1.* We can see from the equation that $A = -2$, so the amplitude is 2.
$$|A| = 2.$$
- *Step 2.* The equation shows that $B = \frac{\pi}{2}$, so the period is
$$P = \frac{2\pi}{\frac{\pi}{2}}$$
$$= 2\pi \cdot \frac{2}{\pi}$$
$$= 4.$$
- *Step 3.* Because A is negative, the graph descends as we move to the right of the origin.
- *Step 4–7.* The x-intercepts are at the beginning of one period, $x = 0$, the horizontal midpoints are at $x = 2$ and at the end of one period at $x = 4$.

The quarter points include the minimum at $x = 1$ and the maximum at $x = 3$. A local minimum will occur 2 units below the midline, at $x = 1$, and a local maximum will occur at 2 units above the midline, at $x = 3$. Figure 20 shows the graph of the function.

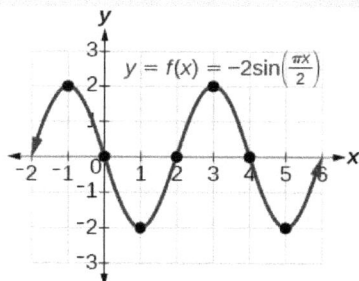

Figure 20: A graph of $y = 2\sin\left(\frac{\pi x}{2}\right).$

The graph has range of [-2,2], period of 4, and amplitude of 2.

TRY IT #8

Sketch a graph of $g(x) = -0.8\cos(2x)$. Determine the midline, amplitude, period, and horizontal shift.

Answer

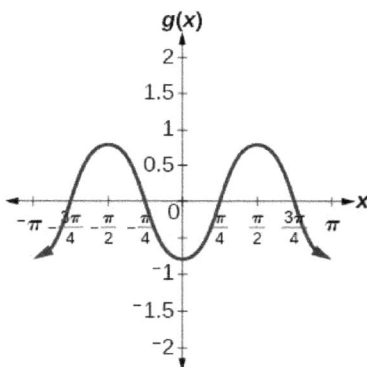

Figure 21: A graph of $y = -0.8\cos(2x).$

midline: $y = 0$; amplitude: $|A| = 0.8$; period: $P = \frac{2\pi}{|B|} = \pi$; horizontal shift: $h = 0$ or none

Given a sinusoidal function with a horizontal shift and a vertical shift, sketch its graph.

1. Express the function in the general form $y = A\sin\left(B\left(x - h\right)\right) + k$ or $y = A\cos\left(B\left(x - h\right)\right) + k$

2. Identify the amplitude, $\left|A\right|$.

3. Identify the period, $P = \frac{2\pi}{\left|B\right|}$.

4. Identify the horizontal shift, h, and the vertical shift, k.

5. Draw the graph of $f\left(x\right) = A\sin\left(Bx\right)$ or $f\left(x\right) = A\cos\left(Bx\right)$ shifted to the right or left by h and up or down by k.

EXAMPLE 9: GRAPHING A TRANSFORMED SINUSOID

Sketch a graph of $f\left(x\right) = 3\sin\left(\frac{\pi}{4}x - \frac{\pi}{4}\right)$.

Answer

- *Step 1.* First rewrite the equation in the form: $y = 3\sin\left(\frac{\pi}{4}\left(x - 1\right)\right)$. This graph will have the shape of a sine function, starting at the midline and increasing to the right.
- *Step 2.* $\left|A\right| = \left|3\right| = 3$. The amplitude is 3.
- *Step 3.* Since $\left|B\right| = \left|\frac{\pi}{4}\right| = \frac{\pi}{4}$, we determine the period as follows.
$$P = \frac{2\pi}{\left|B\right|} = \frac{2\pi}{\frac{\pi}{4}} = 2\pi \cdot \frac{4}{\pi} = 8$$

  The period is 8.

- *Step 4.* Since $h = 1$, the horizontal shift is 1 unit right. Since $k = 0$ there is no vertical shift.

- *Step 5.* Figure 22 shows the graph of the function.

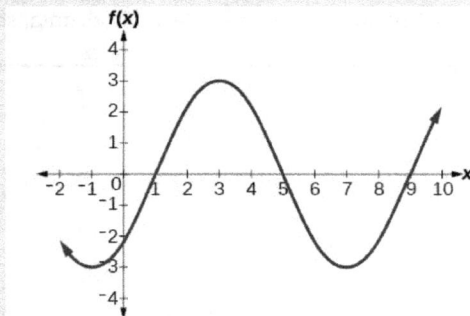

Figure 22: A horizontally stretched, vertically stretched, and horizontally shifted sinusoid graph of
$$y = 3\sin\left(\frac{\pi}{4}x - \frac{\pi}{4}\right).$$

TRY IT #9

Draw a graph of $g\left(x\right) = -2\cos\left(\frac{\pi}{3}x + \frac{\pi}{6}\right)$. Determine the midline, amplitude, period, and horizontal shift.

Answer

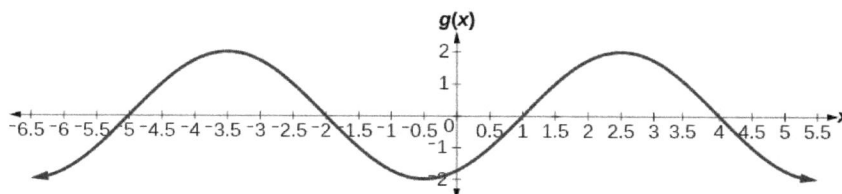

Figure 23: A graph of $y = -2\cos\left(\left(\frac{\pi}{3}\right)x + \left(\frac{\pi}{6}\right)\right)$.

First rewrite the equation in the form $g\left(x\right) = -2\cos\left(\frac{\pi}{3}\left(x + \frac{1}{2}\right)\right)$.

midline: $y = 0$; amplitude: $\left|A\right| = 2$; period: $P = \frac{2\pi}{\left|B\right|} = 6$; horizontal shift: $h = -\frac{1}{2}$ or $\frac{1}{2}$ units left

---

EXAMPLE 10: IDENTIFYING THE PROPERTIES OF A SINUSOIDAL FUNCTION

Given $y = -2\cos\left(\frac{\pi}{2}x + \pi\right) + 3$, determine the amplitude, period, horizontal shift and vertical shift. Then graph the function.

Answer

- *Step 1.* First rewrite the equation in the form: $y = -2\cos\left(\frac{\pi}{2}\left(x + 2\right)\right) + 3$.
- *Step 2.* Since $A = -2$, the amplitude is $\left|A\right| = 2$.
- *Step 3.* $\left|B\right| = \frac{\pi}{2}$, so the period is $P = \frac{2\pi}{\left|B\right|} = \frac{2\pi}{\frac{\pi}{2}} = 2\pi \cdot \frac{2}{\pi} = 4$. The period is 4.
- *Step 4.* $h = -2$, so the horizontal shift is 2 units to the left. $k = 3$, so the midline is $y = 3$, and the vertical shift is up 3.

Since A is negative, the graph of the cosine function has been reflected about the x-axis before the vertical shift is done.

Figure 24 shows one cycle of the graph of the function.

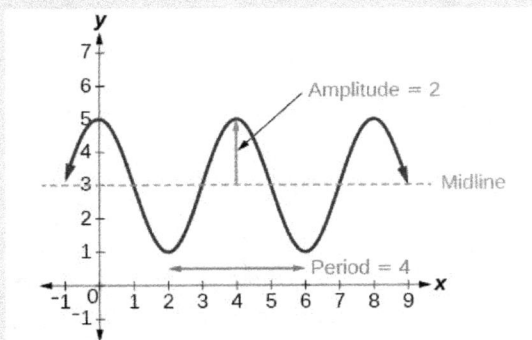

Figure 24: A graph of $\cos\left(\frac{\pi}{2}x + \pi\right) + 3$.

Using Five Key Points to Graph Sinusoidal Functions

One method of graphing sinusoidal functions is to find five key points. These points will correspond to intervals of equal length representing $\frac{1}{4}$ of the period. The key points will indicate the location of maximum and minimum values. If there is no vertical shift, they will also indicate x-intercepts. For example, suppose we want to graph the function $y = \cos\theta$. We know that the period is 2π, so we find the interval between key points as follows.

$$\frac{2\pi}{4} = \frac{\pi}{2}$$

Starting with $\theta = 0$, we calculate the first y-value, add the length of the interval $\frac{\pi}{2}$ to 0, and calculate the second y-value. We then add $\frac{\pi}{2}$ repeatedly until the five key points are determined. The last value should equal the first value, as the calculations cover one full period. Making a table similar to Table 3, we can see these key points clearly on the graph shown in Figure 25.

Table 3

θ	0	$\frac{\pi}{2}$	π	$\frac{3\pi}{2}$	2π
$y = \cos\theta$	1	0	−1	0	1

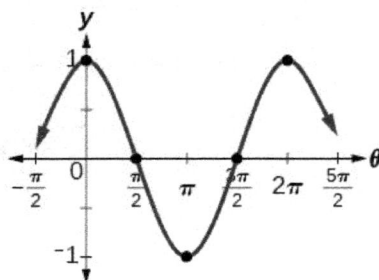

Figure 25

EXAMPLE 11: GRAPHING SINUSOIDAL FUNCTIONS USING KEY POINTS

Graph the function $y = -4\cos(\pi x)$ using amplitude, period, and key points.

Answer

The amplitude is $|-4| = 4$. The period is $\frac{2\pi}{B} = \frac{2\pi}{\pi} = 2$. One cycle of the graph can be drawn over the interval $[0, 2]$. To find the key points, we divide the period by 4 which means we will increment x by $\frac{2}{4} = \frac{1}{2}$. Make a table similar to Table 4, starting with $x = 0$ and then adding $\frac{1}{2}$ successively to x and calculate y. See the graph in Figure 26.

Table 4

x	0	$\frac{1}{2}$	1	$\frac{3}{2}$	2
$y = -4\cos(\pi x)$	-4	0	4	0	-4

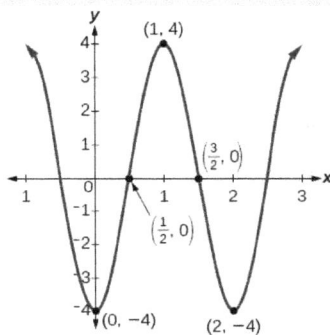

Figure 26

TRY IT #10

Graph the function $y = 3\sin(3x)$ using the amplitude, period, and five key points.

Answer

Table 5

x	$y = 3\sin(3x)$
0	0
$\frac{\pi}{6}$	3
$\frac{\pi}{3}$	0

x	$y = 3\sin(3x)$
$\frac{\pi}{2}$	-3
$\frac{2\pi}{3}$	0

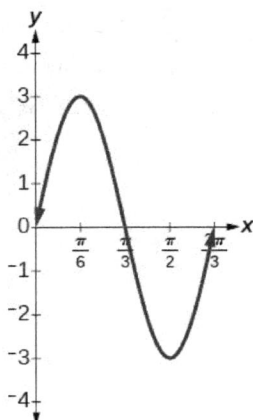

Using Transformations of Sine and Cosine Functions

We can use the transformations of sine and cosine functions in numerous applications. As mentioned at the beginning of the chapter, circular motion can be modeled using either the sine or cosine function.

EXAMPLE 12: FINDING THE VERTICAL COMPONENT OF CIRCULAR MOTION

A point rotates around a circle of radius 3 centered at the origin. Sketch a graph of the y-coordinate of the point as a function of the angle of rotation.

Answer

Recall that we discussed in Section 3.3 that a point on a circle of radius r will have a y coordinate of $y = r\sin(\theta)$. In this case, we get the equation $y(\theta) = 3\sin(\theta)$.

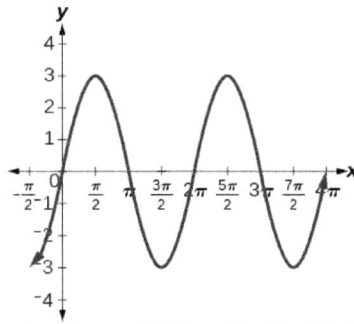

Figure 27: A graph of 3sin(θ).

Notice that the period of the function is still 2π; as we travel around the circle, we return to the point $(3, 0)$ for $x = 2\pi, 4\pi, 6\pi, \ldots$.

From our study of transformations of trigonometric functions, we also know that the constant 3 causes a vertical stretch of the sine function by a factor of 3, which we can see in the graph in Figure 27.

Because the outputs of the graph will now oscillate between -3 and 3, the amplitude of the sine wave is 3. This means that the radius of the circle centered at the origin will correspond to the amplitude of the sine function.

TRY IT #11

What is the radius of the circle whose y-coordinate corresponds to the function $f(x) = 7\cos(x)$? Sketch a graph of this function.

Answer

The radius and amplitude are 7.

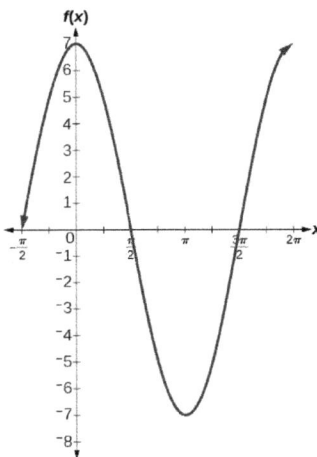

Figure 26: A graph of $y = 7\cos(x)$.

EXAMPLE 13: FINDING THE VERTICAL COMPONENT OF CIRCULAR MOTION

A circle with radius 3 ft is mounted with its center 4 ft off the ground. The point closest to the ground is labeled *P*, as shown in Figure 27. Sketch a graph of the height above the ground of the point P as the circle is rotated; then find a function that gives the height in terms of the angle of rotation.

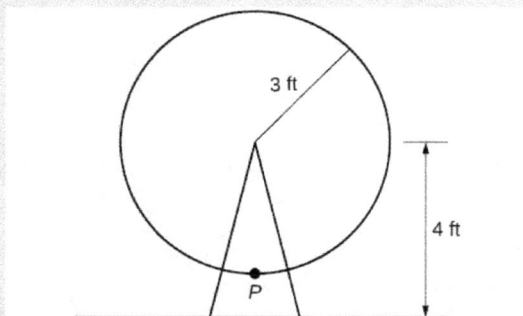

Figure 27: An illustration of a circle lifted 4 feet off the ground.

Answer

Sketching the height, we note that it will start 1 ft above the ground, then increase up to 7 ft above the ground, and continue to oscillate 3 ft above and below the center value of 4 ft, as shown in Figure 28.

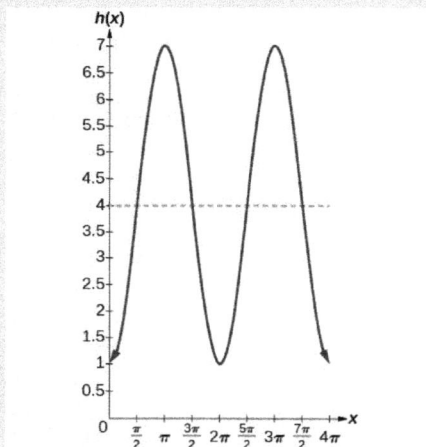

Figure 28: A graph of $y = -3\cos(x) + 4.$

Although we could use a transformation of either the sine or cosine function, we start by looking for characteristics that would make one function easier to use than the other. Let's use a cosine function because it starts at the highest or lowest value, while a sine function starts at the middle value. A standard cosine starts at the highest value, and this graph starts at the lowest value, so we need to incorporate a vertical reflection.

Second, we see that the graph oscillates 3 feet above and below the center, while a basic cosine has an amplitude of 1, so this graph has been vertically stretched by 3, as in the last example.

Finally, to move the center of the circle up to a height of 4, the graph has been vertically shifted up by 4. Putting these transformations together, we find that

$$y = -3\cos(x) + 4.$$

TRY IT #12

A weight is attached to a spring that is then hung from a board, as shown in Figure 29. As the spring oscillates up and down, the position y of the weight relative to the board ranges from -1 in. (at time $x = 0$) to -7 in. (at time $x = \pi$) below the board. Assume the position of y is given as a sinusoidal function of x. Sketch a graph of the function, and then find a cosine function that gives the position y in terms of x.

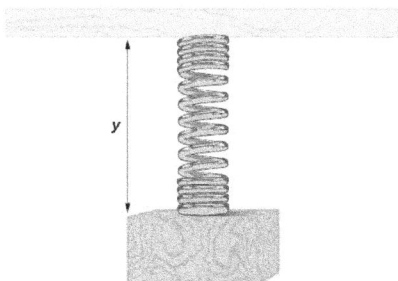

Figure 29: An illustration of a spring with length y.

Answer

$$y = 3\cos(x) - 4$$

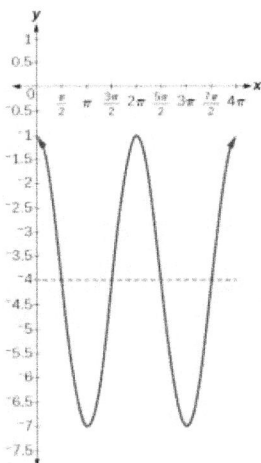

Figure 30

EXAMPLE 14: DETERMINING A RIDER'S HEIGHT ON A FERRIS WHEEL

The London Eye is a huge Ferris wheel with a diameter of 135 meters (443 feet). It completes one rotation every 30 minutes. Riders board from a platform 2 meters above the ground. Express a rider's height above ground as a function of time in minutes.

Answer

With a diameter of 135 m, the wheel has a radius of 67.5 m. The height will oscillate with amplitude 67.5 m above and below the center.

Passengers board 2 m above ground level, so the center of the wheel must be located $67.5 + 2 = 69.5$ m above ground level. The midline of the oscillation will be at 69.5 m.

The wheel takes 30 minutes to complete 1 revolution, so the height will oscillate with a period of 30 minutes.

Lastly, because the rider boards at the lowest point, the height will start at the smallest value and increase, following the shape of a vertically reflected cosine curve.

- Amplitude: 67.5, so $A = 67.5$
- Midline: 69.5, so $k = 69.5$
- Period: 30, so $B = \frac{2\pi}{30} = \frac{\pi}{15}$
- Shape: $-\cos(t)$

An equation for the rider's height would be

$$y = -67.5\cos\left(\tfrac{\pi}{15}t\right) + 69.5$$

where t is in minutes and y is measured in meters.

Access these online resources for additional instruction and practice with graphs of sine and cosine functions.

- Amplitude and Period of Sine and Cosine
- Translations of Sine and Cosine
- Graphing Sine and Cosine Transformations
- Graphing the Sine Function

Key Equations

Sinusoidal functions	$y = A\sin(B(x-h)) + k$ or $y = A\cos(B(x-h)) + k$

KEY CONCEPTS

- Periodic functions repeat after a given value. The smallest such value is the period. The basic sine and cosine functions have a period of 2π.
- The function $\sin x$ is odd, so its graph is symmetric about the origin. The function $\cos x$ is even, so its graph is symmetric about the y-axis.
- The graph of a sinusoidal function has the same general shape as a sine or cosine function.
- In the general formula for a sinusoidal function, the period is $P = \frac{2\pi}{|B|}$.
- In the general formula for a sinusoidal function, $|A|$ represents amplitude. If $|A| > 1$, the function is stretched, whereas if $|A| < 1$, the function is compressed.
- The value h in the general formula for a sinusoidal function indicates the horizontal shift.

- The value k in the general formula for a sinusoidal function indicates the vertical shift.
- Combinations of variations of sinusoidal functions can be detected from an equation.
- The equation for a sinusoidal function can be determined from a graph.
- A function can also be graphed by identifying its amplitude, period, vertical shift, and horizontal shift.
- Sinusoidal functions can be used to solve real-world problems.

GLOSSARY

amplitude
the vertical height of a function; the constant A appearing in the definition of a sinusoidal function

midline
the horizontal line $y = k$, where k appears in the general form of a sinusoidal function

periodic function
a function $f(x)$ that satisfies $f(x + P) = f(x)$ for a specific constant P and any value of x

sinusoidal function
any function that can be expressed in the form $f(x) = A\sin(B(x - h)) + k$ or
$$f(x) = A\cos(B(x - h)) + k$$

3.5 The Other Trigonometric Functions

LEARNING OBJECTIVES

In this section, you will:

- Find exact values of the trigonometric functions secant, cosecant, tangent, and cotangent of $\frac{\pi}{3}$, $\frac{\pi}{4}$, and $\frac{\pi}{6}$.
- Use reference angles to evaluate the trigonometric functions secant, cosecant, tangent, and cotangent.
- Use properties of even and odd trigonometric functions.
- Recognize and use fundamental identities.
- Evaluate trigonometric functions with a calculator.
- Describe the graphical properties of the other trigonometric functions.
- Sketch the tangent function.

A wheelchair ramp that meets the standards of the Americans with Disabilities Act must make an angle with the ground whose tangent is $\frac{1}{12}$ or less, regardless of its length. A tangent represents a ratio, so this means that for every 1 inch of rise, the ramp must have 12 inches of run. Trigonometric functions allow us to specify the shapes and proportions of objects independent of exact dimensions. Though sine and cosine are the trigonometric functions most often used, we know from our work with right triangles that there are six trigonometric functions altogether. In this section, we will investigate the remaining functions in terms of using ideas from the unit circle.

Finding Exact Values of the Trigonometric Functions Secant, Cosecant, Tangent, and Cotangent

Recall the following information that was covered in Section 3.1.

Consider a right triangle $\triangle$ ABC, with the right angle at C and with lengths a, b, and c, as in the Figure 1 below. For the acute angle A, call the leg BC its **opposite side**, and call the leg AC its **adjacent side**. Recall that the **hypotenuse** of the triangle is always opposite the right angle. In the triangle below, this is the side AB. The ratios of sides of a right triangle occur often enough in practical applications to warrant their own names, so we define the **six trigonometric functions** of A as follows:

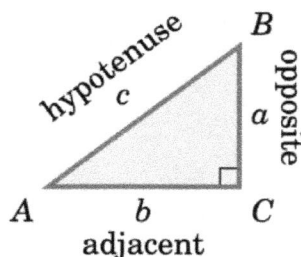

Figure 1: Sides of a right triangle with respect to angle A.

Table 1 The six trigonometric functions of A

Name of function	Abbreviation	Definition	
sine(A)	$sin(A)$	$= \dfrac{\text{opposite side}}{\text{hypotenuse}}$	$= \dfrac{a}{c}$
cosine(A)	$cos(A)$	$= \dfrac{\text{adjacent side}}{\text{hypotenuse}}$	$= \dfrac{b}{c}$
tangent(A)	$tan(A)$	$= \dfrac{\text{opposite side}}{\text{adjacent side}}$	$= \dfrac{a}{b}$
cosecant(A)	$csc(A)$	$= \dfrac{\text{hypotenuse}}{\text{opposite side}}$	$= \dfrac{c}{a}$
secant(A)	$sec(A)$	$= \dfrac{\text{hypotenuse}}{\text{adjacent side}}$	$= \dfrac{c}{b}$
cotangent(A)	$cot(A)$	$= \dfrac{\text{adjacent side}}{\text{opposite side}}$	$= \dfrac{b}{a}$

We will usually use the abbreviated names of the functions. Notice from Table 1 that the pairs sin(A) and csc(A), cos(A) and sec(A), and tan(A) and cot(A) are reciprocals:

$$\csc(A) = \frac{1}{\sin(A)} \qquad \sec(A) = \frac{1}{\cos(A)} \qquad \cot(A) = \frac{1}{\tan(A)}$$

$$\sin(A) = \frac{1}{\csc(A)} \qquad \cos(A) = \frac{1}{\sec(A)} \qquad \tan(A) = \frac{1}{\cot(A)}$$

Also recall the work we did in section 3.3 when we defined the sine and cosine functions in terms of the unit circle. For any angle t, we labeled the intersection of the terminal side and the unit circle as by its coordinates, (x, y). We considered an acute angle in the first quadrant and dropped a perpendicular line to the x- axis to create a right triangle. The sides of the right triangle were then x and y. When we used our right trigonometric definitions above, we saw that $\cos(t) = \frac{x}{1}$ and $\sin(t) = \frac{y}{1}$. This means the ordered pair $(x, y) = (\cos(t), \sin(t))$. See Figure 2 below.

As with the sine and cosine, we can use the (x, y) coordinates to find the other functions.

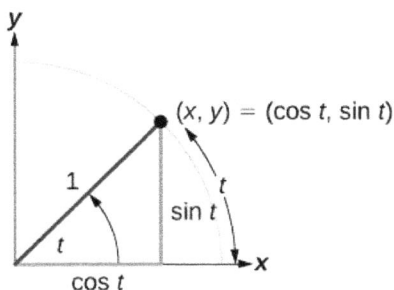

Figure 2: Unit circle where the central angle is in radians

- In right triangle trigonometry, the tangent of an angle is the ratio of the opposite side over the adjacent side with respect to the angle. In Figure 2, the tangent of angle t is equal to $\frac{y}{x}$, where $x \neq 0$.
- Because the y-value is equal to the sine of t, and the x-value is equal to the cosine of t, the tangent of angle t can also be defined as $\dfrac{\sin(t)}{\cos(t)}$, where $\cos(t) \neq 0$.
- The remaining three functions can also all be expressed as functions of a point on the unit circle.
- When we change the y-value to the sine of t, and the x-value to the cosine of t, we can express the functions in terms of the sine and cosine functions. When we do this, we typically refer to these statements as **basic trignometric identities**.

See the definition box below for details.

DEFINITION

Tangent, Secant, Cosecant, and Cotangent Functions and Basic Identities

If t is a real number and (x, y) is a point where the terminal side of an angle of t radians intercepts the unit circle, then we can create the equations below and their corresponding identities since we know that $x = \cos(t)$ and $y = \sin(t)$.

Definition	Trigonometric Identity
$\tan(t) = \frac{y}{x}, \ x \neq 0$	$\tan(t) = \frac{\sin(t)}{\cos(t)}, \ \cos(t) \neq 0$
$\sec(t) = \frac{1}{x}, \ x \neq 0$	$\sec(t) = \frac{1}{\cos(t)}, \ \cos(t) \neq 0$
$\csc(t) = \frac{1}{y}, \ y \neq 0$	$\csc(t) = \frac{1}{\sin(t)}, \ \sin(t) \neq 0$
$\cot(t) = \frac{x}{y}, \ y \neq 0$	$\cot(t) = \frac{\cos(t)}{\sin(t)}, \ \sin(t) \neq 0$

EXAMPLE 1: FINDING TRIGONOMETRIC FUNCTIONS FROM A POINT ON THE UNIT CIRCLE

The point $\left(-\frac{\sqrt{3}}{2}, \frac{1}{2}\right)$ is on the unit circle, as shown in Figure 3. Find $\sin(t), \ \cos(t), \ \tan(t), \ \sec(t), \ \csc(t),$ and $\cot(t).$

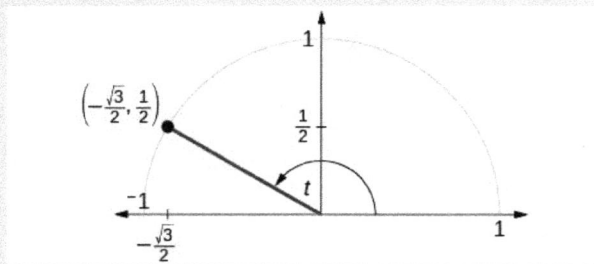

Figure 3: Graph of circle with angle of t inscribed.

Answer

Because we know the (x, y) coordinates of the point on the unit circle indicated by angle t, we can use those coordinates to find the six functions. First we know

$$\sin(t) = y = \frac{1}{2}$$

$$\cos(t) = x = -\frac{\sqrt{3}}{2}$$

Since $\tan(t) = \frac{y}{x}$ or $\tan(t) = \frac{\sin(t)}{\cos(t)}$ using the values for sine and cosine, we have

$$\tan(t) = \frac{y}{x} = \frac{\frac{1}{2}}{-\frac{\sqrt{3}}{2}} = \frac{1}{2}\left(-\frac{2}{\sqrt{3}}\right) = -\frac{1}{\sqrt{3}} = -\frac{\sqrt{3}}{3}.$$

Since $\sec(t) = \frac{1}{x}$ or $\sec(t) = \frac{1}{\cos(t)}$ and we know the value for cosine, we have

$$\sec(t) = \frac{1}{-\frac{\sqrt{3}}{2}} = -\frac{2}{\sqrt{3}} = -\frac{2\sqrt{3}}{3}.$$

Since $\csc(t) = \frac{1}{y}$ or $\sec(t) = \frac{1}{\sin(t)}$ and we know the value for sine, we have

$$\csc(t) = \frac{1}{\frac{1}{2}} = 2.$$

Finally, since $\cot(t) = \frac{x}{y}$ or $\tan(t) = \frac{\cos(t)}{\sin(t)}$ using the values for sine and cosine, we have

$$\tan(t) = \frac{\frac{1}{2}}{-\frac{\sqrt{3}}{2}} = \frac{1}{2}\left(-\frac{2}{\sqrt{3}}\right) = -\frac{1}{\sqrt{3}} = -\frac{\sqrt{3}}{3}.$$

TRY IT #1

The point $\left(\frac{\sqrt{2}}{2}, -\frac{\sqrt{2}}{2}\right)$ is on the unit circle, as shown in Figure 4.

Find $\sin(t)$, $\cos(t)$, $\tan(t)$, $\sec(t)$, $\csc(t)$, and $\cot(t)$.

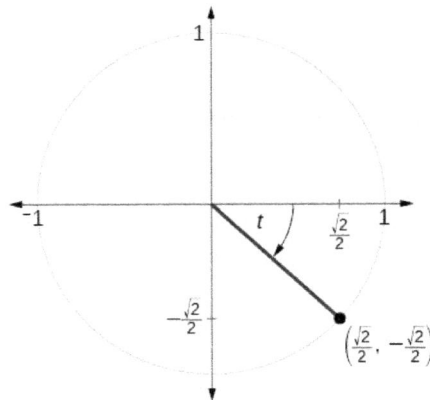

Figure 4: Graph of circle with angle of t inscribed.

Answer

$\sin(t) = -\frac{\sqrt{2}}{2}$, $\cos(t) = \frac{\sqrt{2}}{2}$, $\tan(t) = -1$, $\sec(t) = \sqrt{2}$, $\csc(t) = -\sqrt{2}$, $\cot(t) = -1$

TRY IT #2

Find the values of the six trigonometric functions of angle t based on Figure 5.

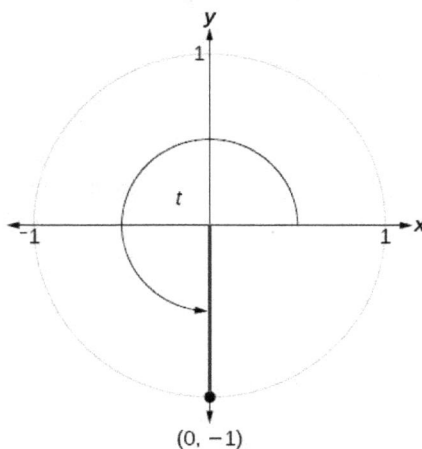

Figure 5: Graph of circle with angle of t inscribed.

Answer

$\sin(t) = -1,$

$\cos(t) = 0,$

$\tan(t) = \text{Undefined}$

$\sec(t) = \text{ Undefined},$

$\csc(t) = -1,$

$\cot(t) = 0$

EXAMPLE 2: FINDING THE TRIGONOMETRIC FUNCTIONS OF AN ANGLE

Find $\sin(t)$, $\cos(t)$, $\tan(t)$, $\sec(t)$, $\csc(t)$, and $\cot(t)$ when $t = \frac{\pi}{4}$.

Answer

We have previously used the relationship of the sides of our special right triangles to demonstrate that $\sin\left(\frac{\pi}{4}\right) = \frac{\sqrt{2}}{2}$ and $\cos\left(\frac{\pi}{4}\right) = \frac{\sqrt{2}}{2}$. We can use these values and the definitions of tangent, secant, cosecant, and cotangent as functions of sine and cosine to find the remaining function values.

$$\tan\left(\tfrac{\pi}{4}\right) = \frac{\sin\left(\frac{\pi}{4}\right)}{\cos\left(\frac{\pi}{4}\right)} = \frac{\frac{\sqrt{2}}{2}}{\frac{\sqrt{2}}{2}} = 1$$

$$\sec\left(\tfrac{\pi}{4}\right) = \frac{1}{\cos\left(\frac{\pi}{4}\right)} = \frac{1}{\frac{\sqrt{2}}{2}} = \frac{2}{\sqrt{2}} = \frac{2\sqrt{2}}{2} = \sqrt{2}$$

$$\csc\left(\tfrac{\pi}{4}\right) = \frac{1}{\sin\left(\frac{\pi}{4}\right)} = \frac{1}{\frac{\sqrt{2}}{2}} = \frac{2}{\sqrt{2}} = \frac{2\sqrt{2}}{2} = \sqrt{2}$$

$$\cot\left(\frac{\pi}{4}\right) = \frac{\cos\left(\frac{\pi}{4}\right)}{\sin\left(\frac{\pi}{4}\right)} = \frac{\frac{\sqrt{2}}{2}}{\frac{\sqrt{2}}{2}} = 1$$

TRY IT #3

Find $\sin(t)$, $\cos(t)$, $\tan(t)$, $\sec(t)$, $\csc(t)$, and $\cot(t)$ when $t = \frac{\pi}{3}$.

Answer

$$\sin\left(\frac{\pi}{3}\right) = \frac{\sqrt{3}}{2}$$

$$\cos\left(\frac{\pi}{3}\right) = \frac{1}{2}$$

$$\tan\left(\frac{\pi}{3}\right) = \sqrt{3}$$

$$\sec\left(\frac{\pi}{3}\right) = 2$$

$$\csc\left(\frac{\pi}{3}\right) = \frac{2\sqrt{3}}{3}$$

$$\cot\left(\frac{\pi}{3}\right) = \frac{\sqrt{3}}{3}$$

Because we know the sine and cosine values for the common first-quadrant angles, we can find the other function values for those angles as well by setting x equal to the cosine and y equal to the sine and then using the definitions of tangent, secant, cosecant, and cotangent. The results are shown in Table 2.

Table 2

Angle	0	$\frac{\pi}{6}$, or 30	$\frac{\pi}{4}$, or 45	$\frac{\pi}{3}$, or 60	$\frac{\pi}{2}$, or 90
Cosine	1	$\frac{\sqrt{3}}{2}$	$\frac{1}{\sqrt{2}}$ or $\frac{\sqrt{2}}{2}$	$\frac{1}{2}$	0
Sine	0	$\frac{1}{2}$	$\frac{1}{\sqrt{2}}$ or $\frac{\sqrt{2}}{2}$	$\frac{\sqrt{3}}{2}$	1
Tangent	0	$\frac{1}{\sqrt{3}}$ or $\frac{\sqrt{3}}{3}$	1	$\sqrt{3}$	Undefined
Secant	1	$\frac{2}{\sqrt{3}}$ or $\frac{2\sqrt{3}}{3}$	$\sqrt{2}$	2	Undefined
Cosecant	Undefined	2	$\sqrt{2}$	$\frac{2}{\sqrt{3}}$ or $\frac{2\sqrt{3}}{3}$	1
Cotangent	Undefined	$\sqrt{3}$	1	$\frac{1}{\sqrt{3}}$ or $\frac{\sqrt{3}}{3}$	0

Using Reference Angles to Evaluate Tangent, Secant, Cosecant, and Cotangent

We can evaluate trigonometric functions of angles outside the first quadrant using reference angles as we have already done with the sine and cosine functions. The procedure is the same: Find the reference angle formed by the terminal side of the given angle with the horizontal axis. The trigonometric function values for the original angle will be the same as those for the reference angle, except for the positive or negative sign, which is determined by x and y-values in the original quadrant. Figure 6 shows which functions are positive in which quadrant.

To help us remember which of the six trigonometric functions are positive in each quadrant, we can use the mnemonic phrase "A Smart Trig Class." Each of the four words in the phrase corresponds to one of the four quadrants, starting with quadrant I and rotating counterclockwise. In quadrant I, which is "**A**," **a**ll of the six trigonometric functions are positive. In quadrant II, "**S**mart," only **s**ine and its reciprocal function, cosecant, are positive. In quadrant III, "**T**rig," only **t**angent and its reciprocal function, cotangent, are positive. Finally, in quadrant IV, "**C**lass," only **c**osine and its reciprocal function, secant, are positive.

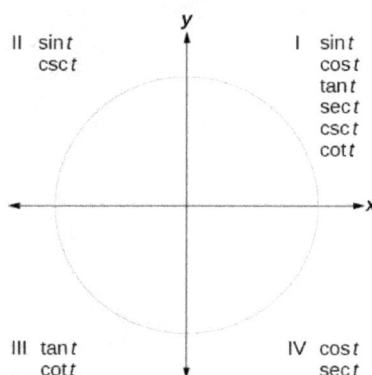

Figure 6: Quadrants with trig functions that are positive.

HOW TO

Given an angle not in the first quadrant, use reference angles to find all six trigonometric functions.

1. If the angle is not between 0 and 2π or $0°$ and $360°$, first add or subtract as many full revolutions as necessary in order to find a coterminal angle that is within these boundaries.
2. Measure the angle formed by the terminal side of your angle and the horizontal axis. This is the reference angle.
3. Evaluate the function at the reference angle.
4. Observe the quadrant where the terminal side of the original angle is located. Based on the quadrant, determine whether the output is positive or negative.

EXAMPLE 3: USING REFERENCE ANGLES TO FIND TRIGONOMETRIC FUNCTIONS

Use reference angles to find all six trigonometric functions of $-\frac{5\pi}{6}$.

Answer

A coterminal angle within 0 and 2π will be $\frac{7\pi}{6}$. The angle between this angle's terminal side and the x-axis is $\frac{\pi}{6}$, so that is the reference angle. Since $-\frac{5\pi}{6}$ is in the third quadrant, where both x and y are negative, cosine, sine, secant, and cosecant will be negative, while tangent and cotangent will be positive.

$$\cos\left(-\frac{5\pi}{6}\right) = -\frac{\sqrt{3}}{2},$$

$$\sin\left(-\frac{5\pi}{6}\right) = -\frac{1}{2},$$

$$\tan\left(-\frac{5\pi}{6}\right) = \frac{\sqrt{3}}{3}$$

$$\sec\left(-\frac{5\pi}{6}\right) = -\frac{2\sqrt{3}}{3},$$

$$\csc\left(-\frac{5\pi}{6}\right) = -2,$$

$$\cot\left(-\frac{5\pi}{6}\right) = \sqrt{3}$$

TRY IT #4

Use reference angles to find all six trigonometric functions of $-\frac{7\pi}{4}$.

Answer A co-terminal angle is $\frac{\pi}{4}$ and since this angle is in quadrant 1, it is also the reference angle.

$\sin\left(\frac{-7\pi}{4}\right) = \frac{\sqrt{2}}{2}$, $\cos\left(\frac{-7\pi}{4}\right) = \frac{\sqrt{2}}{2}$, $\tan\left(\frac{-7\pi}{4}\right) = 1$,

$\sec\left(\frac{-7\pi}{4}\right) = \sqrt{2}$, $\csc\left(\frac{-7\pi}{4}\right) = \sqrt{2}$, $\cot\left(\frac{-7\pi}{4}\right) = 1$

Using Even and Odd Trigonometric Functions

To be able to use our six trigonometric functions freely with both positive and negative angle inputs, we should examine how each function treats a negative input. As it turns out, there is an important difference among the functions in this regard.

Recall that:

- An even function is one in which $f(-x) = f(x)$.
- An odd function is one in which $f(-x) = -f(x)$.

We can test whether a trigonometric function is even or odd by drawing a unit circle with a positive and a negative angle, as in Figure 7. The sine of the positive angle is y. The sine of the negative angle is $-y$. The sine function, then, is an odd function. We can test each of the six trigonometric functions in this fashion. The results are shown in Table 3.

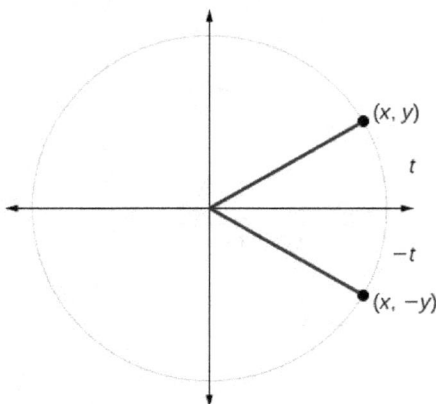

Figure 7: Graph of circle with angle of t and -t inscribed.

Table 3

$\sin(t) = y$	$\cos(t) = x$	$\tan(t) = \dfrac{y}{x}$
$\sin(-t) = -y$	$\cos(-t) = x$	$\tan(-t) = -\dfrac{y}{x}$
$\sin(-t) = -\sin(t)$	$\cos(-t) = \cos(t)$	$\tan(-t) = -\tan(t)$
$\csc(t) = \dfrac{1}{y}$	$\sec(t) = \dfrac{1}{x}$	$\cot(t) = \dfrac{x}{y}$
$\csc(-t) = \dfrac{1}{-y}$	$\sec(-t) = \dfrac{1}{x}$	$\cot(-t) = \dfrac{x}{-y}$
$\csc(-t) = -\csc(t)$	$\sec(-t) = \sec(t)$	$\cot(-t) = -\cot(t)$

Therefore, we can see that cosine and secant are even and sine, tangent, cosecant, and cotangent are odd.

EXAMPLE 4: USING EVEN AND ODD PROPERTIES OF TRIGONOMETRIC FUNCTIONS

If the secant of angle t is 2, what is the secant of $-t$?

Answer

Secant is an even function. The secant of an angle is the same as the secant of its opposite. So if the secant of angle t is 2, the secant of $-t$ is also 2.

TRY IT #5

If the cotangent of angle t is $\sqrt{3}$, what is the cotangent of $-t$?

Answer

$-\sqrt{3}$

Recognizing and Using Fundamental Identities

We have now explored a number of definitions and properties of trigonometric functions and can use them to help us find values for other trigonometric function values for a specific angle. We can also use the definitions to help simplify trigonometric expressions. As you continue on to Calculus, you will see that it is oftentimes advantageous to work with the simplest expression possible.

EXAMPLE 5: USING IDENTITIES TO SIMPLIFY TRIGONOMETRIC EXPRESSIONS

Simplify $\dfrac{\sec(t)}{\tan(t)}$.

Answer

We can simplify this by rewriting both functions in terms of sine and cosine.

$$\frac{\sec{(t)}}{\tan{(t)}} = \frac{\frac{1}{\cos(t)}}{\frac{\sin(t)}{\cos(t)}}$$

$$= \frac{1}{\cos{(t)}} \frac{\cos{(t)}}{\sin{(t)}} \quad \text{To divide the functions, multiply by the reciprocal.}$$

$$= \frac{1}{\sin{(t)}} \quad \text{Divide out the cosines.}$$

$$= \csc{(t)} \quad \text{Simplify and use the identity.}$$

By showing that $\frac{\sec(t)}{\tan(t)}$ can be simplified to $\csc{(t)}$, we have, in fact, established a new identity.

$$\frac{\sec(t)}{\tan(t)} = \csc{(t)}.$$

Analysis

When simplifying trigonometric expressions, we need to consider the domain restrictions of the original expression before any algebra is done. The equivalent expression will only be valid on the original expression's domain. The expression $\frac{\sec(t)}{\tan(t)}$ has restrictions which include those of the domain of the secant function, the domain of the tangent function and finally those where the tangent function is zero.

The domain of the tangent and secant function both exclude values where $\cos{(t)} = 0$ or where $t = \frac{\pi}{2}$ or $t = \frac{3\pi}{2}$ and any coterminal angles. This list of exceptions is often written as $\frac{\pi}{2} + \pi k$ where k is an integer. Further, tangent is zero where $\sin{(t)} = 0$ or where $t = 0$ or π and any coterminal angles. These domain exceptions can be written as $t = k\pi$ where k is an integer. Therefore, our final equality

$$\frac{\sec{(t)}}{\tan{(t)}} = \csc{(t)}$$

is only valid where $\sin{(t)} \neq 0$ and $\cos{(t)} \neq 0$ or where $t \neq \frac{\pi}{2}k$ for k an integer.

TRY IT #6

Simplify $(\tan{(t)})(\cos{(t)})$.

Answer

$\sin{(t)}$

The domain of the tangent function excludes values where $\cos{(t)} = 0$ or where $t = \frac{\pi}{2}$ or $t = \frac{3\pi}{2}$ and any coterminal angles. This list of exceptions is often written as $\frac{\pi}{2} + \pi k$ where k is an integer.

The Pythagorean Identity

You should recall that as a direct result of our definition of the sine and cosine functions in terms of the coordinates of points on the unit circle, we were able to create the Pythagorean Identity given below:

$$\cos^2(\theta) + \sin^2(\theta) = 1.$$

This identity can often be used to find the sine or cosine function value if we know one of these values for a given angle. By using the other basic identities, we can then find the values of all of the trigonometric functions for a given angle.

EXAMPLE 6: USING IDENTITIES TO RELATE TRIGONOMETRIC FUNCTIONS

If $\cos(t) = \frac{12}{13}$ and t is in quadrant IV, as shown in Figure 8, find the values of the other five trigonometric functions.

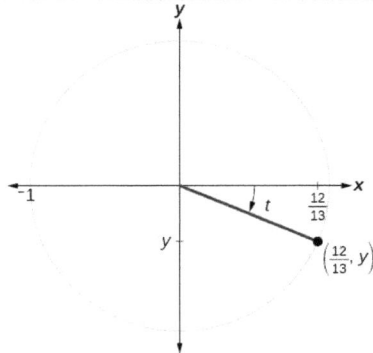

Figure 8: Graph of circle with angle of t inscribed.

Answer

We can find the sine using the Pythagorean Identity, $\cos^2(t) + \sin^2(t) = 1$, and the remaining functions by relating them to sine and cosine.

$$\left(\frac{12}{13}\right)^2 + \sin^2(t) = 1$$

$$\sin^2(t) = 1 - \left(\frac{12}{13}\right)^2$$

$$\sin^2(t) = 1 - \frac{144}{169}$$

$$\sin^2(t) = \frac{25}{169}$$

$$\sin(t) = \pm\sqrt{\frac{25}{169}}$$

$$\sin(t) = \pm\frac{\sqrt{25}}{\sqrt{169}}$$

$$\sin(t) = \pm\frac{5}{13}$$

The sign of the sine depends on the *y*-values in the quadrant where the angle is located. Since the angle is in quadrant IV, where the *y*-values are negative, its sine is negative, $-\frac{5}{13}$.

The remaining functions can be calculated using identities relating them to sine and cosine.

$$\tan(t) = \frac{\sin(t)}{\cos(t)} = \frac{-\frac{5}{13}}{\frac{12}{13}} = -\frac{5}{12}$$

$$\sec(t) = \frac{1}{\cos(t)} = \frac{1}{\frac{12}{13}} = \frac{13}{12}$$

$$\csc(t) = \frac{1}{\sin(t)} = \frac{1}{-\frac{5}{13}} = -\frac{13}{5}$$

$$\cot(t) = \frac{1}{\tan(t)} = \frac{1}{-\frac{5}{12}} = -\frac{12}{5}$$

TRY IT #7

If $\sec(t) = -\frac{17}{8}$ and $0 < t < \pi$, find the values of the other five functions. Hint: First find $\cos(t)$ using a reciprocal relationship.

Answer

$\cos(t) = -\frac{8}{17}$, $\sin(t) = \frac{15}{17}$, $\tan(t) = -\frac{15}{8}$

$\csc(t) = \frac{17}{15}$, $\cot(t) = -\frac{8}{15}$

Evaluating Trigonometric Functions with a Calculator

We have learned how to evaluate the six trigonometric functions for the common first-quadrant angles and to use them as reference angles for angles in other quadrants. To evaluate trigonometric functions of other angles, we use a scientific or graphing calculator or computer software. If the calculator has a degree mode and a radian mode, confirm the correct mode is chosen before making a calculation.

Evaluating a tangent function with a scientific calculator as opposed to a graphing calculator or computer algebra system is like evaluating a sine or cosine: Enter the value and press the TAN key. For the reciprocal functions, there may not be any dedicated keys that say CSC, SEC, or COT. In that case, the function must be evaluated as the reciprocal of a sine, cosine, or tangent.

If we need to work with degrees and our calculator or software does not have a degree mode, we can enter the degrees multiplied by the conversion factor $\frac{\pi}{180}$ to convert the degrees to radians. To find the secant of 30, we could press

(for a scientific calculator): $\dfrac{1}{30\frac{\pi}{180}}\text{COS}$

or

(for a graphing calculator): $\dfrac{1}{\cos\left(\frac{30\pi}{180}\right)}$

HOW TO

Given an angle measure in radians, use a <u>scientific</u> calculator to find the cosecant.

1. If the calculator has degree mode and radian mode, set it to radian mode.
2. Enter: $1 \; /$
3. Enter the value of the angle inside parentheses.
4. Press the SIN key.
5. Press the = key.

Given an angle measure in radians, use a <u>graphing utility</u>/calculator to find the cosecant.

1. If the graphing utility has degree mode and radian mode, set it to radian mode.
2. Enter: $1 \; /$
3. Press the SIN key.
4. Enter the value of the angle inside parentheses.
5. Press the ENTER key.

EXAMPLE 7: EVALUATING THE COSECANT USING TECHNOLOGY

Evaluate the cosecant of $\frac{5\pi}{7}$.

Answer

For a scientific calculator, enter information as follows:

$1 \; / \; (\; 5 \; \pi \; / \; 7 \;) \; \text{SIN} =$

$\csc\left(\frac{5\pi}{7}\right) \approx 1.279$

For a graphing calculator, enter the information as follows:

$1/$ and now press the SIN key. Then enter $\frac{5\pi}{7}$ followed by a closing parenthesis. Now press ENTER.

TRY IT #8

Evaluate the cotangent of $-\frac{\pi}{8}$.

Answer

≈ -2.414

Analyzing the Graph of y = tan(x)

We will begin with the graph of the tangent function, plotting points as we did for the sine and cosine functions. Recall that

$$\tan(x) = \frac{\sin(x)}{\cos(x)}.$$

Remember that there are some values of x for which $\cos(x) = 0$. For example, $\cos\left(\frac{\pi}{2}\right) = 0$ and $\cos\left(\frac{3\pi}{2}\right) = 0$. At these values, the tangent function is undefined, so the graph of $y = \tan(x)$ has discontinuities at $x = \frac{\pi}{2}$ and $\frac{3\pi}{2}$. We will examine the function from a numerical point of view to see if there is evidence that there are vertical asymptotes at these points of discontinuity.

We have already shown, using the ideas from the unit circle, that the tangent function is odd.

We can further analyze the numerical behavior of the tangent function by looking at values for some of the special angles, as listed in Table 4.

Table 4

x	$-\frac{\pi}{2}$	$-\frac{\pi}{3}$	$-\frac{\pi}{4}$	$-\frac{\pi}{6}$	0	$\frac{\pi}{6}$	$\frac{\pi}{4}$	$\frac{\pi}{3}$	$\frac{\pi}{2}$
$\tan(x)$	undefined	$-\sqrt{3}$	-1	$-\frac{\sqrt{3}}{3}$	0	$\frac{\sqrt{3}}{3}$	1	$\sqrt{3}$	undefined

These points will help us draw our graph, but we need to determine how the graph behaves where the function is undefined. If we look more closely at values when $\frac{\pi}{3} < x < \frac{\pi}{2}$, we can use a table to look for a trend. Because $\frac{\pi}{3} \approx 1.05$ and $\frac{\pi}{2} \approx 1.5707$, we will evaluate x at radian measures $1.05 < x < 1.5707$ as shown in Table 5.

Table 5

x	1.3	1.5	1.55	1.56	1.57
$\tan(x)$	3.6	14.1	48.1	92.6	1255.8

As x approaches $\frac{\pi}{2}$ from the left hand side, the outputs of the function get larger and larger or as $x \to \frac{\pi}{2}^{-}$, $\tan(x) \to \infty$. This provides us with evidence that there is a vertical asymptote at $\frac{\pi}{2}$.

Because $y = \tan(x)$ is an odd function, we see the corresponding table of negative values in Table 6.

Table 6

x		-1.3	-1.5	-1.55	-1.56	-1.57
$\tan(x)$		-3.6	-14.1	-48.1	-92.6	-1255.8

We can see that, as x approaches $-\frac{\pi}{2}$ from the right hand side, the outputs get more and more negative $x \to \frac{-\pi}{2}^{+}$, $\tan(x) \to -\infty$. Again, this gives us evidence that these is a vertical asymptote at $-\frac{\pi}{2}$.

Figure 9 represents the graph of $y = \tan(x)$. The tangent is positive from 0 to $\frac{\pi}{2}$ and from π to $\frac{3\pi}{2}$, corresponding to quadrants I and III of the unit circle.

We could create more points for other intervals and we will see that the tangent function repeats its behavior every π units. We therefore conclude that the tangent function has a period of π.

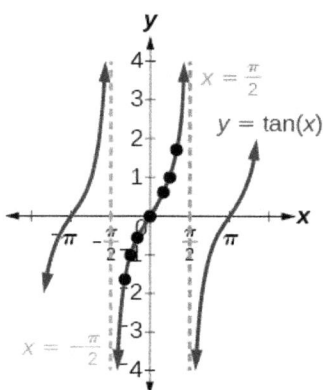

Figure 9: Graph of the tangent function.

Analyzing the Graphs of *y* = sec(x) and *y* = csc(x)

The secant was defined by the reciprocal identity $\sec\left(x\right) = \frac{1}{\cos(x)}$. Notice that the function is undefined when the cosine is 0, leading to vertical asymptotes at $\frac{\pi}{2}, \frac{3\pi}{2}$, etc. Because the cosine is never more than 1 in absolute value, the secant, being the reciprocal, will never be less than 1 in absolute value.

We can graph $y = \sec\left(x\right)$ by observing the graph of the cosine function because these two functions are reciprocals of one another. See Figure 10. The graph of the cosine is shown as a dashed orange wave so we can see the relationship. Where the graph of the cosine function decreases, the graph of the secant function increases. Where the graph of the cosine function increases, the graph of the secant function decreases. When the cosine function is zero, the secant is undefined.

The secant graph has vertical asymptotes at each value of x where the cosine graph crosses the *x*-axis; we show these in the graph below with dashed vertical lines, but will not show all the asymptotes explicitly on all later graphs involving the secant and cosecant.

Note that, because cosine is an even function, secant is also an even function. That is, $\sec\left(-x\right) = \sec\left(x\right)$

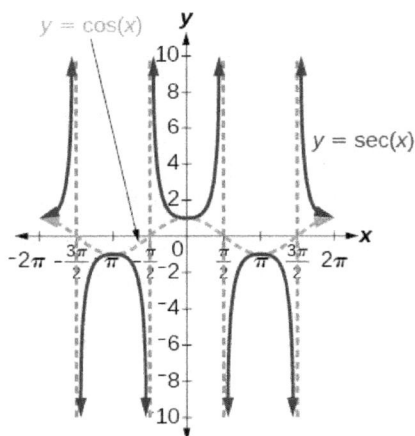

Figure 10: Graph of the secant function,
$$f\left(x\right) = \sec\left(x\right) = \frac{1}{\cos(x).}$$

Similar to the secant, the cosecant is defined by the reciprocal identity $\csc\left(x\right) = \frac{1}{\sin}\left(x\right).$ Notice that the function is undefined when the sine is 0, leading to a vertical asymptote in the graph at $0, \pi$, etc. Since the sine is never more than 1 in absolute value, the cosecant, being the reciprocal, will never be less than 1 in absolute value.

We can graph $y = \csc\left(x\right)$ by observing the graph of the sine function because these two functions are reciprocals of one another. See Figure 11. The graph of sine is shown as a dashed orange wave so we can see the relationship. Where the graph of the sine function decreases, the graph of the cosecant function increases. Where the graph of the sine function increases, the graph of the cosecant function decreases.

The cosecant graph has vertical asymptotes at each value of x where the sine graph crosses the x-axis; we show these in the graph below with dashed vertical lines.

Note that, since sine is an odd function, the cosecant function is also an odd function. That is, $\csc\left(-x\right) = -\csc\left(x\right)$.

The graph of cosecant, which is shown in Figure 9, is similar to the graph of secant.

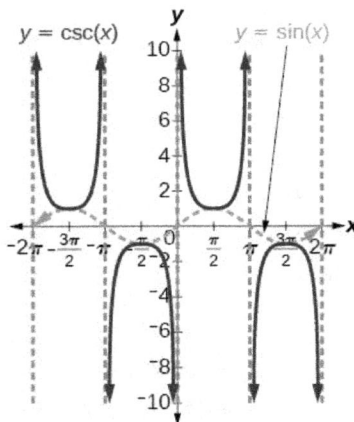

Figure 11: The graph of the cosecant function,
$$f\left(x\right) = \csc\left(x\right) = \frac{1}{\sin(x).}$$

Analyzing the Graph of *y* = cot(*x*)

The last trigonometric function we need to explore is cotangent. The cotangent is defined by the reciprocal identity $\cot\left(x\right) = \frac{1}{\tan(x)}$. Notice that the function is undefined when the tangent function is 0, leading to a vertical asymptote in the graph at $0, \pi$, etc. Since the output of the tangent function is all real numbers, the output of the cotangent function is also all real numbers.

We can graph $y = \cot\left(x\right)$ by observing the graph of the tangent function because these two functions are reciprocals of one another. See Figure 12. Where the graph of the tangent function decreases, the graph of the cotangent function increases. Where the graph of the tangent function increases, the graph of the cotangent function decreases.

The cotangent graph has vertical asymptotes at each value of x where $\tan\left(x\right) = 0$; we show these in the graph below with dashed lines. Since the cotangent is the reciprocal of the tangent, $\cot\left(x\right)$ has vertical asymptotes at all values of x where $\tan\left(x\right) = 0$, and $\cot\left(x\right) = 0$ at all values of x where $\tan\left(x\right)$ has its vertical asymptotes.

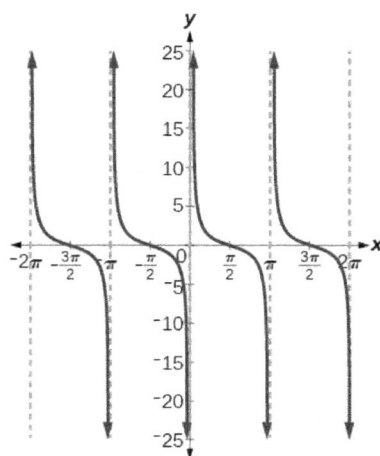

Figure 12: The cotangent function.

Period of a Function

As we have previously discussed, a function that repeats its values in regular intervals is known as a periodic function. For the four trigonometric functions, sine, cosine, cosecant and secant, a revolution of one circle, or 2π, will result in the same outputs for these functions. And for tangent and cotangent, only a half a revolution will result in the same outputs.

Remember, the period P of a repeating function f is the number representing the interval such that $f(x + P) = f(x)$ for any value of x.

The period of the cosine, sine, secant, and cosecant functions is 2π.

The period of the tangent and cotangent functions is π.

Other functions can also be periodic. For example, the lengths of months repeat every four years. If x represents the length time, measured in years, and $f(x)$ represents the number of days in February, then $f(x + 4) = f(x)$. This pattern repeats over and over through time. In other words, every four years, February is guaranteed to have the same number of days as it did 4 years earlier. The positive number 4 is the smallest positive number that satisfies this condition and is called the period. A **period** is the shortest interval over which a function completes one full cycle—in this example, the period is 4 and represents the time it takes for us to be certain February has the same number of days.

Access these online resources for additional instruction and practice with other trigonometric functions.

- Determining Trig Function Values
- More Examples of Determining Trig Functions
- Pythagorean Identities
- Trig Functions on a Calculator

Key Equations

Tangent function $\quad \tan(t) = \dfrac{\sin(t)}{\cos(t)}$

Secant function $\quad \sec(t) = \dfrac{1}{\cos(t)}$

Cosecant function $\quad \csc(t) = \dfrac{1}{\sin(t)}$

Cotangent function	$\cot(t) = \dfrac{1}{\tan(t)} = \dfrac{\cos(t)}{\sin(t)}$

KEY CONCEPTS

- The tangent of an angle is the ratio of the *y*-value to the *x*-value of the corresponding point on the unit circle.
- Secant, cotangent, and cosecant are all reciprocals of other functions. The secant function is the reciprocal of the cosine function, the cotangent function is the reciprocal of the tangent function, and the cosecant function is the reciprocal of the sine function.
- The six trigonometric functions can be found from a point on the unit circle.
- Trigonometric functions can also be found from an angle.
- Trigonometric functions of angles outside the first quadrant can be determined using reference angles.
- A function is said to be even if $f(-x) = f(x)$ and odd if $f(-x) = -f(x)$.

- Cosine and secant are even; sine, tangent, cosecant, and cotangent are odd.
- Even and odd properties can be used to evaluate trigonometric functions.
- The Pythagorean Identity makes it possible to find a cosine from a sine or a sine from a cosine.
- Identities can be used to evaluate trigonometric functions.
- Fundamental identities such as the Pythagorean Identity can be manipulated algebraically to produce new identities.
- The trigonometric functions repeat at regular intervals.
- The period P of a repeating function f is the smallest interval such that $f(x+P) = f(x)$ for any value of x.
- The values of trigonometric functions of special angles can be found by mathematical analysis.
- To evaluate trigonometric functions of other angles, we can use a calculator or computer software.

GLOSSARY

cosecant

the reciprocal of the sine function: on the unit circle, $\csc(t) = \frac{1}{y}, y \neq 0$

cotangent

the reciprocal of the tangent function: on the unit circle, $\cot(t) = \frac{x}{y}, y \neq 0$

identities

statements that are true for all values of the input on which they are defined

period

the smallest interval P of a repeating function f such that $f(x+P) = f(x)$

secant

the reciprocal of the cosine function: on the unit circle, $\sec(t) = \frac{1}{x}, x \neq 0$

tangent

the quotient of the sine and cosine: on the unit circle, $\tan(t) = \frac{y}{x}, x \neq 0$

3.6 Inverse Trigonometric Functions

LEARNING OBJECTIVES

In this section, you will:

- Understand and use the inverse sine, cosine, and tangent functions.
- Find the exact value of expressions involving the inverse sine, cosine, and tangent functions.
- Use a calculator to evaluate inverse trigonometric functions.
- Find exact values of composite functions with inverse trigonometric functions.

For any right triangle, given one other angle and the length of one side, we can figure out what the other angles and sides are. But what if we are given only two sides of a right triangle? We need a procedure that leads us from a ratio of sides to an angle. This is where the notion of an inverse to a trigonometric function comes into play. In this section, we will explore the inverse trigonometric functions.

Understanding and Using the Inverse Sine, Cosine, and Tangent Functions

In order to use inverse trigonometric functions, we need to understand that an inverse trigonometric function "undoes" what the original trigonometric function "does," as is the case with any other function and its inverse. In other words, the domain of the inverse function is the range of the original function, and vice versa, as summarized in Figure 1.

Trig Functions	Inverse Trig Functions
Domain: Measure of an angle	Domain: Ratio
Range: Ratio	Range: Measure of an angle

Figure 1

For example, if $f(x) = \sin(x)$, then we would write $f^{-1}(x) = \sin^{-1}(x)$. Be aware that $\sin^{-1}(x)$ does not mean $\frac{1}{\sin(x)}$. The following examples illustrate the inverse trigonometric functions:

- Since $\sin\left(\frac{\pi}{6}\right) = \frac{1}{2}$, then $\frac{\pi}{6} = \sin^{-1}\left(\frac{1}{2}\right)$.
- Since $\cos(\pi) = -1$, then $\pi = \cos^{-1}(-1)$.
- Since $\tan\left(\frac{\pi}{4}\right) = 1$, then $\frac{\pi}{4} = \tan^{-1}(1)$.

In previous sections, we evaluated the trigonometric functions at various angles, but at times we need to know what angle would yield a specific sine, cosine, or tangent value. For this, we need inverse functions. Recall that, for a one-to-one function, if $f(a) = b$, then an inverse function would satisfy $f^{-1}(b) = a$.

Bear in mind that the sine, cosine, and tangent functions are not one-to-one functions. The graph of each function would fail the horizontal line test. In fact, no periodic function can be one-to-one because each output in its range corresponds to at least one input in every period, and there are an infinite number of periods. As with other functions that are not one-to-one, we will need to restrict the domain of each function to yield a new function that is one-to-one. We choose a domain for each function that includes the number 0. Figure 2 shows the graph of the sine function limited to $\left[-\frac{\pi}{2}, \frac{\pi}{2}\right]$ and the graph of the cosine function limited to $[0, \pi]$.

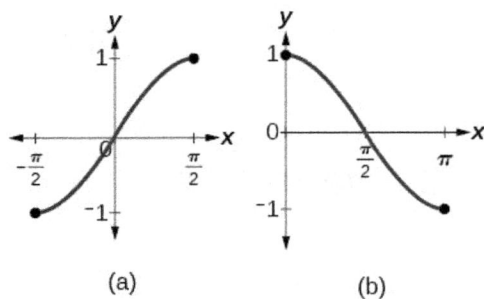

(a)   (b)

Figure 2: (a) Sine function on a restricted domain of $\left[-\frac{\pi}{2}, \frac{\pi}{2}\right]$; (b) Cosine function on a restricted domain of $[0, \pi]$

Figure 3 shows the graph of the tangent function limited to $\left(-\frac{\pi}{2}, \frac{\pi}{2}\right)$.

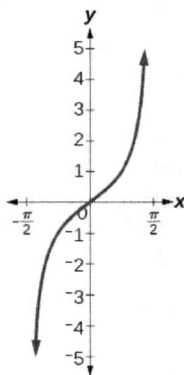

Figure 3: Tangent function on a restricted domain of $\left(-\frac{\pi}{2}, \frac{\pi}{2}\right)$

These conventional choices for the restricted domain are somewhat arbitrary, but they have important, helpful characteristics. Each domain includes the origin and some positive values, and most importantly, each results in a one-to-one function that is invertible. The conventional choice for the restricted domain of the tangent function also has the useful property that it extends from one vertical asymptote to the next instead of being divided into two parts by an asymptote.

On these restricted domains, we can define the inverse trigonometric functions.

- The inverse sine function $y = \sin^{-1}(x)$ means $x = \sin(y)$. The inverse sine function is sometimes called the arcsine function, and notated $\arcsin(x)$.
$$y = \sin^{-1}(x) \text{ has domain } [-1, 1] \text{ and range } \left[-\frac{\pi}{2}, \frac{\pi}{2}\right].$$
- The inverse cosine function $y = \cos^{-1}(x)$ means $x = \cos(y)$. The inverse cosine function is sometimes called the arccosine function, and notated $\arccos(x)$.
$$y = \cos^{-1}(x) \text{ has domain } [-1, 1] \text{ and range } [0, \pi].$$
- The inverse tangent function $y = \tan^{-1}(x)$ means $x = \tan(y)$. The inverse tangent function is sometimes called the arctangent function, and notated $\arctan(x)$.
$$y = \tan^{-1}(x) \text{ has domain } (-\infty, \infty) \text{ and range } \left(-\frac{\pi}{2}, \frac{\pi}{2}\right).$$

The graphs of the inverse functions are shown in Figure 4, Figure 5, and Figure 6. Notice that the output of each of these inverse functions is a *number*, an angle in radian measure. We see that $\sin^{-1}(x)$ has domain $[-1, 1]$ and range $\left[-\frac{\pi}{2}, \frac{\pi}{2}\right]$,

$\cos^{-1}(x)$ has domain $[-1, 1]$ and range $[0, \pi]$, and $\tan^{-1}(x)$ has domain of all real numbers and range $\left(-\frac{\pi}{2}, \frac{\pi}{2}\right)$. To find the domain and range of inverse trigonometric functions, switch the restricted domain and range of the original functions. Each graph of the inverse trigonometric function is a reflection of the graph of the original function about the line $y = x$.

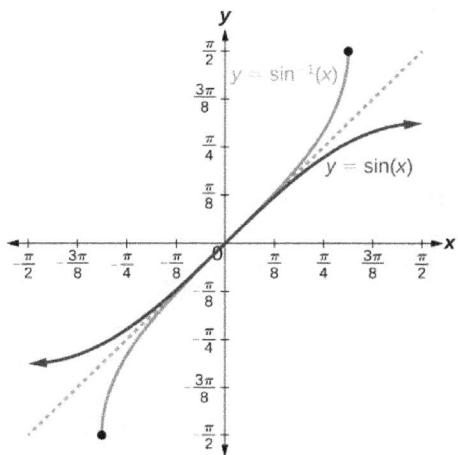

Figure 4: The sine function and inverse sine (or arcsine) function.

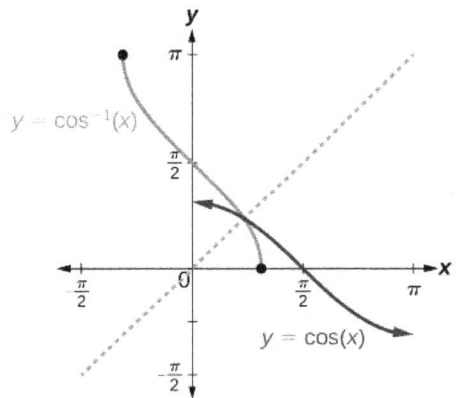

Figure 5: The cosine function and inverse cosine (or arccosine) function.

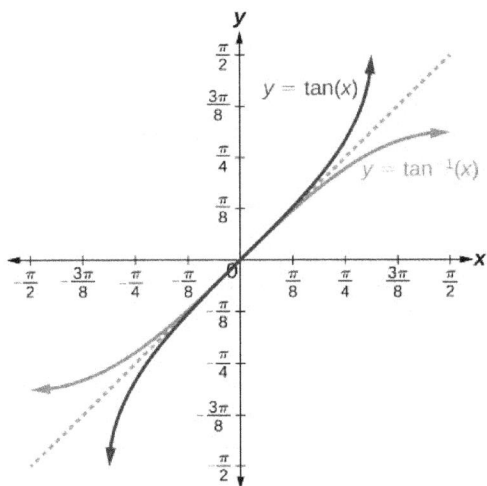

Figure 6: The tangent function and inverse tangent (or arctangent) function.

Relations for Inverse Sine, Cosine, and Tangent Functions

For angles in the interval $\left[-\frac{\pi}{2}, \frac{\pi}{2}\right]$, if $\sin(x) = y$, then $\sin^{-1}(y) = x$.

For angles in the interval $[0, \pi]$, if $\cos(x) = y$, then $\cos^{-1}(y) = x$.

For angles in the interval $\left(-\frac{\pi}{2}, \frac{\pi}{2}\right)$, if $\tan(x) = y$, then $\tan^{-1}(y) = x$.

EXAMPLE 1: WRITING A RELATION FOR AN INVERSE FUNCTION

Given $\sin\left(\frac{5\pi}{12}\right) \approx 0.96593$, write a relation involving the inverse sine.

Answer

Use the relation for the inverse sine. If $\sin(x) = y$, then $\sin^{-1}(y) = x$.

In this problem, $x = \frac{5\pi}{12}$, and $y = 0.96593$

$\sin^{-1}(0.96593) \approx \frac{5\pi}{12}$.

TRY IT #1

Given $\cos(0.5) \approx 0.8776$, write a relation involving the inverse cosine.

Answer

$\arccos(0.8776) \approx 0.5$

Finding the Exact Value of Expressions Involving the Inverse Sine, Cosine, and Tangent Functions

Now that we can identify inverse functions, we will learn to evaluate them. For most values in their domains, we must evaluate the inverse trigonometric functions by using a calculator, interpolating from a table, or using some other numerical technique. Just as we did with the original trigonometric functions, we can give exact values for the inverse functions when we are using the special angles, specifically $\frac{\pi}{6}$ (30°), $\frac{\pi}{4}$ (45°), and $\frac{\pi}{3}$ (60°), and the reflections into other quadrants.

HOW TO

Given a "special" input value, evaluate an inverse trigonometric function.

1. Find angle x for which the original trigonometric function has an output equal to the given input for the inverse trigonometric function.

---

2. If x is not in the defined range of the inverse, find another angle that is in the defined range and has the same sine, cosine, or tangent as x, depending on which corresponds to the given inverse function.

EXAMPLE 2: EVALUATING INVERSE TRIGONOMETRIC FUNCTIONS FOR SPECIAL INPUT VALUES

Evaluate each of the following.

a. $\sin^{-1}\left(\frac{1}{2}\right)$

b. $\sin^{-1}\left(-\frac{\sqrt{2}}{2}\right)$

c. $\cos^{-1}\left(-\frac{\sqrt{3}}{2}\right)$

d. $\tan^{-1}(1)$

Answer

a. Evaluating $\sin^{-1}\left(\frac{1}{2}\right)$ is the same as determining the angle that would have a sine value of $\frac{1}{2}$. In other words, what angle x would satisfy $\sin(x) = \frac{1}{2}$? There are multiple values that would satisfy this relationship, such as $\frac{\pi}{6}$ and $\frac{5\pi}{6}$, but we know we need the angle in the interval $\left[-\frac{\pi}{2}, \frac{\pi}{2}\right]$, so the answer will be $\sin^{-1}\left(\frac{1}{2}\right) = \frac{\pi}{6}$. Remember that the inverse is a function, so for each input, we will get exactly one output.

b. To evaluate $\sin^{-1}\left(-\frac{\sqrt{2}}{2}\right)$, we know that $\frac{5\pi}{4}$ and $\frac{7\pi}{4}$ both have a sine value of $-\frac{\sqrt{2}}{2}$, but neither is in the interval $\left[-\frac{\pi}{2}, \frac{\pi}{2}\right]$. For that, we need the negative angle coterminal with $\frac{7\pi}{4}$: $\sin^{-1}\left(-\frac{\sqrt{2}}{2}\right) = -\frac{\pi}{4}$.

c. To evaluate $\cos^{-1}\left(-\frac{\sqrt{3}}{2}\right)$, we are looking for an angle in the interval $[0, \pi]$ with a cosine value of $-\frac{\sqrt{3}}{2}$. The angle that satisfies this is $\cos^{-1}\left(-\frac{\sqrt{3}}{2}\right) = \frac{5\pi}{6}$.

d. Evaluating $\tan^{-1}(1)$, we are looking for an angle in the interval $\left(-\frac{\pi}{2}, \frac{\pi}{2}\right)$ with a tangent value of 1. The correct angle is $\tan^{-1}(1) = \frac{\pi}{4}$.

TRY IT #2

Evaluate each of the following.

a. $\sin^{-1}(-1)$

b. $\tan^{-1}(-1)$

c. $\cos^{-1}(-1)$

d. $\cos^{-1}\left(\frac{1}{2}\right)$

Answer

a. $-\frac{\pi}{2}$; b. $-\frac{\pi}{4}$; c. π; d. $\frac{\pi}{3}$

Using a Calculator to Evaluate Inverse Trigonometric Functions

To evaluate inverse trigonometric functions that do not involve the special angles discussed previously, we will need to use a calculator or other type of technology. Most scientific calculators and calculator-emulating applications have specific keys or buttons for the inverse sine, cosine, and tangent functions. These may be labeled, for example, SIN^{-1}, ARCSIN, or ASIN.

In Section 3.1, we worked with trigonometry on a right triangle to solve for the sides of a triangle given one side and an additional angle. Using the inverse trigonometric functions, we can solve for the angles of a right triangle given two sides, and we can use a calculator to find the values to several decimal places.

In these examples and exercises, the answers will be interpreted as angles and we will use θ as the independent variable. The value displayed on the calculator may be in degrees or radians, so be sure to set the mode appropriate to the application.

EXAMPLE 3: EVALUATING THE INVERSE SINE ON A CALCULATOR

Evaluate $\sin^{-1}(0.97)$ using a calculator.

Answer

Because the output of the inverse function is an angle, the calculator will give us a degree value if in degree mode and a radian value if in radian mode. Calculators also use the same domain restrictions on the angles as we are using.

In radian mode, $\sin^{-1}(0.97) \approx 1.3252$. In degree mode, $\sin^{-1}(0.97) \approx 75.93$. Note that in calculus and beyond we will use radians in almost all cases.

TRY IT #3

Evaluate $\cos^{-1}(-0.4)$ using a calculator.

Answer

1.9823 or 113.578°

HOW TO

Given two sides of a right triangle like the one shown in Figure 7, find an angle.

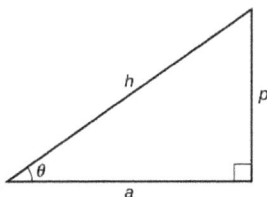

Figure 7

1. If one given side is the hypotenuse of length h and the side of length a adjacent to the desired angle is given, use the equation $\theta = \cos^{-1}\left(\frac{a}{h}\right)$.

2. If one given side is the hypotenuse of length h and the side of length p opposite to the desired angle is given, use the equation $\theta = \sin^{-1}\left(\frac{p}{h}\right)$.

3. If the two legs (the sides adjacent to the right angle) are given, then use the equation $\theta = \tan^{-1}\left(\frac{p}{a}\right)$.

EXAMPLE 4: APPLYING AN INVERSE FUNCTION TO A RIGHT TRIANGLE

Solve the triangle in Figure 8 for the angle θ.

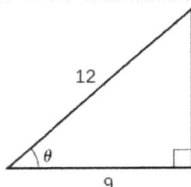

Figure 8

Answer

Because we know the hypotenuse and the side adjacent to the angle, it makes sense for us to use the cosine function.

$$\cos \theta = \frac{9}{12}$$

$$\theta = \cos^{-1}\left(\frac{9}{12}\right) \qquad \text{Apply definition of the inverse.}$$

$$\theta \approx 0.7227 \text{ or about } 41.4096 \qquad \text{Evaluate.}$$

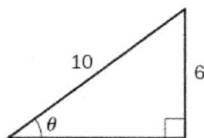
Finding Exact Values of Composite Functions with Inverse Trigonometric Functions

There are times when we need to compose a trigonometric function with an inverse trigonometric function. In these cases, we can usually find exact values for the resulting expressions without resorting to a calculator. Even when the input to the composite function is a variable or an expression, we can often find an expression for the output.

Evaluating Compositions of the Form $f\left(f^{-1}(y)\right)$ and $f^{-1}(f(x))$

For any trigonometric function, $f\left(f^{-1}(y)\right) = y$ for all y in the proper domain for the given function. This follows from the definition of the inverse and from the fact that the range of f was defined to be identical to the domain of f^{-1}. However, we have to be a little more careful with expressions of the form $f^{-1}(f(x))$.

Compositions of a trigonometric function and its inverse

$$\sin\left(\sin^{-1}(x)\right) = x \text{ for } -1 \leq x \leq 1$$
$$\cos\left(\cos^{-1}(x)\right) = x \text{ for } -1 \leq x \leq 1$$
$$\tan\left(\tan^{-1}(x)\right) = x \text{ for } -\infty < x < \infty$$
$$\sin^{-1}(\sin(x)) = x \text{ only for } -\frac{\pi}{2} \leq x \leq \frac{\pi}{2}$$
$$\cos^{-1}(\cos(x)) = x \text{ only for } 0 \leq x \leq \pi$$
$$\tan^{-1}(\tan(x)) = x \text{ only for } -\frac{\pi}{2} < x < \frac{\pi}{2}$$

Is it correct that $\sin^{-1}\left(\sin\left(x\right)\right) = x$**?**

No. This equation is correct if x belongs to the restricted domain $\left[-\frac{\pi}{2}, \frac{\pi}{2}\right]$, but sine is defined for all real input values, and for x outside the restricted interval, the equation is not correct because its inverse always returns a value in $\left[-\frac{\pi}{2}, \frac{\pi}{2}\right]$. The situation is similar for cosine and tangent and their inverses. For example, $\sin^{-1}\left(\sin\left(\frac{3\pi}{4}\right)\right) = \frac{\pi}{4}$.

HOW TO

Given an expression of the form $f^{-1}\left(f\left(\theta\right)\right)$ **where** $f\left(\theta\right) = \sin\left(\theta\right),\ \cos\left(\theta\right),\ \text{or}\ \tan\left(\theta\right)$, **evaluate.**

1. If θ is in the restricted domain of $f,$ then $f^{-1}\left(f\left(\theta\right)\right) = \theta.$
2. If not, then find an angle φ within the restricted domain of f such that $f\left(\varphi\right) = f\left(\theta\right)$. Then
   $f^{-1}\left(f\left(\theta\right)\right) = \varphi.$

EXAMPLE 5: USING INVERSE TRIGONOMETRIC FUNCTIONS

Evaluate the following:

1. $\sin^{-1}\left(\sin\left(\frac{\pi}{3}\right)\right)$
2. $\sin^{-1}\left(\sin\left(\frac{2\pi}{3}\right)\right)$
3. $\cos^{-1}\left(\cos\left(\frac{2\pi}{3}\right)\right)$
4. $\cos^{-1}\left(\cos\left(-\frac{\pi}{3}\right)\right)$

Answer

a. $\frac{\pi}{3}$ is in $\left[-\frac{\pi}{2}, \frac{\pi}{2}\right]$, so $\sin^{-1}\left(\sin\left(\frac{\pi}{3}\right)\right) = \frac{\pi}{3}$.

b. $\frac{2\pi}{3}$ is not in $\left[-\frac{\pi}{2}, \frac{\pi}{2}\right]$, but $\sin\left(\frac{2\pi}{3}\right) = \sin\left(\frac{\pi}{3}\right)$, so $\sin^{-1}\left(\sin\left(\frac{2\pi}{3}\right)\right) = \frac{\pi}{3}$.

c. $\frac{2\pi}{3}$ is in $\left[0, \pi\right]$, so $\cos^{-1}\left(\cos\left(\frac{2\pi}{3}\right)\right) = \frac{2\pi}{3}$.

d. $-\frac{\pi}{3}$ is not in $\left[0, \pi\right]$, but $\cos\left(-\frac{\pi}{3}\right) = \cos\left(\frac{\pi}{3}\right)$ because cosine is an even function. $\frac{\pi}{3}$ is in $\left[0, \pi\right]$, so $\cos^{-1}\left(\cos\left(-\frac{\pi}{3}\right)\right) = \frac{\pi}{3}$.

Evaluate $\tan^{-1}\left(\tan\left(\frac{\pi}{8}\right)\right)$ and $\tan^{-1}\left(\tan\left(\frac{11\pi}{9}\right)\right)$.

Answer

$\frac{\pi}{8}$; $\frac{2\pi}{9}$

Evaluating Compositions of the Form $f^{-1}\left(g\left(x\right)\right)$ (Optional)

Now that we can compose a trigonometric function with its inverse, we can explore how to evaluate a composition of a trigonometric function and the inverse of another trigonometric function. We will begin with compositions of the form $f^{-1}\left(g\left(x\right)\right)$. For special values of x, we can exactly evaluate the inner function and then the outer, inverse function. However, we can find a more general approach by considering the relation between the two acute angles of a right triangle where one is θ, making the other $\frac{\pi}{2}-\theta$. Consider the sine and cosine of each angle of the right triangle in Figure 9.

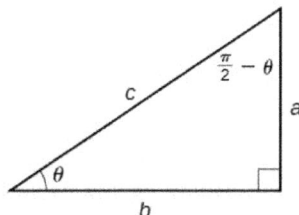

Figure 9: Right triangle illustrating the cofunction relationships

Because $\cos\left(\theta\right)=\frac{b}{c}=\sin\left(\frac{\pi}{2}-\theta\right)$, we have $\sin^{-1}\left(\cos\left(\theta\right)\right)=\frac{\pi}{2}-\theta$ if $0\leq\theta\leq\pi$. If θ is not in this domain, then we need to find another angle that has the same cosine as θ and does belong to the restricted domain; we then subtract this angle from $\frac{\pi}{2}$. Similarly, $\sin\left(\theta\right)=\frac{a}{c}=\cos\left(\frac{\pi}{2}-\theta\right)$, so $\cos^{-1}\left(\sin\left(\theta\right)\right)=\frac{\pi}{2}-\theta$ if $-\frac{\pi}{2}\leq\theta\leq\frac{\pi}{2}$. These are just the function-cofunction relationships presented in another way.

HOW TO

Given functions of the form $\sin^{-1}\left(\cos\left(x\right)\right)$ **and** $\cos^{-1}\left(\sin\left(x\right)\right)$, **evaluate them.**

1. If x is in $\left[0,\pi\right]$, then $\sin^{-1}\left(\cos\left(x\right)\right)=\frac{\pi}{2}-x$.
2. If x is not in $\left[0,\pi\right]$ then find another angle y in $\left[0,\pi\right]$ such that $\cos\left(y\right)=\cos\left(x\right)$.
$$\sin^{-1}\left(\cos\left(x\right)\right)=\frac{\pi}{2}-y.$$
3. If x is in $\left[-\frac{\pi}{2},\frac{\pi}{2}\right]$, then $\cos^{-1}\left(\sin\left(x\right)\right)=\frac{\pi}{2}-x$.
4. If x is not in $\left[-\frac{\pi}{2},\frac{\pi}{2}\right]$, then find another angle y in $\left[-\frac{\pi}{2},\frac{\pi}{2}\right]$ such that $\sin\left(y\right)=\sin\left(x\right)$.
$$\cos^{-1}\left(\sin\left(x\right)\right)=\frac{\pi}{2}-y.$$

EXAMPLE 6: EVALUATING THE COMPOSITION OF AN INVERSE SINE WITH A COSINE

Evaluate $\sin^{-1}\left(\cos\left(\frac{13\pi}{6}\right)\right)$

 a. by direct evaluation.
 b. by the method described previously.

Answer

 a. Here, we can directly evaluate the inside of the composition.

$$\cos\left(\frac{13\pi}{6}\right) = \cos\left(\frac{\pi}{6} + 2\pi\right)$$
$$= \cos\left(\frac{\pi}{6}\right)$$
$$= \frac{\sqrt{3}}{2}$$

Now, we can evaluate the inverse function as we did earlier.

$$\sin^{-1}\left(\frac{\sqrt{3}}{2}\right) = \frac{\pi}{3}$$

 b. We have $x = \frac{13\pi}{6}$, which is outside the restricted domain of the cosine function. Therefore, we have to consider another angle within the restricted domain of $0 \le x \le 2\pi$ which has the same cosine value. We will use $\frac{\pi}{6}$.

$$\sin^{-1}\left(\cos\left(\frac{13\pi}{6}\right)\right) = \frac{\pi}{2} - \frac{\pi}{6}$$
$$= \frac{\pi}{3}$$

TRY IT #6

Evaluate $\cos^{-1}\left(\sin\left(-\frac{11\pi}{4}\right)\right)$.

Answer

$\frac{3\pi}{4}$

Evaluating Compositions of the Form $f\left(g^{-1}\left(x\right)\right)$

To evaluate compositions of the form $f\left(g^{-1}\left(x\right)\right)$, where f and g are any two of the functions sine, cosine, or tangent and x is any input in the domain of g^{-1}, we have exact formulas, such as $\sin\left(\cos^{-1}\left(x\right)\right) = \sqrt{1-x^2}$. When we need to use them, we can derive these formulas by using the trigonometric relations between the angles and sides of a right triangle, together with the use of the Pythagorean relation between the lengths of the sides. We can use the Pythagorean identity, $\sin^2\left(x\right) + \cos^2\left(x\right) = 1$, to solve for one when given the other. We can also use the inverse trigonometric functions to find compositions involving algebraic expressions.

EXAMPLE 7: EVALUATING THE COMPOSITION OF A SINE WITH AN INVERSE COSINE

Find an exact value for $\sin\left(\cos^{-1}\left(\frac{4}{5}\right)\right)$.

Answer

Beginning with the inside, we can say there is some angle such that $\theta = \cos^{-1}\left(\frac{4}{5}\right)$, which means $\cos(\theta) = \frac{4}{5}$, and we are looking for $\sin(\theta)$. We can use the Pythagorean identity to do this.

$$\sin^2(\theta) + \cos^2(\theta) = 1 \qquad \text{Use our known value for cosine.}$$

$$\sin^2(\theta) + \left(\frac{4}{5}\right)^2 = 1 \qquad \text{Solve for sine.}$$

$$\sin^2(\theta) = 1 - \frac{16}{25}$$

$$\sin(\theta) = \sqrt{\frac{9}{25}} = \frac{3}{5}$$

Since $\theta = \cos^{-1}\left(\frac{4}{5}\right)$ is in quadrant I, $\sin(\theta)$ must be positive, so the solution is $\frac{3}{5}$. See Figure 10.

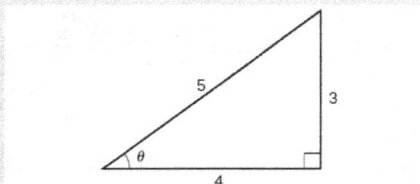

Figure 10: Right triangle illustrating that if $\cos(\theta) = \frac{4}{5}$, then $\sin(\theta) = \frac{3}{5}$.

We know that the inverse cosine always gives an angle on the interval $[0, \pi]$, so we know that the sine of that angle must be positive; therefore $\sin\left(\cos^{-1}\left(\frac{4}{5}\right)\right) = \sin(\theta) = \frac{3}{5}$.

TRY IT #7

Evaluate $\cos\left(\tan^{-1}\left(\frac{5}{12}\right)\right)$.

Answer

$\frac{12}{13}$

EXAMPLE 8: EVALUATING THE COMPOSITION OF A SINE WITH AN INVERSE TANGENT

Find an exact value for $\sin\left(\tan^{-1}\left(\frac{7}{4}\right)\right)$.

Answer

While we could use a similar technique as in Figure 10, we will demonstrate a different technique here. From the inside, we know there is an angle such that $\tan(\theta) = \frac{7}{4}$. We can envision this as the opposite and adjacent sides on a right triangle, as shown in Figure 11.

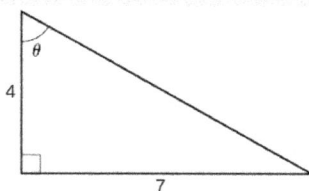

Figure 11: A right triangle with two sides known

Using the Pythagorean Theorem, we can find the hypotenuse of this triangle.

$$4^2 + 7^2 = \text{hypotenuse}^2$$
$$\text{hypotenuse} = \sqrt{65}$$

Now, we can evaluate the sine of the angle as the opposite side divided by the hypotenuse.

$$\sin(\theta) = \frac{7}{\sqrt{65}}$$

This gives us our desired composition.

$$\sin\left(\tan^{-1}\left(\frac{7}{4}\right)\right) = \sin(\theta)$$
$$= \frac{7}{\sqrt{65}}$$
$$= \frac{7\sqrt{65}}{65}$$

TRY IT #8

Evaluate $\cos\left(\sin^{-1}\left(\frac{7}{9}\right)\right)$.

Answer

$\frac{4\sqrt{2}}{9}$

EXAMPLE 9: FINDING THE COSINE OF THE INVERSE SINE OF AN ALGEBRAIC EXPRESSION

Find a simplified expression for $\cos\left(\sin^{-1}\left(\frac{x}{3}\right)\right)$ for $-3 \le x \le 3$.

Answer

We know there is an angle θ such that $\sin(\theta) = \frac{x}{3}$.

$$\sin^2(\theta) + \cos^2(\theta) = 1 \qquad \text{Use the Pythagorean Theorem.}$$

$$\left(\frac{x}{3}\right)^2 + \cos^2(\theta) = 1 \qquad \text{Solve for cosine.}$$

$$\cos^2(\theta) = 1 - \frac{x^2}{9}$$

$$\cos(\theta) = \sqrt{\frac{9 - x^2}{9}}$$

$$= \frac{\sqrt{9 - x^2}}{3}$$

Because we know that the inverse sine must give an angle on the interval $\left[-\frac{\pi}{2}, \frac{\pi}{2}\right]$, we can deduce that the cosine of that angle must be positive.

$$\cos\left(\sin^{-1}\left(\frac{x}{3}\right)\right) = \frac{\sqrt{9-x^2}}{3}$$

TRY IT #9

Find a simplified expression for $\sin\left(\tan^{-1}(4x)\right)$ for $-\frac{1}{4} \le x \le \frac{1}{4}$.

Answer

$$\frac{4x}{\sqrt{16x^2 + 1}}$$

Access this online resource for additional instruction and practice with inverse trigonometric functions.

- Evaluate Expressions Involving Inverse Trigonometric Functions

Visit this website for additional practice questions from Learningpod.

KEY CONCEPTS

- An inverse function is one that "undoes" another function. The domain of an inverse function is the range of the original function and the range of an inverse function is the domain of the original function.
- Because the trigonometric functions are not one-to-one on their natural domains, inverse trigonometric functions are defined for restricted domains.
- For any trigonometric function $f(x)$, if $x = f^{-1}(y)$, then $f(x) = y$. However, $f(x) = y$ only implies $x = f^{-1}(y)$ if x is in the restricted domain of f.
- Special angles are the outputs of inverse trigonometric functions for special input values; for example, $\frac{\pi}{4} = \tan^{-1}(1)$ and $\frac{\pi}{6} = \sin^{-1}\left(\frac{1}{2}\right)$.
- A calculator will return an angle within the restricted domain of the original trigonometric function.
- Inverse functions allow us to find an angle when given two sides of a right triangle.
- In function composition, if the inside function is an inverse trigonometric function, then there are exact expressions; for example, $\sin\left(\cos^{-1}(x)\right) = \sqrt{1 - x^2}$.
- If the inside function is a trigonometric function, then the only possible combinations are $\sin^{-1}(\cos(x)) = \frac{\pi}{2} - x$ if $0 \le x \le \pi$ and $\cos^{-1}(\sin(x)) = \frac{\pi}{2} - x$ if $-\frac{\pi}{2} \le x \le \frac{\pi}{2}$.
- When evaluating the composition of a trigonometric function with an inverse trigonometric function, draw a reference triangle to assist in determining the ratio of sides that represents the output of the trigonometric function.
- When evaluating the composition of a trigonometric function with an inverse trigonometric function, you may use trig identities to assist in determining the ratio of sides.

GLOSSARY

arccosine
    another name for the inverse cosine; $\arccos(x) = \cos^{-1}(x)$
arcsine
    another name for the inverse sine; $\arcsin(x) = \sin^{-1}(x)$
arctangent
    another name for the inverse tangent; $\arctan(x) = \tan^{-1}(x)$
inverse cosine function
    the function $\cos^{-1}(x)$, which is the inverse of the cosine function and the angle that has a cosine equal to a given number
inverse sine function
    the function $\sin^{-1}(x)$, which is the inverse of the sine function and the angle that has a sine equal to a given number
inverse tangent function
    the function $\tan^{-1}(x)$, which is the inverse of the tangent function and the angle that has a tangent equal to a given number

3.7 Trigonometric Identities

LEARNING OBJECTIVES

In this section, you will:

- Verify the fundamental trigonometric identities.
- Simplify trigonometric expressions using algebra and the identities.

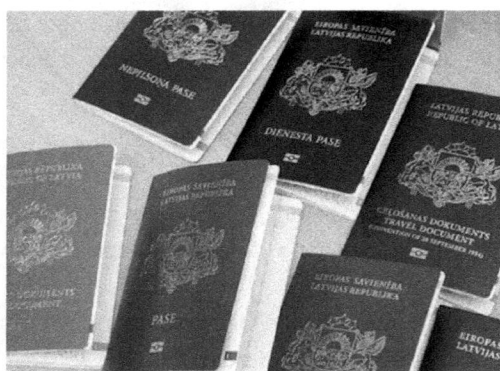

Figure 1. International passports and travel documents

In espionage movies, we see international spies with multiple passports, each claiming a different identity. However, we know that each of those passports represents the same person. The trigonometric identities act in a similar manner to multiple passports—there are many ways to represent the same trigonometric expression. Just as a spy will choose an Italian passport when traveling to Italy, we choose the identity that applies to the given scenario when solving a trigonometric equation.

In this section, we will review some trigonometric identities that we have already seen in earlier sections, create some new ones and show how we can use identities, along with basic tools of algebra, to simplify trigonometric expressions.

Some Fundamental Trigonometric Identities

We have previously discussed the set of **even-odd identities.** The **even-odd identities** relate the value of a trigonometric function at a given angle to the value of the function at the opposite angle and determine whether the identity is odd or even. (See Table 1).

Table 1

Even-Odd Identities		
$\tan(-\theta) = -\tan(\theta)$	$\sin(-\theta) = -\sin(\theta)$	$\cos(-\theta) = \cos(\theta)$
$\cot(-\theta) = -\cot(\theta)$	$\csc(-\theta) = -\csc(\theta)$	$\sec(-\theta) = \sec(\theta)$

The next set of fundamental identities is the set of **reciprocal identities**, which, as their name implies, relate trigonometric functions that are reciprocals of each other. See Table 2.

Table 2

Reciprocal Identities

$$\sin\left(\theta\right) = \frac{1}{\csc\left(\theta\right)} \quad \csc\left(\theta\right) = \frac{1}{\sin\left(\theta\right)}$$

$$\cos\left(\theta\right) = \frac{1}{\sec\left(\theta\right)} \quad \sec\left(\theta\right) = \frac{1}{\cos\left(\theta\right)}$$

$$\tan\left(\theta\right) = \frac{1}{\cot\left(\theta\right)} \quad \cot\left(\theta\right) = \frac{1}{\tan\left(\theta\right)}$$

Another set of identities is the set of **quotient identities**, which define relationships among certain trigonometric functions and can be very helpful in verifying other identities. See Table 3.

Table 3

Quotient Identities

$$\tan\left(\theta\right) = \frac{\sin(\theta)}{\cos(\theta)} \quad \cot\left(\theta\right) = \frac{\cos(\theta)}{\sin(\theta)}$$

Alternate Forms of the Pythagorean Identity

We can use these fundamental identities to derive alternative forms of the Pythagorean Identity,

$$\cos^2\left(\theta\right) + \sin^2\left(\theta\right) = 1.$$

You should recall that the identity shown above is a direct result of our definition of the sine and cosine functions in terms of the coordinates of points on the unit circle. We can derive two more identities using the methods shown below.

The identity $1 + \cot^2\left(\theta\right) = \csc^2\left(\theta\right)$ is found by dividing each term of the first identity by $\sin^2\left(\theta\right)$, and then rewriting each part of the equation using the identities we have already discussed in earlier sections.

$$\frac{\sin^2\left(\theta\right)}{\sin^2\left(\theta\right)} + \frac{\cos^2\left(\theta\right)}{\sin^2\left(\theta\right)} = \frac{1}{\sin^2\left(\theta\right)}$$

We can then use our earlier quotient and reciprocal identities to rewrite the expression in this equation as shown below.

$$1 + \cot^2\left(\theta\right) = \csc^2\left(\theta\right)$$

Similarly, $1 + \tan^2\left(\theta\right) = \sec^2\left(\theta\right)$ can be obtained by dividing all terms in the first identity by $\cos^2\left(\theta\right)$, and then rewriting each part of the equation using the quotient and reciprocal identities .

$$\frac{\sin^2\left(\theta\right)}{\cos^2\left(\theta\right)} + \frac{\cos^2\left(\theta\right)}{\cos^2\left(\theta\right)} = \frac{1}{\cos^2\left(\theta\right)}$$

$$\tan^2\left(\theta\right) + 1 = \sec^2\left(\theta\right)$$

We now have the three **Pythagorean identities** shown in Table 4.

Table 4

Pythagorean Identities

$$\sin^2(\theta) + \cos^2(\theta) = 1 \quad 1 + \cot^2(\theta) = \csc^2(\theta) \quad 1 + \tan^2(\theta) = \sec^2(\theta)$$

EXAMPLE 1: GRAPHING THE EQUATIONS OF AN IDENTITY

Graph both sides of the identity $\cot(\theta) = \frac{1}{\tan(\theta)}$. In other words, on the graphing calculator, graph $y = \cot(\theta)$ and $y = \frac{1}{\tan(\theta)}$.

Answer

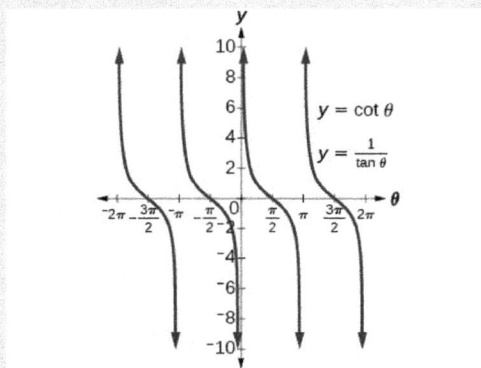

Figure 1.

Analysis

We see only one graph because both expressions generate the same image. One is on top of the other. This is a good way to confirm an identity verified with analytical means. If both expressions give the same graph, then they are most likely identities.

HOW TO:

Given a trigonometric identity, verify that it is true.

1. Work on one side of the equation. It is usually better to start with the more complex side, as it is easier to simplify than to build.
2. Look for opportunities to factor expressions, square a binomial, or add fractions.
3. Noting which functions are in the final expression, look for opportunities to use the identities and make the proper substitutions.
4. If these steps do not yield the desired result, try converting all terms to sines and cosines.
5. Note the values not in the domain of the expression on the left and right as the identity does not hold for those values.

EXAMPLE 2: VERIFYING A TRIGONOMETRIC IDENTITY

Verify $\tan\left(\theta\right)\cos\left(\theta\right) = \sin\left(\theta\right)$.

Answer

We will start on the left side, as it is the more complicated side:

$$\tan\left(\theta\right)\cos\left(\theta\right) = \left(\frac{\sin\left(\theta\right)}{\cos\left(\theta\right)}\right)\cos\left(\theta\right)$$
$$= \sin\left(\theta\right)$$

Analysis

This identity was fairly simple to verify, as it only required writing $\tan\left(\theta\right)$ in terms of $\sin\left(\theta\right)$ and $\cos\left(\theta\right)$. When determining identities we need to also consider the domains of the expressions on the left and right sides of the equation. The identity is only valid where both expressions are defined. For this problem, the domain of $\tan\left(\theta\right)\cos\left(\theta\right)$ is all real numbers except $\frac{\pi}{2} \pm n\pi$ where n is an integer. The expression $\frac{\pi}{2} \pm n\pi$ means that we will have breaks in the domain starting at $\frac{\pi}{2}$ and then every time we add or subtract π from there. This means the identity will not be valid at these points. We also need to consider the left hand side of the equation. However, since the domain of $\sin\left(\theta\right)$ is all real numbers there are no additional places that the identity is not valid.

TRY IT #1

Verify the identity $\csc\left(\theta\right)\,\cos\left(\theta\right)\,\tan\left(\theta\right) = 1$.

Answer

$$\csc\left(\theta\right)\cos\left(\theta\right)\tan\left(\theta\right) = \left(\frac{1}{\sin\left(\theta\right)}\right)\cos\left(\theta\right)\left(\frac{\sin\left(\theta\right)}{\cos\left(\theta\right)}\right)$$
$$= \frac{\cos\left(\theta\right)}{\sin\left(\theta\right)}\left(\frac{\sin\left(\theta\right)}{\cos\left(\theta\right)}\right)$$
$$= \frac{\sin\left(\theta\right)\cos\left(\theta\right)}{\sin\left(\theta\right)\cos\left(\theta\right)}$$
$$= 1$$

This identity is not valid where $\csc\left(\theta\right)$ or $\tan\left(\theta\right)$ are not defined. This means that it is not valid at
$\ldots - \pi, \frac{-\pi}{2}, 0, \frac{\pi}{2}, \pi, \ldots$.
We can capture these values with the expression $\frac{\pi}{2} \pm \frac{n\pi}{2}$ where n is an integer.

EXAMPLE 3: VERIFYING THE EQUIVALENCY USING THE EVEN-ODD IDENTITIES

Verify the following equivalency using the even-odd identities:

$$(1 + \sin(x))(1 + \sin(-x)) = \cos^2(x).$$

Answer

Working on the left side of the equation, we have

$$
\begin{aligned}
(1 + \sin(x))(1 + \sin(-x)) &= (1 + \sin(x))(1 - \sin(x)) && \text{Since } \sin(-x) = -\sin(x). \\
&= 1 - \sin^2(x) && \text{Difference of squares.} \\
&= \cos^2(x) && \cos^2(x) = 1 - \sin^2(x) \text{ from Pythagorean Identity.}
\end{aligned}
$$

This identity is valid for all real numbers since $\sin(\theta)$ and $\cos(\theta)$ have a domain of all real numbers.

EXAMPLE 4: VERIFYING A TRIGONOMETRIC IDENTITY INVOLVING $SEC^2(\Theta)$

Verify the identity $\dfrac{\sec^2(\theta) - 1}{\sec^2(\theta)} = \sin^2(\theta).$

Answer

As the left side is more complicated, let's begin there.

$$
\begin{aligned}
\frac{\sec^2(\theta) - 1}{\sec^2(\theta)} &= \frac{(\tan^2(\theta) + 1) - 1}{\sec^2(\theta)} && \text{Because } \sec^2(\theta) = \tan^2(\theta) + 1. \\
&= \frac{\tan^2(\theta)}{\sec^2(\theta)} \\
&= \tan^2(\theta)\left(\frac{1}{\sec^2(\theta)}\right) \\
&= \tan^2(\theta)\cos^2(\theta) && \text{Because } \cos^2(\theta) = \frac{1}{\sec^2(\theta)}. \\
&= \frac{\sin^2(\theta)}{\cos^2(\theta)}\cos^2(\theta) && \text{Because } \tan^2(\theta) = \frac{\sin^2(\theta)}{\cos^2(\theta)}. \\
&= \sin^2(\theta)
\end{aligned}
$$

There is more than one way to verify an identity. Here is another possibility. Again, we can start with the left side.

$$\frac{\sec^2(\theta) - 1}{\sec^2(\theta)} = \frac{\sec^2(\theta)}{\sec^2(\theta)} - \frac{1}{\sec^2(\theta)}$$

$$= 1 - \cos^2(\theta)$$

$$= \sin^2(\theta)$$

This identity is valid on the domain of $\sec(\theta)$. This means that we need to exclude values where $\cos(\theta) = 0$. This would give us all real numbers except $\frac{\pi}{2} \pm n\pi$ where n is an integer.

Analysis

In the first method, we used the identity $\sec^2(\theta) = \tan^2(\theta) + 1$ and continued to simplify. In the second method, we split the fraction, putting both terms in the numerator over the common denominator. This problem illustrates that there are multiple ways we can verify an identity. Employing some creativity can sometimes simplify a procedure. As long as the substitutions are correct, the answer will be the same.

TRY IT #2

Show that $\frac{\cot(\theta)}{\csc(\theta)} = \cos(\theta)$.

Answer

$$\frac{\cot(\theta)}{\csc(\theta)} = \frac{\frac{\cos(\theta)}{\sin(\theta)}}{\frac{1}{\sin(\theta)}} \qquad \text{This is a complex fraction.}$$

$$= \frac{\cos(\theta)}{\sin(\theta)} \cdot \frac{\sin(\theta)}{1} \qquad \text{Multiple the numerator by the reciprocal of the denominator.}$$

$$= \cos(\theta)$$

This identity is valid on the domain of $\cot(\theta)$ and $\csc(\theta)$. This means that the identity is not valid for the values $\ldots -2\pi, -\pi, 0, \pi, 2\pi, \ldots$

This would give us all real numbers except $\pi \pm n\pi$ where n is an integer.

EXAMPLE 5: CREATING AND VERIFYING AN IDENTITY

Create an identity for the expression $2\tan(\theta)\sec(\theta)$ by rewriting strictly in terms of sine.

Answer

There are a number of ways to begin, but here we will use the quotient and reciprocal identities to rewrite the expression:

$$2\tan(\theta)\sec(\theta) = 2\left(\frac{\sin(\theta)}{\cos(\theta)}\right)\left(\frac{1}{\cos(\theta)}\right)$$

$$= \frac{2\sin(\theta)}{\cos^2(\theta)}$$

$$= \frac{2\sin(\theta)}{1-\sin^2(\theta)} \qquad \text{Substitute } 1-\sin^2(\theta) \text{ for } \cos^2(\theta).$$

$$2\tan(\theta)\sec(\theta) = \frac{2\sin(\theta)}{1-\sin^2(\theta)}$$

This identity holds on values in both of the domains of $\tan(\theta)$ and $\sec(\theta)$. This means it holds for all real numbers except $\ldots -\frac{3\pi}{2}, -\frac{\pi}{2}, \frac{\pi}{2}, \frac{3\pi}{2} \ldots$

This would give us all real numbers except $\frac{\pi}{2} \pm n\pi$ where n is an integer.

The expression on the left also needs to be considered for restrictions but the denominator is zero exactly when the left side of the expression is undefined so no more values need to be excluded.

EXAMPLE 6: VERIFYING AN IDENTITY USING ALGEBRA AND EVEN/ODD IDENTITIES

Verify the identity:

$$\frac{\sin^2(-\theta)-\cos^2(-\theta)}{\sin(-\theta)-\cos(-\theta)} = \cos(\theta) - \sin(\theta)$$

Answer

Let's start with the left side and simplify:

$$\frac{\sin^2(-\theta) - \cos^2(-\theta)}{\sin(-\theta) - \cos(-\theta)} = \frac{(\sin(-\theta) - \cos(-\theta))(\sin(-\theta) + \cos(-\theta))}{\sin(-\theta) - \cos(-\theta)} \qquad \text{Difference of squares.}$$

$$= \frac{(-\sin(\theta) - \cos(\theta))(-\sin(\theta) + \cos(\theta))}{-\sin(\theta) - \cos(\theta)} \qquad \text{Odd-Even Identities.}$$

$$= -\sin(\theta) + \cos(\theta) \qquad \text{Cancel the common factor.}$$

$$= \cos(\theta) - \sin(\theta) \qquad \text{Reorder terms.}$$

This identity will be valid when $\sin(-\theta) - \cos(-\theta) \neq 0$. Solving this type of equation is discussed in section 3.8 Solving Trigonometric Equations.

TRY IT #3

Verify the identity $\dfrac{\sin^2(\theta)-1}{\tan(\theta)\sin(\theta)-\tan(\theta)} = \dfrac{\sin(\theta)+1}{\tan(\theta)}$.

Answer

$$\frac{\sin^2(\theta)-1}{\tan(\theta)\sin(\theta)-\tan(\theta)} = \frac{(\sin(\theta)+1)(\sin(\theta)-1)}{\tan(\theta)(\sin(\theta)-1)}$$
$$= \frac{\sin(\theta)+1}{\tan(\theta)}$$

This equation is valid on the domain of tangent and where we don't get a divide by zero. Determining when the denominator in the expression on the left requires equation solving discussed in section 3.8 Solving Trigonometric Equations.

EXAMPLE 7: VERIFYING AN IDENTITY INVOLVING COSINES AND COTANGENTS

Verify the identity: $\left(1 - \cos^2(x)\right)\left(1 + \cot^2(x)\right) = 1$.

Answer

We will work on the left side of the equation.

$$\left(1 - \cos^2(x)\right)\left(1 + \cot^2(x)\right)$$

$$= \left(1 - \cos^2(x)\right)\left(1 + \frac{\cos^2(x)}{\sin^2(x)}\right) \qquad \text{Rewrite } \cot^2(x).$$

$$= \left(1 - \cos^2(x)\right)\left(\frac{\sin^2(x)}{\sin^2 x} + \frac{\cos^2 x}{\sin^2(x)}\right) \qquad \text{Find the common denominator.}$$

$$= \left(1 - \cos^2(x)\right)\left(\frac{\sin^2(x) + \cos^2(x)}{\sin^2(x)}\right)$$

$$= \left(\sin^2(x)\right)\left(\frac{\sin^2(x) + \cos^2(x)}{\sin^2(x)}\right) \qquad \text{Because } 1 - \cos^2(x) = \sin^2(x).$$

$$= \left(\sin^2(x)\right)\left(\frac{1}{\sin^2(x)}\right) \qquad \text{Because } \sin^2(x) + \cos^2(x) = 1.$$

$$= 1$$

This identity is valid on the domain of cotangent. This would give us all real numbers except $\pi \pm n\pi$ where n is an integer.

An alternate way to verify this identity is to begin by multiplying out the expression on the left and simplifying.

$$\left(1 - \cos^2\left(x\right)\right)\left(1 + \cot^2\left(x\right)\right) = 1 - \cos^2\left(x\right) + \cot^2\left(x\right) - \cot^2\left(x\right)\cos^2\left(x\right)$$
$$= 1 - \cos^2\left(x\right) + \cot^2\left(x\right)\left(1 - \cos^2\left(x\right)\right)$$
$$= 1 - \cos^2\left(x\right) + \frac{\cos^2\left(x\right)}{\sin^2\left(x\right)}\sin^2\left(x\right)$$
$$= 1 - \cos^2\left(x\right) + \cos^2\left(x\right)$$
$$= 1$$

Using Algebra to Simplify Trigonometric Expressions

We have seen that algebra is very important in verifying trigonometric identities, but it is just as critical in simplifying trigonometric expressions when we are solving equations. Being familiar with the basic properties and formulas of algebra, such as the difference of squares formula, the perfect square formula, or substitution, will simplify the work involved with trigonometric expressions and equations.

An example is the difference of squares formula, $a^2 - b^2 = \left(a - b\right)\left(a + b\right)$, which is widely used in many areas other than mathematics, such as engineering, architecture, and physics

For example, the expression $\sin^2\left(x\right) - 1$ resembles the difference of squares $x^2 - 1$. Recognizing that $x^2 - 1$ can be factored as $\left(x + 1\right)\left(x - 1\right)$ helps us quickly recognize that $\sin^2\left(x\right) - 1$ can be factored as $\left(\sin\left(x\right) + 1\right)\left(\sin\left(x\right) - 1\right)$.

We can also create our own identities by continually expanding an expression and making the appropriate substitutions. Using algebraic properties and formulas makes many trigonometric expressions and equations easier to work with.

EXAMPLE 8: WRITING THE TRIGONOMETRIC EXPRESSION AS AN ALGEBRAIC EXPRESSION

Write the following trigonometric expression as an algebraic expression: $2\cos^2\left(\theta\right) + \cos\left(\theta\right) - 1$.

Answer

Notice that the pattern displayed has the same form as a standard quadratic expression, $ax^2 + bx + c$. Letting $\cos\left(\theta\right) = x$, we can rewrite the expression as follows:

$$2x^2 + x - 1$$

This expression can be factored as $\left(2x - 1\right)\left(x + 1\right)$. If it were set equal to zero and we wanted to solve the equation, we would use the zero factor property and solve each factor for x. At this point, we would replace x with $\cos\left(\theta\right)$ and solve for θ.

EXAMPLE 9: REWRITING A TRIGONOMETRIC EXPRESSION USING THE DIFFERENCE OF SQUARES

Rewrite the trigonometric expression: $4\cos^2(\theta) - 1$.

Answer

Notice that both the coefficient and the trigonometric expression in the first term are squared, and the square of the number 1 is 1. This is the difference of squares. Thus,

$$4\cos^2(\theta) - 1 = (2\cos(\theta) - 1)(2\cos(\theta) + 1)$$

Analysis

If this expression were written in the form of an equation set equal to zero, we could solve each factor using the zero factor property. We could also use substitution like we did in the previous problem and let $\cos(\theta) = x$, rewrite the expression as $4x^2 - 1$, and factor $(2x - 1)(2x + 1)$. Then replace x with $\cos(\theta)$ and solve for the angle.

TRY IT #4

Rewrite the trigonometric expression: $25 - 9\sin^2(\theta)$.

Answer

This is a difference of squares formula: $25 - 9\sin^2(\theta) = (5 - 3\sin(\theta))(5 + 3\sin(\theta))$.

EXAMPLE 10: SIMPLIFY BY REWRITING AND USING SUBSTITUTION

Simplify the expression by rewriting and using identities:

$$\csc^2(\theta) - \cot^2(\theta)$$

Answer

We can start with the Pythagorean identity. We know: $1 + \cot^2(\theta) = \csc^2(\theta)$

$$\csc^2(\theta) - \cot^2(\theta) = 1 + \cot^2(\theta) - \cot^2(\theta) \quad \text{Substitute expression in for } \csc^2(\theta).$$
$$= 1$$

This identity is valid on the domains of cotangent and cosecant. This would give us all real numbers except $\pi \pm n\pi$ where n is an integer.

TRY IT #5

Use algebraic techniques to verify the identity: $\frac{\cos(\theta)}{1+\sin(\theta)} = \frac{1-\sin(\theta)}{\cos(\theta)}$.

(Hint: Multiply the numerator and denominator on the left side by $1 - \sin(\theta)$.)

Answer

$$\frac{\cos(\theta)}{1+\sin(\theta)}\left(\frac{1-\sin(\theta)}{1-\sin(\theta)}\right) = \frac{\cos(\theta)(1-\sin(\theta))}{1-\sin^2(\theta)}$$

$$= \frac{\cos(\theta)(1-\sin(\theta))}{\cos^2(\theta))} \quad \text{We know } 1 - \sin^2(\theta) = \cos^2(\theta).$$

$$= \frac{1-\sin(\theta)}{\cos(\theta)} \quad \text{Cancel a factor of } \cos(\theta).$$

MEDIA

Access these online resources for additional instruction and practice with the fundamental trigonometric identities.

- Fundamental Trigonometric Identities
- Verifying Trigonometric Identities

Key Equations

Pythagorean identities	$\sin^2(\theta) + \cos^2(\theta) = 1$ $1 + \cot^2(\theta) = \csc^2(\theta)$ $1 + \tan^2(\theta) = \sec^2(\theta)$
Even-odd identities	$\tan(-\theta) = -\tan(\theta)$ $\cot(-\theta) = -\cot(\theta)$ $\sin(-\theta) = -\sin(\theta)$ $\csc(-\theta) = -\csc(\theta)$ $\cos(-\theta) = \cos(\theta)$ $\sec(-\theta) = \sec(\theta)$

Reciprocal identities

$$\sin(\theta) = \frac{1}{\csc(\theta)}$$

$$\cos(\theta) = \frac{1}{\sec(\theta)}$$

$$\tan(\theta) = \frac{1}{\cot(\theta)}$$

$$\csc(\theta) = \frac{1}{\sin(\theta)}$$

$$\sec(\theta) = \frac{1}{\cos(\theta)}$$

$$\cot(\theta) = \frac{1}{\tan(\theta)}$$

Quotient identities

$$\tan(\theta) = \frac{\sin(\theta)}{\cos(\theta)}$$

$$\cot(\theta) = \frac{\cos(\theta)}{\sin(\theta)}$$

Shift Identities

$$\sin(\theta) = \cos\left(\theta - \frac{\pi}{2}\right)$$

$$\cos(\theta) = \sin\left(\theta - \frac{\pi}{2}\right)$$

KEY CONCEPTS

- There are multiple ways to represent a trigonometric expression. Verifying the identities illustrates how expressions can be rewritten to simplify a problem.
- Graphing both sides of an identity will verify it.
- Simplifying one side of the equation to equal the other side is another method for verifying an identity.
- The approach to verifying an identity depends on the nature of the identity. It is often useful to begin on the more complex side of the equation.
- We can create an identity by simplifying an expression and then verifying it.
- Verifying an identity may involve algebra with the fundamental identities.
- Algebraic techniques can be used to simplify trigonometric expressions. We use algebraic techniques throughout this text, as they consist of the fundamental rules of mathematics.

GLOSSARY

even-odd identities

set of equations involving trigonometric functions such that if $f(-x) = -f(x)$, the identity is odd, and if $f(-x) = f(x)$, the identity is even

Pythagorean identities

set of equations involving trigonometric functions based on the right triangle properties

quotient identities

pair of identities based on the fact that tangent is the ratio of sine and cosine, and cotangent is the ratio of cosine and sine

reciprocal identities

set of equations involving the reciprocals of basic trigonometric definitions

3.8 Solving Trigonometric Equations

> **LEARNING OBJECTIVES**
>
> In this section, you will:
>
> - Solve linear trigonometric equations in sine and cosine.
> - Solve equations involving a single trigonometric function.
> - Solve trigonometric equations using a calculator.
> - Solve trigonometric equations that are quadratic in form.
> - Solve trigonometric equations using fundamental identities.
> - Solve trigonometric equations with multiple angles.

Egyptian pyramids standing near a modern city. (credit: Oisin Mulvihill)

Thales of Miletus (circa 625–547 BC) is known as the founder of geometry. The legend is that he calculated the height of the Great Pyramid of Giza in Egypt using the theory of *similar triangles*, which he developed by measuring the shadow of his staff. Based on proportions, this theory has applications in a number of areas, including fractal geometry, engineering, and architecture. Often, the angle of elevation and the angle of depression are found using similar triangles.

In earlier sections of this chapter, we looked at trigonometric identities. Identities are true for all values in the domain of the variable. In this section, we begin our study of trigonometric equations to study real-world scenarios such as the finding the dimensions of the pyramids.

Solving Linear Trigonometric Equations in Sine and Cosine

Trigonometric equations are, as the name implies, equations that involve trigonometric functions. Similar in many ways to solving polynomial equations or rational equations, only specific values of the variable will be solutions, if there are solutions at all. Additionally, like rational equations, the domain of the function must be considered before we assume that any solution is valid.

Often we will solve a trigonometric equation over a specified interval. However, just as often, we will be asked to find all possible solutions, and as trigonometric functions are periodic, solutions are repeated within each period. In other words, trigonometric equations may have an infinite number of solutions. The period of both the sine function and the cosine function is 2π. In other words, every 2π units, the y-values repeat, so $\sin\left(\theta\right) = \sin\left(\theta \pm 2k\pi\right)$. If we need to find all possible solutions, then we must add $2\pi k$, where k is an integer, to the initial solution.

There are similar rules for indicating all possible solutions for the other trigonometric functions. Solving trigonometric equations requires the same techniques as solving algebraic equations. We read the equation from left to right, horizontally, like a sentence. We look for known patterns, factor, find common denominators, and substitute certain expressions with a variable to make solving a more straightforward process. However, with trigonometric equations, we also have the advantage of using the identities we developed in the previous sections.

EXAMPLE 1: SOLVING A LINEAR TRIGONOMETRIC EQUATION INVOLVING THE COSINE FUNCTION

Find all possible exact solutions for the equation $\cos(\theta) = \frac{1}{2}$.

Answer

From the unit circle, we know that there will be two angles where $\cos(\theta) = \frac{1}{2}$ in one complete revolution, i.e. $0 \le \theta \le 2\pi$. They will occur in the first and fourth quadrants. We recognize $\frac{1}{2}$ as a value from one of our special right triangles. We can identify the acute angle as $\frac{\pi}{3}$. We can then use this as the reference angle to find the angle in the fourth quadrant by computing $2\pi - \frac{\pi}{3}$.

$$\cos(\theta) = \frac{1}{2}$$

$$\theta = \frac{\pi}{3}, \frac{5\pi}{3}$$

These are the solutions in the interval $[0, 2\pi]$. All possible solutions are given by $\frac{\pi}{3} \pm 2k\pi$ and $\frac{5\pi}{3} \pm 2k\pi$ where k is an integer.

EXAMPLE 2: SOLVING A LINEAR EQUATION INVOLVING THE SINE FUNCTION

Find all possible exact solutions for the equation $\sin(t) = \frac{1}{2}$.

Answer

Solving for all possible values of t means that solutions include angles beyond the period of 2π. From previous work with the unit circle, we know that there are two solutions in one revolution. Since the value is positive, we know these solutions are in the first and second quadrants. From our special right triangles, we know that the angle in the first quadrant is $\frac{\pi}{6}$ and using this as the reference angle, the solution in the second quadrant is $\frac{5\pi}{6}$. But the problem is asking for all possible values that solve the equation.

Therefore, the answer is $\frac{\pi}{6} \pm 2\pi k$ and $\frac{5\pi}{6} \pm 2\pi k$ where k is an integer.

EXAMPLE 3: SOLVE THE TRIGONOMETRIC EQUATION IN LINEAR FORM

Solve the equation exactly: $2 \cos(\theta) - 3 = -5, \ \ 0 \le \theta < 2\pi$.

Answer

Use algebraic techniques to solve the equation.

$$2 \cos(\theta) - 3 = -5$$
$$2\cos(\theta) = -2 \qquad \text{Added 3 to both sides.}$$
$$\cos(\theta) = -1 \qquad \text{Divided both sides by 2.}$$
$$\theta = \pi$$

Thinking of the unit circle, we can see that there is only one place where the cosine value is -1 in one complete revolution.

TRY IT #1

Solve exactly the following linear equation on the interval $[0, 2\pi) : \ 2 \sin(x) + 1 = 0$.

Answer

$x = \frac{7\pi}{6}, \ \frac{11\pi}{6}$

Solving Equations Involving a Single Trigonometric Function

When we are given equations that involve only one of the six trigonometric functions, their solutions involve using algebraic techniques and information we know from the unit circle. We need to make several considerations when the equation involves trigonometric functions other than sine and cosine. Problems involving the reciprocals of the primary trigonometric functions need to be viewed from an algebraic perspective. In other words, we will write the equation in terms of the reciprocal function, and solve for the angles using the functions we are most familiar with. Also, an equation involving the tangent function is slightly different from one containing a sine or cosine function. First, as we know, the period of tangent is π, not 2π. Further, the domain of tangent is all real numbers with the exception of odd integer multiples of $\frac{\pi}{2}$, unless, of course, a problem places its own restrictions on the domain.

EXAMPLE 4: SOLVING A TRIGONOMETRIC EQUATION INVOLVING COSECANT

Solve the following equation exactly: $\csc(\theta) = -2, \ 0 \le \theta < 4\pi$.

Answer

We want all values of θ for which $\csc(\theta) = -2$ over the interval $0 \leq \theta < 4\pi$.

$$\csc(\theta) = -2$$

$$\frac{1}{\sin(\theta)} = -2$$

$$\sin(\theta) = -\frac{1}{2}$$

$$= \frac{7\pi}{6}, \frac{11\pi}{6}, \frac{19\pi}{6}, \frac{23\pi}{6}$$

As $\sin(\theta) = -\frac{1}{2}$, notice that all four solutions are in the third and fourth quadrants.

EXAMPLE 5: SOLVING AN EQUATION INVOLVING TANGENT

Solve the equation exactly: $\tan\left(\theta - \frac{\pi}{2}\right) = 1$, $0 \leq \theta < 2\pi$.

Answer

Recall that the tangent function has a period of π. On the interval $[0, \pi)$, and at the angle of $\frac{\pi}{4}$, the tangent has a value of 1. However, the angle we want is $\left(\theta - \frac{\pi}{2}\right)$. Thus, if $\tan\left(\frac{\pi}{4}\right) = 1$, then

$$\theta - \frac{\pi}{2} = \frac{\pi}{4}$$

$$\theta = \frac{3\pi}{4}\pi$$

Over the interval $[0, 2\pi)$, we have two solutions:

$$\frac{3\pi}{4} \text{ and } \frac{3\pi}{4} + \pi = \frac{7\pi}{4}$$

TRY IT #2

Find all solutions for $\tan(x) = \sqrt{3}$.

Answer

$\frac{\pi}{3} \pm \pi k$

EXAMPLE 6: IDENTIFY ALL SOLUTIONS TO THE EQUATION INVOLVING TANGENT

Identify all exact solutions to the equation $2\left(\tan\left(x\right)+3\right)=5+\tan\left(x\right),\ 0\le x<2\pi$.

Answer

We can solve this equation using only algebra. Isolate the expression $\tan\left(x\right)$ on the left side of the equals sign.

$$2\ \left(\tan\left(x\right)\right)+2\ \left(3\right)=5+\tan\left(x\right)\quad\text{Distribute the 2 on the left hand side.}$$
$$2\tan\left(x\right)+6=5+\tan\left(x\right)$$
$$2\tan\left(x\right)-\tan\left(x\right)=5-6\qquad\text{Isolate the tangent on one side.}$$
$$\tan\left(x\right)=-1$$

There are two angles on the unit circle that have a tangent value of -1 : $\theta=\frac{3\pi}{4}$ and $\theta=\frac{7\pi}{4}$.

Solve Trigonometric Equations Using a Calculator

Not all functions can be solved exactly using only the unit circle. When we must solve an equation involving an angle other than one of the special angles, we will need to use a calculator. Make sure it is set to the proper mode, either degrees or radians, depending on the criteria of the given problem.

EXAMPLE 7: USING A CALCULATOR TO SOLVE A TRIGONOMETRIC EQUATION INVOLVING SINE

Use a calculator to solve the equation $\sin\left(\theta\right)=0.8,$ where θ is in radians.

Answer

Make sure mode is set to radians. To find θ, use the inverse sine function. On most calculators, you will need to push the 2ND button and then the SIN button to bring up the $\sin^{-1}$ function. What is shown on the screen is $\sin^{-1}\left(\ \right)$. The calculator is ready for the input within the parentheses. For this problem, we enter $\sin^{-1}\left(0.8\right)$, and press ENTER. Thus, to four decimals places,

$$\sin^{-1}\left(0.8\right)\approx0.9273$$

The solution is $0.9273\pm2\pi k$.

Keep in mind that a calculator will only return an angle in quadrants I or IV for the sine function, since that is the range of the inverse sine. Since the sine value is positive, there will be another solution in quadrant 2. The other angle is obtained by using $\pi-\theta$.

This gives us a second set of values in the form $2.2143\pm2\pi k$.

The angle measurements in degrees are based on the first revolution angles of

$$\theta\approx53.1°\text{ or }\theta\approx180°-53.1°\approx126.9°$$

EXAMPLE 8: USING A CALCULATOR TO SOLVE A TRIGONOMETRIC EQUATION INVOLVING SECANT

Use a calculator to solve the equation $\sec(\theta) = -4$, giving your answer in radians.

Answer

We can begin with some algebra.

$$\sec(\theta) = -4$$

$$\frac{1}{\cos(\theta)} = -4$$

$$\cos(\theta) = -\frac{1}{4}$$

Check that the MODE is in radians. Now use the inverse cosine function.

$$\cos^{-1}\left(-\frac{1}{4}\right) \approx 1.8235$$

$$\theta \approx 1.8235 + 2\pi k$$

Since $\frac{\pi}{2} \approx 1.57$ and $\pi \approx 3.14$, we know that 1.8235 is between these two numbers, thus $\theta \approx 1.8235$ is in quadrant II. Cosine is also negative in quadrant III. Note that a calculator will only return an angle in quadrants I or II for the cosine function, since that is the range of the inverse cosine. See Figure 1.

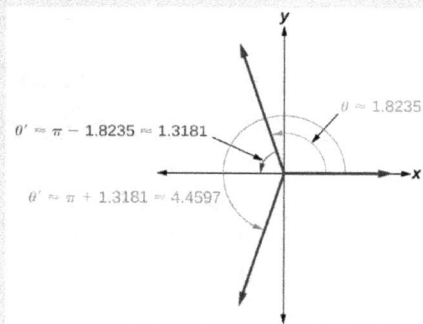

Figure 1

So, we also need to find the measure of the angle in quadrant III. In quadrant III, the reference angle is $\theta' \approx \pi - 1.8235 \approx 1.3181$. The other solution in quadrant III is $\pi + 1.3181 \approx 4.4597$.

The solutions are $1.8235 \pm 2\pi k$ and $4.4597 \pm 2\pi k$.

TRY IT #3

Solve $\csc(\theta) = 3$.

Answer

$\theta \approx 0.33984 \pm 2\pi k$ and $\theta \approx 2.80176 \pm 2\pi k$

Solving Trigonometric Equations in Quadratic Form

Solving a quadratic equation may be more complicated, but once again, we can use algebra as we would for any quadratic equation. Look at the pattern of the equation. Is there more than one trigonometric function in the equation, or is there only one? Which trigonometric function is squared? If there is only one function represented and one of the terms is squared, think about the standard form of a quadratic. Replace the trigonometric function with a variable such as x or u. If substitution makes the equation look like a quadratic equation, then we can use the same methods for solving quadratics to solve the trigonometric equations.

HOW TO

Given a trigonometric equation, solve using algebra.

1. Look for a pattern that suggests an algebraic property, such as the difference of squares or a factoring opportunity.
2. Substitute the trigonometric expression with a single variable, such as x or u.
3. Solve the equation the same way an algebraic equation would be solved.
4. Substitute the trigonometric expression back in for the variable in the resulting expressions.
5. Solve for the angle.

EXAMPLE 9: SOLVING A TRIGONOMETRIC EQUATION IN QUADRATIC FORM USING THE SQUARE ROOT PROPERTY

Solve the problem exactly: $2\sin^2(\theta) - 1 = 0, \ 0 \le \theta < 2\pi$.

Answer

As this problem is not easily factored, we will solve using the square root property. First, we use algebra to isolate $\sin(\theta)$. Then we will find the angles.

$$2\sin^2(\theta) - 1 = 0$$

$$2\sin^2(\theta) = 1 \qquad \text{Add 1 to both sides.}$$

$$\sin^2(\theta) = \frac{1}{2} \qquad \text{Divide both sides by 2.}$$

$$\sqrt{\sin^2(\theta)} = \pm\sqrt{\frac{1}{2}} \qquad \text{Consider both positive and negative square root values.}$$

$$\sin(\theta) = \pm\frac{1}{\sqrt{2}} \qquad \text{Recognize the value from a special right triangle.}$$

$$\theta = \frac{\pi}{4}, \ \frac{3\pi}{4}, \ \frac{5\pi}{4}, \ \frac{7\pi}{4} \qquad \text{Since we have + and -, we have answers in all 4 quadrants}$$

EXAMPLE 10: SOLVING A TRIGONOMETRIC EQUATION IN QUADRATIC FORM USING THE QUADRATIC EQUATION

Solve the equation exactly: $\cos^2(\theta) + 3\cos(\theta) - 1 = 0, \; 0 \leq \theta < 2\pi$.

Answer

We begin by using substitution and replacing $\cos(\theta)$ with u. It is not necessary to use substitution, but it may make the problem easier to solve visually. Let $\cos(\theta) = u$. We have

$$u^2 + 3u - 1 = 0$$

The equation cannot be factored, so we will use the quadratic formula $u = \frac{-b \pm \sqrt{b^2 - 4ac}}{2a}$, where $a = 1, \; b = 3$ and $c = -1$.

$$u = \frac{-3 \pm \sqrt{(3)^2 - 4(1)(-1)}}{2}$$

$$= \frac{-3 \pm \sqrt{13}}{2}$$

Replace u with $\cos(\theta)$, and solve. Thus,

$$\cos(\theta) = \frac{-3 \pm \sqrt{13}}{2}$$

$$\theta = \cos^{-1}\left(\frac{-3 + \sqrt{13}}{2}\right)$$

Note that only the + sign is used. This is because we get an error when we solve $\theta = \cos^{-1}\left(\frac{-3-\sqrt{13}}{2}\right)$ on a calculator, since the domain of the inverse cosine function is $[-1, 1]$. Therefore the solution is

$$\cos^{-1}\left(\frac{-3 + \sqrt{13}}{2}\right) \approx 1.26$$

This terminal side of the angle lies in quadrant I. Since cosine is also positive in quadrant IV, the second solution is

$$2\pi - \cos^{-1}\left(\frac{-3 + \sqrt{13}}{2}\right) \approx 5.02$$

EXAMPLE 11: SOLVING A TRIGONOMETRIC EQUATION IN QUADRATIC FORM BY FACTORING

Solve the equation exactly: $2\sin^2(\theta) - 5\sin(\theta) + 3 = 0$, $0 \le \theta \le 2\pi$.

Answer

Using grouping, this quadratic can be factored. Either make the real substitution, $\sin(\theta) = u$ and factor $2u^2 - 5u + 3 = 0$, or imagine it, as we factor:

Without the substitution	With the substitution
$2\sin^2(\theta) - 5\sin(\theta) + 3 = 0$	$2u^2 - 5u + 3 = 0$
$(2\sin(\theta) - 3)(\sin(\theta) - 1) = 0$	$(2u - 3)(u - 1)$

Now set each factor equal to zero and solve the two equations $2\sin(\theta) - 3 = 0$ and $\sin(\theta) - 1 = 0$.

For the first equation,

$$2\sin(\theta) - 3 = 0$$

$$2\sin(\theta) = 3 \qquad \text{Add 3 to both sides.}$$

$$\sin(\theta) = \frac{3}{2} \qquad \text{Divide both sides by 2.}$$

$$\sin(\theta) \ne \frac{3}{2} \qquad \frac{3}{2} \text{ is not in the domain of the sine function.}$$

The first equation did not have any solution. To solve the second equation,

$$\sin(\theta) - 1 = 0$$

$$\sin(\theta) = 1 \qquad \text{Add 1 to both sides.}$$

$$\theta = \frac{\pi}{2} \qquad \text{Sine equals one only for the quadrantal angle.}$$

The only solution for this equation is $\theta = \frac{\pi}{2}$.

Analysis

Make sure to check all solutions on the given domain as some factors have no solution. This is because the range of the sine function is $[-1, 1]$.

EXAMPLE 12: SOLVING AN EQUATION USING AN IDENTITY

Solve the equation exactly using an identity: $3\cos(\theta) + 3 = 2\sin^2(\theta)$, $0 \le \theta < 2\pi$.

Answer

If we rewrite the right side, we can write the equation in terms of cosine. Recall that the Pythagorean Identity is $\sin^2(\theta) + \cos^2(\theta) = 1$ and can be solved for $\sin^2(\theta)$ so $\sin^2(\theta) = 1 - \cos^2(\theta)$.

$$3\cos(\theta) + 3 = 2\sin^2(\theta)$$
$$3\cos(\theta) + 3 = 2\left(1 - \cos^2(\theta)\right) \qquad \text{Substitute for } \sin^2(\theta).$$

Recognize that we now have a quadratic function in $\cos(\theta)$. Factoring can be used to solve this quadratic equation. Below, the same steps used on the left are demonstrated on the right using the substitution method since the quadratic equation in that form may be easier to work with. For the substitution method, let $u = \cos(\theta)$.

Without the substitution	With the substitution
$3\cos(\theta) + 3 = 2 - 2\cos^2(\theta)$	$3u + 3 = 2 - 2u^2$
$2\cos^2(\theta) + 3\cos(\theta) + 1 = 0$	$2u^2 + 3u + 1 = 0$
$(2\cos(\theta) + 1)(\cos(\theta) + 1) = 0$	$(2u + 1)(u + 1) = 0$

Working with the factors using cosine, set each factor equal to zero and solve.

$$2\cos(\theta) + 1 = 0 \qquad\qquad \cos(\theta) + 1 = 0$$
$$\cos(\theta) = -\frac{1}{2} \qquad\qquad \cos(\theta) = -1$$
$$\theta = \frac{2\pi}{3}, \frac{4\pi}{3} \qquad\qquad \theta = \pi$$

Again, remember that there are two quadrants where the $\cos(\theta) = -\frac{1}{2}$. These are quadrants 2 and 3. We can find the acute reference angle of $\frac{\pi}{3}$ and use that to generate the solutions in these quadrants.

Our solutions are $\frac{2\pi}{3}$, $\frac{4\pi}{3}$, π.

TRY IT #4

Solve $\sin^2(\theta) = 2\cos(\theta) + 2$, $0 \le \theta \le 2\pi$ [Hint: Make a substitution to express the equation only in terms of cosine.]

Answer

$$\cos(\theta) = -1, \ \theta = \pi$$

Solving Trigonometric Equations Using Fundamental Identities

While algebra can be used to solve a number of trigonometric equations, we can also use the fundamental identities because they make solving equations simpler. Remember that the techniques we use for solving are not the same as those for verifying identities. The basic rules of algebra apply here, as opposed to rewriting one side of the identity to match the other side. In the next example, we use two identities to simplify the equation.

We will need to use some new identities in this section. They are called the **double-angle** identities. We will not take the time to show where the following identities come from. Keep in mind that there are other trigonometric identities that we have not covered in this material. If you are interested, you can look up the sum and difference formulas for sine and cosine, and use those to generate some other identities, including the ones shown below.

DEFINITION

The double angle identities are:

$$\sin(2\theta) = 2\sin(\theta)\cos(\theta)$$

$$\cos(2\theta) = \cos^2(\theta) - \sin^2(\theta) \text{ or,}$$
$$= 1 - 2\sin^2(\theta) \text{ or,}$$
$$= 2\cos^2(\theta) - 1$$

EXAMPLE 13: SOLVING THE EQUATION USING A DOUBLE-ANGLE FORMULA

Solve the equation exactly using a double-angle formula: $\cos(2\theta) = \cos(\theta).$

Answer

We have three choices of expressions to substitute for the double-angle of cosine. As it is simpler to solve for one trigonometric function at a time, we will choose the double-angle identity involving only cosine:

$$\cos(2\theta) = \cos(\theta)$$

$$2\cos^2(\theta) - 1 = \cos(\theta) \quad \text{Replace left hand side with identity.}$$

$$2\cos^2(\theta) - \cos(\theta) - 1 = 0 \quad \text{Move all terms to one side.}$$

$$(2\cos(\theta) + 1)(\cos(\theta) - 1) = 0 \quad \text{Factor the left hand side.}$$

Set the factors equal to zero and solve.

$$2\cos(\theta) + 1 = 0 \qquad\qquad \cos(\theta) - 1 = 0$$

$$\cos(\theta) = -\frac{1}{2} \qquad\qquad \cos(\theta) = 1$$

$$\theta = \frac{2\pi}{3} \pm 2\pi k \qquad\qquad \theta = 0 \pm 2\pi k$$

$$\theta = \frac{4\pi}{3} \pm 2\pi k$$

Solving Trigonometric Equations with Multiple Angles

Sometimes it is not possible to solve a trigonometric equation with identities that have a multiple angle, such as $\sin(2x)$ or $\cos(3x)$. When confronted with these equations, recall that $y = \sin(2x)$ is a horizontal compression by a factor of 2 of the function $y = \sin(x)$. On an interval of 2π, we can graph two periods of $y = \sin(2x)$, as opposed to one cycle of $y = \sin(x)$. This compression of the graph leads us to believe there may be twice as many x-intercepts or solutions to $\sin(2x) = 0$ compared to $\sin(x) = 0$. This information will help us solve the similar type of equation shown in the example.

EXAMPLE 14: SOLVING A MULTIPLE ANGLE TRIGONOMETRIC EQUATION

Solve exactly: $\cos(2x) = \frac{1}{2}$ on $[0, 2\pi)$.

Answer

We can see that this equation is the standard equation with a multiple of an angle. If $\cos(\theta) = \frac{1}{2}$, we know θ is in quadrants I and IV. While $\theta = \cos^{-1}\left(\frac{1}{2}\right)$ will only yield solutions in quadrants I and II because of the range of the inverse cosine function, we recognize that the solutions to the equation $\cos(\theta) = \frac{1}{2}$ will be in quadrants I and IV using ideas from our unit circle.

Therefore, the possible angles are $\theta = \frac{\pi}{3}$ and $\theta = \frac{5\pi}{3}$. So, $2x = \frac{\pi}{3}$ or $2x = \frac{5\pi}{3}$, which means that $x = \frac{\pi}{6}$ or $x = \frac{5\pi}{6}$. Does this make sense? Yes, because $\cos\left(2\left(\frac{\pi}{6}\right)\right) = \cos\left(\frac{\pi}{3}\right) = \frac{1}{2}$.

Are there any other possible answers? Let us return to our first step.

In quadrant I, $2x = \frac{\pi}{3}$, so $x = \frac{\pi}{6}$ as noted. Let us revolve around the circle again:

$$2x = \frac{\pi}{3} + 2\pi$$

$$2x = \frac{\pi}{3} + \frac{6\pi}{3}$$

$$2x = \frac{7\pi}{3}$$

$$x = \frac{7\pi}{6}$$

One more rotation yields

$$2x = \frac{\pi}{3} + 4\pi$$

$$2x = \frac{\pi}{3} + \frac{12\pi}{3}$$

$$2x = \frac{13\pi}{3}$$

$$x = \frac{13\pi}{6}$$

$x = \frac{13\pi}{6} > 2\pi$, so this value for x is larger than 2π, so it is not a solution on $[0, 2\pi)$.

In quadrant IV, $2x = \frac{5\pi}{3}$, so $x = \frac{5\pi}{6}$ as noted. Let us revolve around the circle again:

$$2x = \frac{5\pi}{3} + 2\pi$$

$$2x = \frac{5\pi}{3} + \frac{6\pi}{3}$$

$$2x = \frac{11\pi}{3}$$

$$x = \frac{11\pi}{6}$$

One more rotation yields

$$2x = \frac{5\pi}{3} + 4\pi$$

$$2x = \frac{5\pi}{3} + \frac{12\pi}{3}$$

$$2x = \frac{17\pi}{3}$$

$$x = \frac{17\pi}{6}$$

$x = \frac{17\pi}{6} > 2\pi$, so this value for x is larger than 2π, which means it is not a solution on $[0, 2\pi)$.

Our solutions are $\frac{\pi}{6}$, $\frac{5\pi}{6}$, $\frac{7\pi}{6}$, and $\frac{11\pi}{6}$. Note that whenever we solve a problem in the form of $\sin(nx) = c$ or $\cos(nx) = c$, we must go around the unit circle n times.

We can see this easily if we first write the solution for $2x$ in the general form. One of these equations is shown below:

$$2x = \frac{\pi}{3} + 2\pi \cdot k$$
$$x = \frac{\pi}{6} + \pi \cdot k \qquad \text{Divide both sides by 2.}$$

We can now see that adding π to $\frac{\pi}{6}$ gives us $\frac{7\pi}{6}$ and that adding $2 \cdot \pi$ would give us $\frac{13\pi}{6}$ which is larger than 2π.

Access these online resources for additional instruction and practice with solving trigonometric equations.

- Solving Trigonometric Equations I
- Solving Trigonometric Equations II
- Solving Trigonometric Equations III
- Solving Trigonometric Equations IV
- Solving Trigonometric Equations V
- Solving Trigonometric Equations VI

KEY CONCEPTS

- When solving linear trigonometric equations, we can use algebraic techniques just as we do solving algebraic equations. Look for patterns, like the difference of squares, quadratic form, or an expression that lends itself well to substitution.
- Equations involving a single trigonometric function can be solved or verified using the unit circle.
- We can also solve trigonometric equations using a graphing calculator.
- Many equations appear quadratic in form. We can use substitution to make the equation appear simpler, and then use the same techniques we use solving an algebraic quadratic: factoring, the quadratic formula, etc.
- We can also use the identities to solve trigonometric equation.
- We can use substitution to solve a multiple-angle trigonometric equation, which is a compression of a standard trigonometric function. We will need to take the compression into account and verify that we have found all solutions on the given interval.

3.9 Modeling with Trigonometric Equations

LEARNING OBJECTIVES

In this section, you will:

- Model equations and graph sinusoidal functions.
- Model periodic behavior.
- Model simple harmonic motion functions.

The hands on a clock are periodic: they repeat positions every twelve hours. (credit: "zoutedrop"/Flickr)

Suppose we charted the average daily temperatures in New York City over the course of one year. We would expect to find the lowest temperatures in January and February and highest in July and August. This familiar cycle repeats year after year, and if we were to extend the graph over multiple years, it would resemble a periodic function.

Many other natural phenomena are also periodic. For example, the phases of the moon have a period of approximately 28 days, and birds know to fly south at about the same time each year.

So how can we model an equation to reflect periodic behavior? First, we must collect and record data. We then find a function that resembles an observed pattern. Finally, we make the necessary alterations to the function to get a model that is dependable. In this section, we will take a deeper look at specific types of periodic behavior and model equations to fit data.

Modeling Periodic Behavior

EXAMPLE 1: MODELING AN EQUATION AND SKETCHING A SINUSOIDAL GRAPH TO FIT CRITERIA

The average monthly temperatures for a small town in Oregon are given. Find a sinusoidal function of the form $y = A\sin\left(B\left(t - h\right)\right) + k$ that fits the data (round to the nearest tenth) and sketch the graph.

Table 1

Month	Temperature,$^{\circ}$F
January	42.5
February	44.5
March	48.5
April	52.5
May	58
June	63
July	68.5
August	69
September	64.5
October	55.5
November	46.5
December	43.5

Answer

Recall that amplitude is found using the formula

$$A = \frac{\text{largest value } - \text{smallest value}}{2}.$$

Thus, the amplitude is

$$|A| = \frac{69 - 42.5}{2}$$

$$= 13.25.$$

The data covers a period of 12 months, so $\frac{2\pi}{B} = 12$ which gives $B = \frac{2\pi}{12} = \frac{\pi}{6}$.

The vertical shift is found using the following equation.

$$k = \frac{\text{highest value} + \text{lowest value}}{2}.$$

Thus, the vertical shift is

$$k = \frac{69 + 42.5}{2}$$

$$= 55.8.$$

So far, we have the equation $y = 13.3\sin\left(\frac{\pi}{6}\left(x - h\right)\right) + 55.8.$

To find the horizontal shift, we can input the x and y values for the first month, which will be $x = 1$ and $y = 42.5$. We can then solve for h as shown below.

$$42.5 = 13.3\sin\left(\frac{\pi}{6}(1-h)\right) + 55.8$$

$$-13.3 = 13.3\sin\left(\frac{\pi}{6}(1-h)\right) \qquad \text{Subtracted 55.8 from both sides.}$$

$$-1 = \sin\left(\frac{\pi}{6}(1-h)\right) \qquad \text{Now divide both sides by 13.3.}$$

We can use the following idea: $\sin(\theta) = -1 \rightarrow \theta = -\frac{\pi}{2}$

$$\frac{\pi}{6}(1-h) = -\frac{\pi}{2} \qquad \text{Set the input expression of sine equal to } -\frac{\pi}{2}.$$

$$\frac{\pi}{6} - \frac{\pi}{6}h = -\frac{\pi}{2} \qquad \text{Distribute on the left hand side.}$$

$$-\frac{\pi}{6}h = -\frac{\pi}{2} - \frac{\pi}{6} \qquad \text{Subtract } \frac{\pi}{6} \text{ from both sides.}$$

$$h = \left(-\frac{\pi}{2} - \frac{\pi}{6}\right)\frac{-6}{\pi} \qquad \text{Multiply both sides by } -\frac{6}{\pi}.$$

$$h = 3 + 1 = 4$$

We have the equation $y = 13.3\sin\left(\frac{\pi}{6}(x-4)\right) + 55.8$. See the graph in Figure 1.

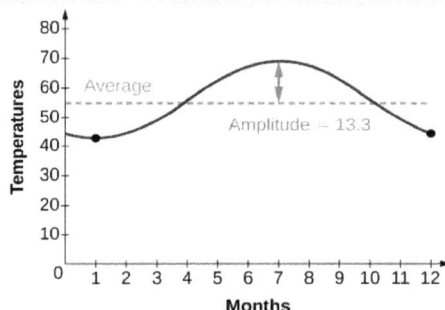

Figure 1

EXAMPLE 2: DESCRIBING PERIODIC MOTION

The hour hand of the large clock on the wall in Union Station measures 24 inches in length. At noon, the tip of the hour hand is 30 inches from the ceiling. At 3 PM, the tip is 54 inches from the ceiling, and at 6 PM, 78 inches. At 9 PM, it is again 54 inches from the ceiling, and at midnight, the tip of the hour hand returns to its original position 30 inches from the ceiling. Let y equal the distance from the tip of the hour hand to the ceiling x hours after noon. Find the equation that models the motion of the clock and sketch the graph.

Answer Begin by making a table of values as shown in Table.

Table 2

x	y	Points to plot
Noon	30 in	$(0, 30)$
3 PM	54 in	$(3, 54)$
6 PM	78 in	$(6, 78)$
9 PM	54 in	$(9, 54)$
Midnight	30 in	$(12, 30)$

To model an equation, we first need to find the amplitude.

$$|A| = |\frac{78 - 30}{2}|$$
$$= 24$$

The clock's cycle repeats every 12 hours. Thus,

$$B = \frac{2\pi}{12}$$
$$= \frac{\pi}{6}$$

The vertical shift is

$$k = \frac{78 + 30}{2}$$
$$= 54$$

Since the function begins with the minimum value of y when $x = 0$ (as opposed to the maximum value), we will use the cosine function with the negative value for A. There is no horizontal shift, so $h = 0$. In the form $y = A\cos\left(B\left(x - h\right)\right) + k$, the equation is

$$y = -24\cos\left(\tfrac{\pi}{6}x\right) + 54$$

See Figure 2.

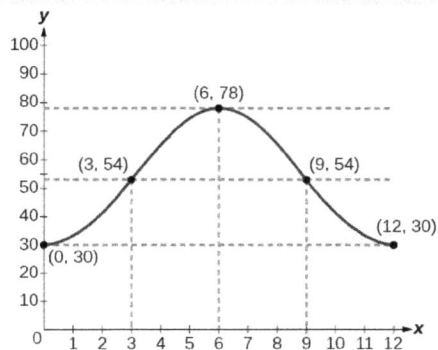

Figure 2

EXAMPLE 3: DETERMINING A MODEL FOR TIDES

The height of the tide in a small beach town is measured along a seawall. Water levels oscillate between 7 feet at low tide and 15 feet at high tide. On a particular day, low tide occurred at 6 AM and high tide occurred at noon. Approximately every 12 hours, the cycle repeats. Find an equation to model the water levels.

Answer

As the water level varies from 7 ft to 15 ft, we can calculate the amplitude as

$$|A| = |\frac{(15-7)}{2}|$$
$$= 4.$$

The cycle repeats every 12 hours; therefore, B is

$$\frac{2\pi}{12} = \frac{\pi}{6}.$$

There is a vertical translation of $\frac{(15+7)}{2} = 11$. Since the value of the function is at a maximum at $t = 0$, we will use the cosine function, with the positive value for A.

$$y = 4\cos\left(\frac{\pi}{6}t\right) + 11$$

See Figure 3.

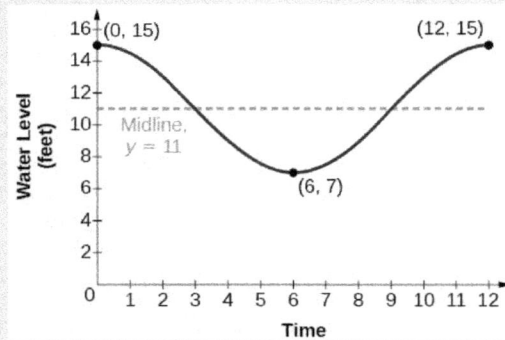

Figure 3

TRY IT #1

The daily temperature in the month of March in a certain city varies from a low of 24F to a high of 40F. Find a sinusoidal function to model daily temperature and sketch the graph. Approximate the time when the temperature reaches the freezing point 32F. Let $t = 0$ correspond to noon.

Answer

$$y = 8\sin\left(\tfrac{\pi}{12}t\right) + 32$$

The temperature reaches freezing at noon and at midnight.

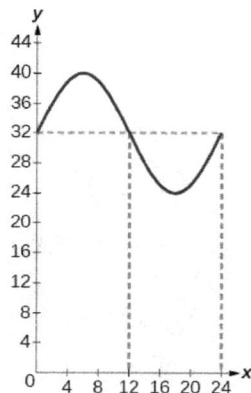

EXAMPLE 4: INTERPRETING THE PERIODIC BEHAVIOR EQUATION

The average person's blood pressure is modeled by the function $f(t) = 20\sin(160\pi t) + 100$, where $f(t)$ represents the blood pressure at time t, measured in minutes. Sketch the graph and find the blood pressure reading.

Answer

The period is given by

$$\frac{2\pi}{B} = \frac{2\pi}{160\pi}$$
$$= \frac{1}{80}.$$

Since the period is $\frac{1}{80}$, we know it takes $\frac{1}{80}^{th}$ of a minute for the blood pressure to cycle through a full range of values.

See the graph in Figure 4.

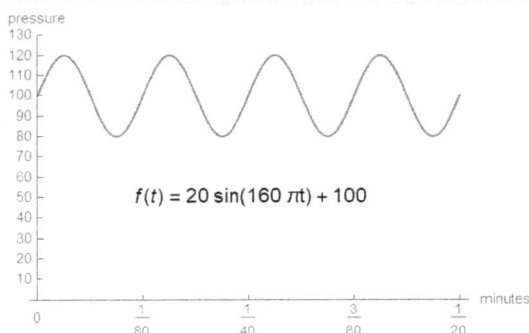

Figure 4

Analysis

Blood pressure of $\frac{120}{80}$ is considered to be normal. The top number is the maximum or systolic reading, which measures the pressure in the arteries when the heart contracts. The bottom number is the minimum or diastolic reading, which measures the pressure in the arteries as the heart relaxes between beats, refilling with blood. Thus, normal blood pressure can be modeled by a periodic function with a maximum of 120 and a minimum of 80. Since the period is $\frac{1}{80}^{th}$ of a minute, we know there are 80 heartbeats in a minute.

Modeling Harmonic Motion Functions

Harmonic motion is a form of periodic motion, but there are factors to consider that differentiate the two types. While general periodic motion applications cycle through their periods with no outside interference, harmonic motion requires a restoring force. Examples of harmonic motion include springs, gravitational force, and magnetic force.

Simple Harmonic Motion

A type of motion described as simple harmonic motion involves a restoring force but assumes that the motion will continue forever. Imagine a weighted object hanging on a spring, When that object is not disturbed, we say that the object is at rest, or in equilibrium. If the object is pulled down and then released, the force of the spring pulls the object back toward equilibrium and harmonic motion begins. The restoring force is directly proportional to the displacement of the object from its equilibrium point. When $t = 0, d = 0$.

Simple Harmonic Motion

We see that simple harmonic motion equations are given in terms of displacement:

$$d = A\cos\left(Bt\right) \ \text{ or } \ d = A\sin\left(Bt\right)$$

where $|A|$ is the amplitude, and $\frac{2\pi}{B}$ is the period.

EXAMPLE 5: FINDING THE DISPLACEMENT, PERIOD, AND FREQUENCY, AND GRAPHING A FUNCTION

For the given functions,

1. Find the maximum displacement of an object.
2. Find the period or the time required for one vibration.
3. Sketch the graph.

    a. $y = 5\sin\left(3t\right)$
    b. $y = 6\cos\left(\pi t\right)$
    c. $y = 5\cos\left(\frac{\pi}{2}t\right)$

Answer

  a. $y = 5\sin\left(3t\right)$

1. The maximum displacement is equal to the amplitude, $|A|$, which is 5.
2. The period is $\frac{2\pi}{B} = \frac{2\pi}{3}$.
3. See Figure 5. The graph indicates the five key points. Remember, you take $\frac{1}{4}$ of the period and starting at 0, add this value repeatedly 4 times. For this problem, you would use an interval of $\frac{1}{4} \times \frac{2\pi}{3} = \frac{\pi}{6}$.

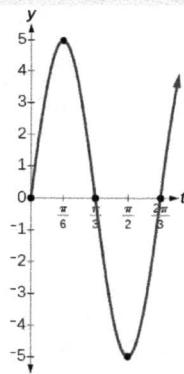

Figure 5

  b. $y = 6\cos\left(\pi t\right)$

1. The maximum displacement is 6.
2. The period is $\frac{2\pi}{B} = \frac{2\pi}{\pi} = 2$.
3. See Figure 6.

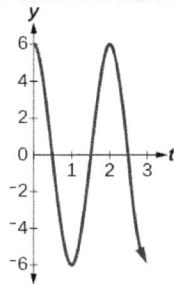

Figure 6

c. $y = 5\cos\left(\frac{\pi}{2}t\right)$

    1. The maximum displacement is 5.

    2. The period is $\frac{2\pi}{B} = \frac{2\pi}{\frac{\pi}{2}} = 4$.

    3. See Figure 7.

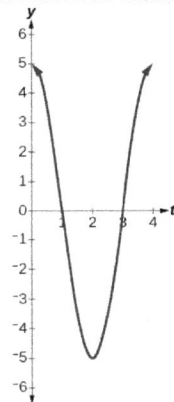

Figure 7

Access these online resources for additional instruction and practice with trigonometric applications.

- Solving Problems Using Trigonometry
- Ferris Wheel Trigonometry
- Daily Temperatures and Trigonometry
- Simple Harmonic Motion

Key Equations

Standard form of sinusoidal equation	$y = A\sin\left(B\left(t - h\right)\right) + k$ or $y = A\cos\left(B\left(t - h\right)\right) + k$
Simple harmonic motion	$d = a\cos\left(Bt\right)$ or $d = a\sin\left(Bt\right)$

KEY CONCEPTS

- Sinusoidal functions are represented by the sine and cosine graphs. In standard form, we can find the amplitude, period, and horizontal and vertical shifts.
- Use key points to graph a sinusoidal function. The five key points include the minimum and maximum values and the midline values.
- Periodic functions can model events that reoccur in set cycles, like the phases of the moon, the hands on a clock, and the seasons in a year.
- Harmonic motion functions are modeled from given data. Similar to periodic motion applications, harmonic motion requires a restoring force. Examples include gravitational force and spring motion activated by weight.

GLOSSARY

simple harmonic motion
    a repetitive motion that can be modeled by periodic sinusoidal oscillation

POLYNOMIAL AND RATIONAL FUNCTIONS

4.1 Power Functions and Polynomial Functions

Figure 1. (credit: Jason Bay, Flickr)

Suppose a certain species of bird thrives on a small island. Its population over the last few years is shown in Table 1.

Table 1

Year	2009	2010	2011	2012	2013
Bird Population	800	897	992	1,083	1,169

The population can be estimated using the function $P(t) = -0.3t^3 + 97t + 800$, where $P(t)$ represents the bird population on the island t years after 2009. We can use this model to estimate the maximum bird population and when it will occur. We can also use this model to predict when the bird population will disappear from the island. In this section, we will examine functions that we can use to estimate and predict these types of changes.

Identifying Power Functions

In order to better understand the bird problem, we need to understand a specific type of function. A **power function** is a function with a single term that is the product of a real number, a **coefficient,** and a variable raised to a fixed real number. (A number that multiplies a variable raised to an exponent is known as a coefficient.)

As an example, consider functions for area or volume. The function for the **area of a circle** with radius r is

$$A(r) = \pi r^2$$

and the function for the **volume of a sphere** with radius r is

$$V(r) = \tfrac{4}{3}\pi r^3$$

Both of these are examples of power functions because they consist of a coefficient, π or $\tfrac{4}{3}\pi$, multiplied by a variable r raised to a power.

DEFINITION

A **power function** is a function that can be represented in the form

$$f(x) = kx^p$$

where k and p are real numbers, and k is known as the **coefficient**.

Q&A

Is $f(x) = 2^x$ a **power function?**

No. A power function contains a variable base raised to a fixed power. This function has a constant base raised to a variable power. This is an exponential function, not a power function.

EXAMPLE 1: IDENTIFYING POWER FUNCTIONS

Which of the following functions are power functions?

$f(x) = 1$	Constant function
$f(x) = x$	Identify function
$f(x) = x^2$	Quadratic function
$f(x) = x^3$	Cubic function
$f(x) = \frac{1}{x}$	Reciprocal function
$f(x) = \frac{1}{x^2}$	Reciprocal squared function
$f(x) = \sqrt{x}$	Square root function
$f(x) = \sqrt[3]{x}$	Cube root function

Answer

All of the listed functions are power functions.

The constant and identity functions are power functions because they can be written as $f(x) = x^0$ and $f(x) = x^1$ respectively.

The quadratic and cubic functions are power functions with whole number powers $f(x) = x^2$ and $f(x) = x^3$.

The **reciprocal** and reciprocal squared functions are power functions with negative whole number powers because they can be written as $f(x) = x^{-1}$ and $f(x) = x^{-2}$.

The square and **cube root** functions are power functions with fractional powers because they can be written as $f(x) = x^{1/2}$ or $f(x) = x^{1/3}$.

TRY IT #1

Which functions are power functions?

$$f(x) = 2x^2 \cdot 4x^3$$
$$g(x) = -x^5 + 5x^3 - 4x$$
$$h(x) = \frac{2x^5 - 1}{3x^2 + 4}$$

Answer

$f(x)$ is a power function because it can be written as $f(x) = 8x^5$. The other functions are not power functions.

Identifying End Behavior of Power Functions

Figure 2 shows the graphs of $f(x) = x^2$, $g(x) = x^4$ and $h(x) = x^6$, which are all power functions with even, whole-number powers. Notice that these graphs have similar shapes, very much like that of the quadratic function in the toolkit. However, as the power increases, the graphs flatten somewhat near the origin and become steeper away from the origin.

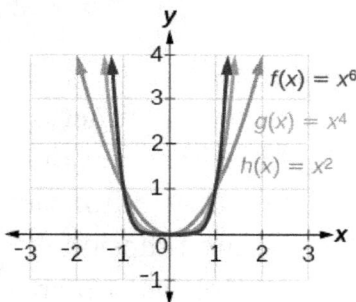

Figure 2. Even-power functions

To describe the behavior as numbers become larger and larger, we use the idea of infinity. We use the symbol ∞ for positive infinity and $-\infty$ for negative infinity. When we symbolically write as $x \to \infty$, we are describing a behavior; we are saying that x is increasing without bound.

With the even-power function, as the input increases or decreases without bound, the output values become very large, positive numbers. Equivalently, we could describe this behavior by saying that as x increases or decreases without bound, the $f(x)$ values increase without bound. In symbolic form, we could write

$$\text{as } x \to \infty, f(x) \to \infty$$

Figure 3 shows the graphs of $f(x) = x^3$, $g(x) = x^5$, and $h(x) = x^7$, which are all power functions with odd, whole-number powers. Notice that these graphs look similar to the cubic function in the toolkit. Again, as the power increases, the graphs flatten near the origin and become steeper away from the origin.

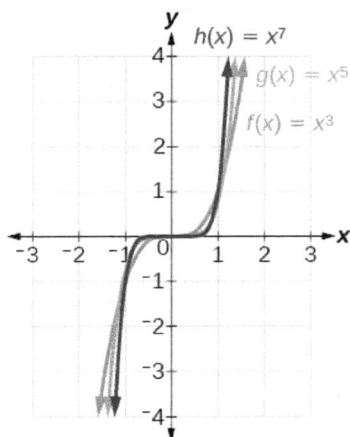

Figure 3. Odd-power function

These examples illustrate that functions of the form $f(x) = x^n$ reveal symmetry of one kind or another. First, in Figure 2 we see that even functions of the form $f(x) = x^n$, n even, are symmetric about the y-axis. In Figure 3 we see that odd functions of the form $f(x) = x^n$, n odd, are symmetric about the origin.

For these odd power functions, as x decreases without bound, $f(x)$ decreases without bound. As x increases without bound, $f(x)$ increases without bound. In symbolic form we write

$$\text{as } x \to -\infty, f(x) \to -\infty$$
$$\text{as } x \to \infty, f(x) \to \infty$$

The behavior of the graph of a function as the input values move far to the left of the origin ($x \to -\infty$) and move far to the right of the origin ($x \to \infty$) is referred to as the **end behavior** of the function. We can use words or symbols to describe end behavior.

Figure 4 shows the end behavior of power functions in the form $f(x) = kx^n$ where n is a non-negative integer depending on the power and the constant.

	Even power	Odd power
Positive constant $k > 0$	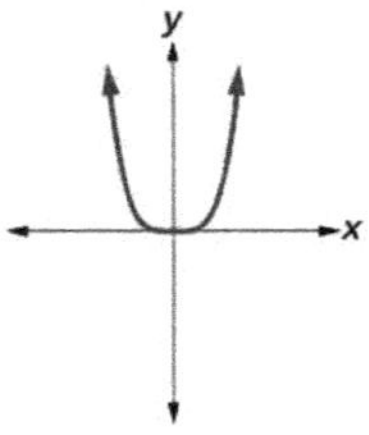 $x \to -\infty, f(x) \to \infty$ and $x \to \infty, f(x) \to \infty$	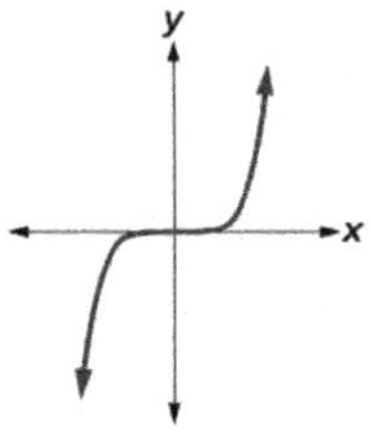 $x \to -\infty, f(x) \to -\infty$ and $x \to \infty, f(x) \to \infty$
Negative constant $k < 0$	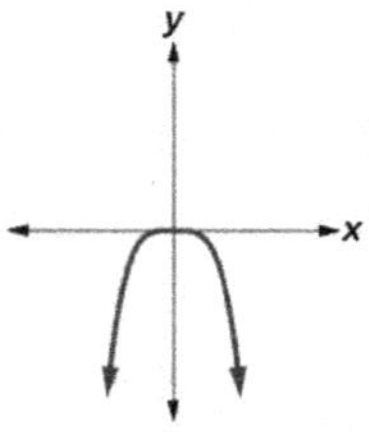 $x \to -\infty, f(x) \to -\infty$ and $x \to \infty, f(x) \to -\infty$	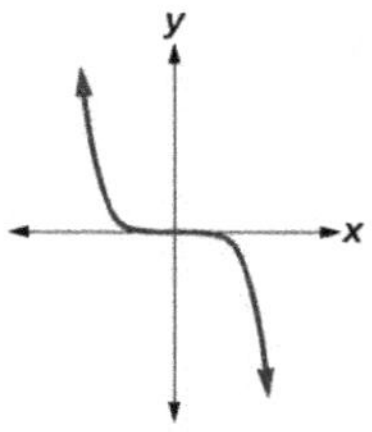 $x \to -\infty, f(x) \to \infty$ and $x \to \infty, f(x) \to -\infty$

Figure 4.

HOW TO

Given a power function $f(x) = kx^n$ where n is a non-negative integer, identify the end behavior.

1. Determine whether the power is even or odd.
2. Determine whether the constant is positive or negative.
3. Use Figure 4 to identify the end behavior.

EXAMPLE 2: IDENTIFYING THE END BEHAVIOR OF A POWER FUNCTION

Describe the end behavior of the graph of $f(x) = x^8$.

Answer

The coefficient is 1 (positive) and the exponent of the power function is 8 (an even number). As x increases without bound, the output (value of $f(x)$) increases without bound. We write as $x \to \infty, f(x) \to \infty$. As x decreases without bound, the output increases without bound. In symbolic form, as $x \to -\infty, f(x) \to \infty$. We can graphically represent the function as shown in Figure 5.

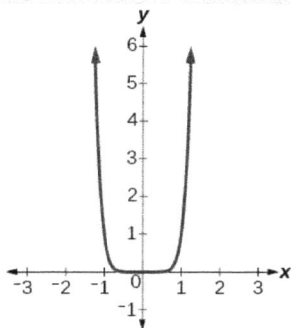

Figure 5.

EXAMPLE 3: IDENTIFYING THE END BEHAVIOR OF A POWER FUNCTION.

Describe the end behavior of the graph of $f(x) = -x^9$.

Answer

The exponent of the power function is 9 (an odd number). Because the coefficient is -1 (negative), the graph is the reflection about the x-axis of the graph of $f(x) = x^9$. Figure 6 shows that as x increases without bound, the output decreases without bound. As x decreases without bound, the output increases without bound. In symbolic form, we would write

$$\text{as } x \to -\infty, f(x) \to \infty$$
$$\text{as } x \to \infty, f(x) \to -\infty$$

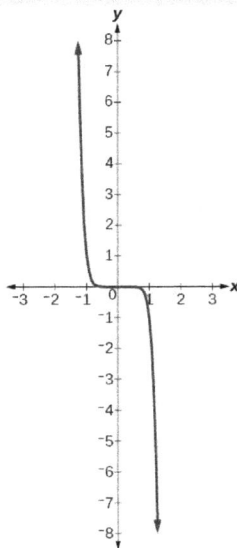

Figure 6.

Analysis

We can check our work by using the table feature on a graphing utility.

Table 2

x	$f(x)$
–10	1,000,000,000
–5	1,953,125
0	0
5	–1,953,125
10	–1,000,000,000

We can see from Table 2 that, when we substitute more and more negative values for x, the output becomes very large, and when we substitute larger positive values for x, the output becomes more negative.

TRY IT #2

Describe in words and symbols the end behavior of $f(x) = -5x^4$.

Answer

As x increases or decreases without bound, $f(x)$ decreases without bound: as $x \to \infty$, $f(x) \to -\infty$ because of the negative coefficient.

Identifying Polynomial Functions

An oil pipeline bursts in the Gulf of Mexico, causing an oil slick in a roughly circular shape. The slick is currently 24 miles in radius, but that radius is increasing by 8 miles each week. We want to write a formula for the area covered by the oil slick by combining two functions. The radius r of the spill depends on the number of weeks w that have passed. This relationship is linear.

$$r(w) = 24 + 8w$$

We can combine this with the formula for the area A of a circle.

$$A(r) = \pi r^2$$

Composing these functions gives a formula for the area in terms of weeks.

$$A(w) = A(r(w))$$
$$= A(24 + 8w)$$
$$= \pi(24 + 8w)^2$$

Multiplying gives the formula

$$A\left(w\right) = 576\pi + 384\pi w + 64\pi w^2.$$

This formula is an example of a **polynomial function**. A polynomial function consists of either zero or the sum of a finite number of non-zero terms, each of which is a product of a number, called the coefficient of the term, and a variable raised to a non-negative integer power.

DEFINITION

Let n be a non-negative integer. A **polynomial function** is a function that can be written in the form
$$f\left(x\right) = a_n x^n + a_{n-1} x^{n-1} + \dots + a_2 x^2 + a_1 x + a_0$$
This is called the general form of a polynomial function. Each a_i is a **coefficient** and can be any real number, but a_n cannot equal 0. Each product $a_i x^i$ is a **term of a polynomial function**.

EXAMPLE 4: IDENTIFYING POLYNOMIAL FUNCTIONS

Which of the following are polynomial functions?

$$f\left(x\right) = 2x^3 \cdot 3x + 4$$
$$g\left(x\right) = -x\left(x^2 - 4\right)$$
$$h\left(x\right) = 5\sqrt{x} + 2$$

Answer

The first two functions are examples of polynomial functions because they can be written in the form $f\left(x\right) = a_n x^n + a_{n-1} x^{n-1} + \dots + a_2 x^2 + a_1 x + a_0$ where the powers are non-negative integers and the coefficients are real numbers.

- $f\left(x\right)$ can be written as $f\left(x\right) = 6x^4 + 4$.
- $g\left(x\right)$ can be written as $g\left(x\right) = -x^3 + 4x$.
- $h\left(x\right)$ cannot be written in this form and is therefore not a polynomial function.

Identifying the Degree and Leading Coefficient of a Polynomial Function

Because of the form of a polynomial function, we can see an infinite variety in the number of terms and the power of the variable. Although the order of the terms in the polynomial function is not important for performing operations, we typically arrange the terms in descending order of power, or in general form. The **degree** of the polynomial is the highest power of the variable that occurs in the polynomial; it is the power of the first variable if the function is in general form. The **leading term** is the term containing the highest power of the variable, or the term with the highest degree. The **leading coefficient** is the coefficient of the leading term. For the constant function, the degree is zero, and for a non-constant linear function, the degree is one.

TERMINOLOGY OF POLYNOMIAL FUNCTIONS

We often rearrange polynomials so that the powers are descending.

Leading coefficient   Degree

$$f(x) = \underline{a_n x^n} + \ldots + a_2 x^2 + a_1 x + a_0$$

Leading term

When a polynomial is written in this way, we say that it is in general form.

HOW TO

Given a polynomial function, identify the degree and leading coefficient.

1. Find the highest power of x to determine the degree of the function.
2. Identify the term containing the highest power of x to find the leading term.
3. Identify the coefficient of the leading term.

EXAMPLE 5: IDENTIFYING THE DEGREE AND LEADING COEFFICIENT OF A POLYNOMIAL FUNCTION

Identify the degree, leading term, and leading coefficient of the following polynomial functions.

$$f(x) = 3 + 2x^2 - 4x^3$$
$$g(t) = 5t^5 - 2t^3 + 7t$$
$$h(p) = 6p - p^3 - 2$$

Answer

For the function $f(x)$, the highest power of x is 3, so the degree is 3. The leading term is the term containing that degree, $-4x^3$. The leading coefficient is the coefficient of that term, -4.

For the function $g(t)$, the highest power of t is 5, so the degree is 5. The leading term is the term containing that degree, $5t^5$. The leading coefficient is the coefficient of that term, 5.

For the function $h(p)$, the highest power of p is 3, so the degree is 3. The leading term is the term containing that degree, $-p^3$; the leading coefficient is the coefficient of that term, -1.

Identify the degree, leading term, and leading coefficient of the polynomial $f\left(x\right) = 4x^2 - x^6 + 2x - 6$.

Answer

The degree is 6. The leading term is $-x^6$. The leading coefficient is -1.

Identifying End Behavior of Polynomial Functions

Knowing the degree of a polynomial function is useful in helping us predict its end behavior. To determine its end behavior, look at the leading term of the polynomial function. Because the power of the leading term is the highest, that term will grow significantly faster than the other terms as x gets more and more positive or negative, so its behavior will dominate the graph. For any polynomial, the end behavior of the polynomial will match the end behavior of the term of highest degree. See Table 3.

Table 3

Polynomial Function	Leading Term	Graph of Polynomial Function
$f\left(x\right) = 5x^4 + 2x^3 - x - 4$	$5x^4$	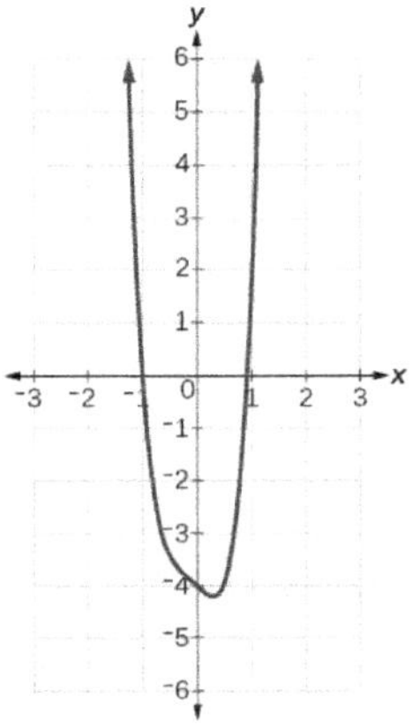
$f\left(x\right) = -2x^6 - x^5 + 3x^4 + x^3$	$-2x^6$	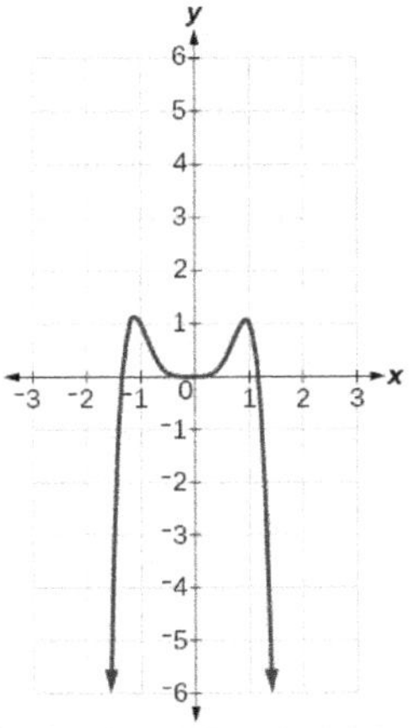

Polynomial Function	Leading Term	Graph of Polynomial Function
$f(x) = 3x^5 - 4x^4 + 2x^2 + 1$	$3x^5$	
$f(x) = -6x^3 + 7x^2 + 3x + 1$	$-6x^3$	

EXAMPLE 6: IDENTIFYING END BEHAVIOR AND DEGREE OF A POLYNOMIAL FUNCTION

Describe the end behavior and determine a possible degree of the polynomial function in Figure 7.

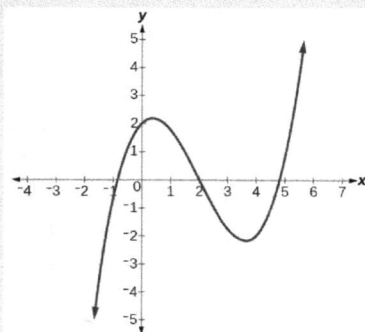

Figure 7.

Answer

As the input values x get very large, the output values $f\left(x\right)$ increase without bound. As the input values x get more and more negative, the output values $f\left(x\right)$ decrease without bound. We can describe the end behavior symbolically by writing

$$\text{as } x \to -\infty, f\left(x\right) \to -\infty$$
$$\text{as } x \to \infty, f\left(x\right) \to \infty.$$

In words, we read this notation, "as x values increase without bound, the function values increase without bound and as x values decrease without bound, the function values decrease without bound."

We can tell this graph has the shape of an odd degree power function that has not been reflected, so the degree of the polynomial creating this graph must be odd and the leading coefficient must be positive.

TRY IT #4

Describe the end behavior, and determine a possible degree of the polynomial function in Figure.

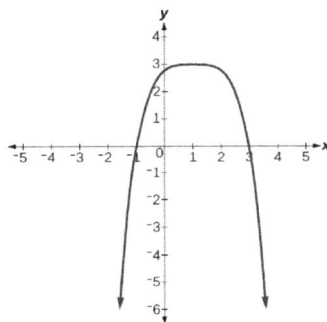

Figure 8.

Answer

As the input increases without bound, the output decreases without bound. As the input decreases without bound, the output decreases without bound. As $x \to \infty, f\left(x\right) \to -\infty$; as $x \to -\infty, f\left(x\right) \to -\infty$. It has the shape of an even degree power function with a negative coefficient.

EXAMPLE 7: IDENTIFYING END BEHAVIOR AND DEGREE OF A POLYNOMIAL FUNCTION

Given the function $f\left(x\right) = -3x^2\left(x-1\right)\left(x+4\right)$, express the function as a polynomial in general form, and determine the leading term, degree, and end behavior of the function.

Answer

Obtain the general form by expanding the given expression for $f\left(x\right)$.

$$f(x) = -3x^2(x-1)(x+4)$$
$$= -3x^2(x^2 + 3x - 4)$$
$$= -3x^4 - 9x^3 + 12x^2$$

The general form is $f(x) = -3x^4 - 9x^3 + 12x^2$. The leading term is $-3x^4$; therefore, the degree of the polynomial is 4. The degree is even (4) and the leading coefficient is negative (-3), so the end behavior is

$$\text{as } x \to -\infty, f(x) \to -\infty$$
$$\text{as } x \to \infty, f(x) \to -\infty.$$

TRY IT #5

Given the function $f(x) = 0.2(x-2)(x+1)(x-5)$, express the function as a polynomial in general form and determine the leading term, degree, and end behavior of the function.

Answer

The function in general form is $f(x) = 0.2x^3 - 1.2x^2 + 0.6x + 2$. The leading term is $0.2x^3$, so it is a degree 3 polynomial. As x increases without bound, $f(x)$ increases without bound; as x decreases without bound, $f(x)$ decreases without bound.

Identifying Local Behavior of Polynomial Functions

In addition to the end behavior of polynomial functions, we are also interested in what happens in the "middle" of the function. In particular, we are interested in locations where graph behavior changes. A **turning point** is a point at which the function values change from increasing to decreasing (**local maximum**) or decreasing to increasing (**local minimum**).

We are also interested in the intercepts. As with all functions, the y-intercept is the point at which the graph intersects the vertical axis. The point corresponds to the coordinate pair in which the input value is zero. Because a polynomial is a function, only one output value corresponds to each input value so there can be only one y-intercept $(0, a_0)$. The x-intercepts occur at the input values that correspond to an output value of zero. It is possible to have more than one x-intercept. See Figure 9.

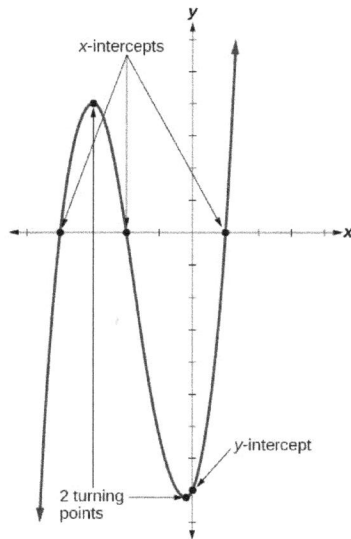

Figure 9.

DEFINITION

A **turning point** of a graph is a point at which the graph changes direction from increasing to decreasing (**local maximum**) or decreasing to increasing (**local minimum**). The **y-intercept** is the point at which the function has an input value of zero. The **x-intercepts** are the points at which the output value is zero.

HOW TO

Given a polynomial function, determine the intercepts.

1. Determine the y-intercept by setting $x = 0$ and finding the corresponding output value.
2. Determine the x-intercepts by solving for the input values that yield an output value of zero.

EXAMPLE 8: DETERMINING THE INTERCEPTS OF A POLYNOMIAL FUNCTION

Given the polynomial function $f(x) = (x - 2)(x + 1)(x - 4)$, written in factored form for your convenience, determine the y-and x-intercepts.

Answer

The y-intercept occurs when the input is zero so substitute 0 for x.

$$f(0) = (0-2)(0+1)(0-4)$$
$$= (-2)(1)(-4)$$
$$= 8$$

The y-intercept is (0, 8).

The x-intercepts occur when the output is zero. To solve $0 = (x-2)(x+1)(x-4)$ set each factor equal to zero and simplify.

$$x - 2 = 0 \qquad \text{or} \quad x + 1 = 0 \qquad \text{or} \quad x - 4 = 0$$
$$x = 2 \qquad\qquad x = -1 \qquad\qquad x = 4$$

The x-intercepts are $(2, 0), (-1, 0)$, and $(4, 0)$.

We can see these intercepts on the graph of the function shown in Figure 10.

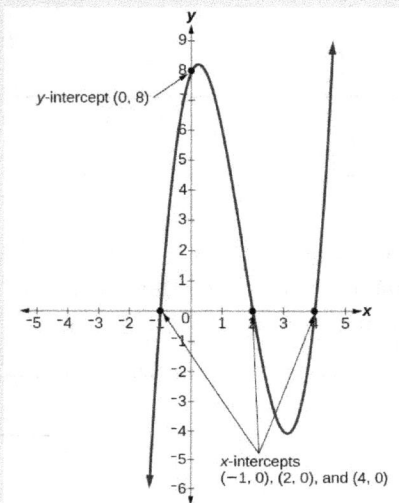

Figure 10.

EXAMPLE 9: DETERMINING THE INTERCEPTS OF A POLYNOMIAL FUNCTION WITH FACTORING

Given the polynomial function $f(x) = x^4 - 4x^2 - 45$, determine the y- and x-intercepts.

Answer The y-intercept occurs when the input is zero.

$$f(0) = (0)^4 - 4(0)^2 - 45$$
$$= -45$$

The y-intercept is $(0, -45)$.

The x-intercepts occur when the output is zero. To determine when the output is zero, we will need to factor the polynomial.

$$f(x) = x^4 - 4x^2 - 45$$
$$= (x^2 - 9)(x^2 + 5)$$
$$= (x - 3)(x + 3)(x^2 + 5)$$
$$0 = (x - 3)(x + 3)(x^2 + 5)$$
$$x - 3 = 0 \quad \text{or} \quad x + 3 = 0 \quad \text{or} \quad x^2 + 5 = 0$$
$$x = 3 \quad \text{or} \quad x = -3 \quad \text{or} \quad \text{(no real solution)}$$

The x-intercepts are $(3, 0)$ and $(-3, 0)$.

We can see these intercepts on the graph of the function shown in Figure 11. We can see that the function is even because $f(x) = f(-x)$.

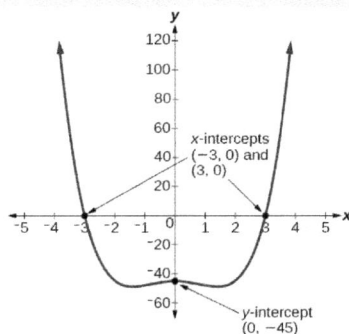

Figure 11.

TRY IT #6

Given the polynomial function $f(x) = 2x^3 - 6x^2 - 20x$, determine the y-and x-intercepts.

Answer

y-intercept $(0, 0)$; x-intercepts $(0, 0)$, $(-2, 0)$, and $(5, 0)$

Comparing Smooth and Continuous Graphs

The degree of a polynomial function helps us to determine the number of x-intercepts and the number of turning points. A polynomial function of $n\text{th}$ degree is the product of at most n factors, so it will have at most n roots or zeros, or x-intercepts. The graph of the polynomial function of degree n must have at most $n - 1$ turning points. This means the graph has at most one fewer turning points than the degree of the polynomial.

A **continuous function** has no breaks in its graph: the graph can be drawn without lifting the pen from the paper. A **smooth curve** is a graph that has no sharp corners. The turning points of a smooth graph must always occur at rounded curves. The graphs of polynomial functions are both continuous and smooth.

INTERCEPTS AND TURNING POINTS OF POLYNOMIALS

A polynomial of degree n will have, at most, n x-intercepts and $n - 1$ turning points.

EXAMPLE 10: DETERMINING THE MAXIMUM POSSIBLE NUMBER OF INTERCEPTS AND TURNING POINTS OF A POLYNOMIAL

Without graphing the function, determine the maximum number of possible x-intercepts and turning points for $f(x) = -3x^{10} + 4x^7 - x^4 + 2x^3$.

Answer

The polynomial has a degree of 10, so there are at most 10 x-intercepts and at most $10 - 1 = 9$ turning points.

TRY IT #7

Without graphing the function, determine the maximum number of possible x-intercepts and turning points for $f(x) = 108 - 13x^9 - 8x^4 + 14x^{12} + 2x^3$.

Answer

There are at most 12 x-intercepts and at most 11 turning points.

EXAMPLE 11: DRAWING CONCLUSIONS ABOUT A POLYNOMIAL FUNCTION FROM THE GRAPH

What can we conclude about the polynomial represented by the graph shown in Figure 12 based on its intercepts and turning points?

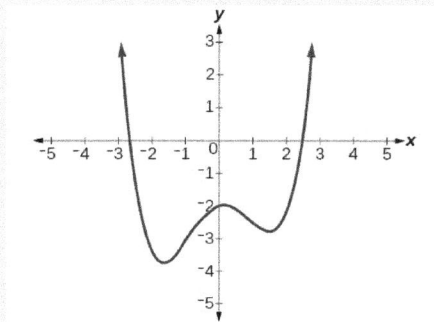

Figure 12.

Answer

The end behavior of the graph tells us this is the graph of an even-degree polynomial. See Figure 13.

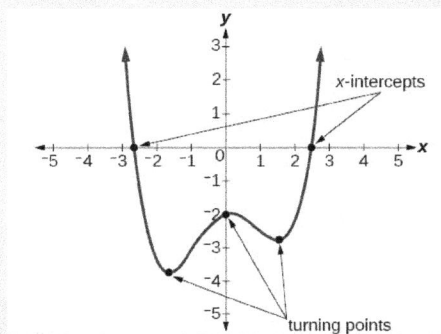

Figure 13.

The graph has 2 x-intercepts, suggesting a degree of 2 or greater, and 3 turning points, suggesting a degree of 4 or greater. Based on this, it would be reasonable to conclude that the degree is even and at least 4.

TRY IT #8

What can we conclude about the polynomial represented by the graph shown in Figure 14 based on its intercepts and turning points?

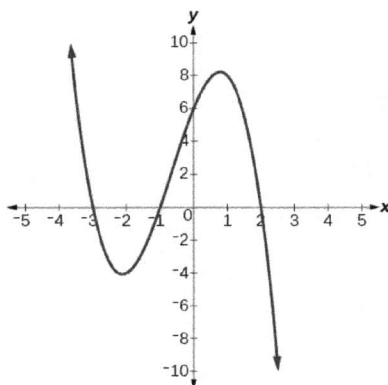

Figure 14.

Answer The end behavior indicates an odd-degree polynomial function; there are 3 x-intercepts and 2 turning points, so the degree is odd and at least 3. Because of the end behavior, we know that the lead coefficient must be negative.

EXAMPLE 12: DRAWING CONCLUSIONS ABOUT A POLYNOMIAL FUNCTION FROM THE FACTORS

Given the function $f(x) = -4x(x+3)(x-4)$, determine the x and y intercepts and the maximum number of turning points possible.

Answer

The y-intercept is found by evaluating $f(0)$.

$$f(0) = -4(0)(0+3)(0-4)$$
$$= 0$$

The y-intercept is $(0,0)$.

The x-intercepts are found by determining the zeros of the function.

$$0 = -4x(x+3)(x-4)$$

$$x = 0 \quad \text{or} \quad x+3 = 0 \quad \text{or} \quad x-4 = 0$$
$$x = 0 \quad \text{or} \quad x = -3 \quad \text{or} \quad x = 4$$

The x-intercepts are $(0,0), (-3,0)$, and $(4,0)$.

The degree is 3 so the graph has at most 2 turning points.

TRY IT #9

Given the function $f(x) = 0.2(x-2)(x+1)(x-5)$, determine the x and y intercepts and the maximum number of turning points possible.

Answer

The x-intercepts are $(2,0)$, $(-1,0)$, and $(5,0)$, the y-intercept is $(0,2)$, and the graph has at most 2 turning points.

MEDIA:

Access these online resources for additional instruction and practice with power and polynomial functions.

- Find Key Information about a Given Polynomial Function
- End Behavior of a Polynomial Function
- Turning Points and x-intercepts of Polynomial Functions
- Least Possible Degree of a Polynomial Functio

Key Equations

general form of a polynomial function
$$f(x) = a_n x^n + a_{n-1} x^{n-1} + \ldots + a_2 x^2 + a_1 x + a_0$$

KEY CONCEPTS

- A power function is a coefficient multiplied by a variable base raised to a number power.
- The behavior of a graph as the input decreases without bound and increases without bound is called the end behavior.
- The end behavior depends on whether the power is even or odd.
- A polynomial function is the sum of terms, each of which consists of a transformed power function with positive whole number power.
- The degree of a polynomial function is the highest power of the variable that occurs in a polynomial. The term containing the highest power of the variable is called the leading term. The coefficient of the leading term is called the leading coefficient.
- The end behavior of a polynomial function is the same as the end behavior of the power function represented by the leading term of the function.
- A polynomial of degree n will have at most n x-intercepts and at most $n-1$ turning points.

GLOSSARY

coefficient
a nonzero real number multiplied by a variable raised to an exponent

continuous function
a function whose graph can be drawn without lifting the pen from the paper because there are no breaks in the graph

degree
the highest power of the variable that occurs in a polynomial

end behavior
the behavior of the graph of a function as the input decreases without bound and increases without bound

leading coefficient
the coefficient of the leading term

leading term
the term containing the highest power of the variable

polynomial function
a function that consists of either zero or the sum of a finite number of non-zero terms, each of which is a product of a number, called the coefficient of the term, and a variable raised to a non-negative integer power.

power function
a function that can be represented in the form $f(x) = kx^p$ where k is a constant, the base is a variable, and the exponent, p, is a constant

smooth curve
a graph with no sharp corners

term of a polynomial function
any $a_i x^i$ of a polynomial function in the form $f(x) = a_n x^n + a_{n-1} x^{n-1} + ... + a_2 x^2 + a_1 x + a_1$

turning point
the location at which the graph of a function changes direction

4.2 Graphs of Polynomial Functions

The revenue in millions of dollars for a fictional cable company from 2006 through 2013 is shown in Table 1.

Table 1

Year	2006	2007	2008	2009	2010	2011	2012	2013
Revenues	52.4	52.8	51.2	49.5	48.6	48.6	48.7	47.1

The revenue can be modeled by the polynomial function

$$R(t) = -0.037t^4 + 1.414t^3 - 19.777t^2 + 118.696t - 205.332$$

where R represents the revenue in millions of dollars and t represents the year, with $t = 6$ corresponding to 2006. Over which intervals is the revenue for the company increasing? Over which intervals is the revenue for the company decreasing? These questions, along with many others, can be answered by examining the graph of the polynomial function. In this section we will explore the behavior of polynomials in general.

Recognizing Characteristics of Graphs of Polynomial Functions

Polynomial functions of degree 2 or more have graphs that do not have sharp corners. These types of graphs are called **smooth curves**. Polynomial functions also display graphs that have no breaks. Curves with no breaks are called **continuous**. Figure 1 shows a graph that represents a polynomial function on the left and a graph that represents a function that is not a polynomial on the right.

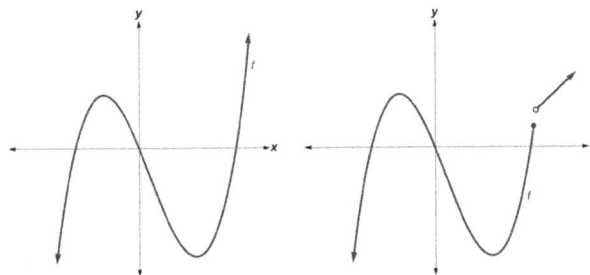

Figure 1

Which of the graphs in Figure 2 and Figure 3 represent a polynomial function?

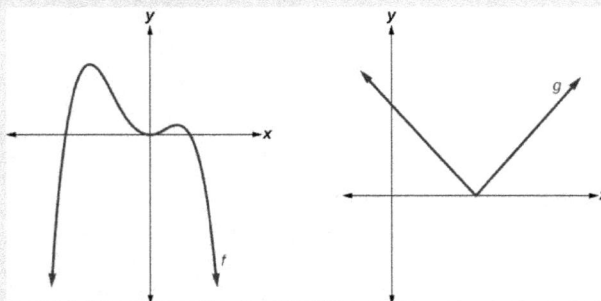

Figure 2

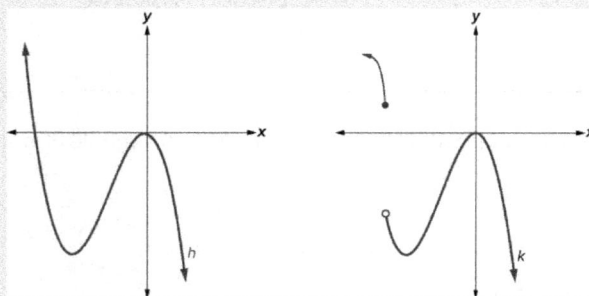

Figure 3

Answer

The graphs of f and h are graphs of polynomial functions. They are smooth and continuous.

The graphs of g and k are graphs of functions that are not polynomials. The graph of function g has a sharp corner. The graph of function k is not continuous.

Q&A

Do all polynomial functions have all real numbers as their domain?

Yes. Any real number is a valid input for a polynomial function.

Using Factoring to Find Zeros of Polynomial Functions

Recall that if f is a polynomial function, the values of x for which $f(x) = 0$ are called zeros of f. If the equation of the polynomial function can be factored, we can set each factor equal to zero and solve for the zeros.

We can use this method to find x-intercepts because at the x-intercepts, we find the input values when the output value is zero. For general polynomials, this can be a challenging prospect. While quadratics can be solved using the relatively simple quadratic formula, the corresponding formulas for cubic and fourth-degree polynomials are not simple enough to remember, and formulas do not exist for general higher-degree polynomials. Consequently, we will limit ourselves to three cases in this section:

1. The polynomial can be factored using known methods: greatest common factor and trinomial factoring.
2. The polynomial is given in factored form.
3. Technology is used to determine the intercepts.

HOW TO

Given a polynomial function f, find the x-intercepts by factoring.

1. Set $f(x) = 0$.
2. If the polynomial function is not given in factored form:
    a. Factor out any common monomial factors.
    b. Factor any factorable binomials or trinomials.
3. Set each factor equal to zero and solve to find the x-intercepts.

EXAMPLE 2: FINDING THE *X*-INTERCEPTS OF A POLYNOMIAL FUNCTION BY FACTORING

Find the x-intercepts of $f(x) = x^6 - 3x^4 + 2x^2$.

Answer

We can attempt to factor this polynomial to find solutions for $f(x) = 0$.

$$x^6 - 3x^4 + 2x^2 = 0$$
$$x^2 \left(x^4 - 3x^2 + 2 \right) = 0 \qquad \text{Factor out the greatest common factor.}$$
$$x^2 \left(x^2 - 1 \right) \left(x^2 - 2 \right) = 0 \qquad \text{Factor the trinomial.}$$

Set each factor equal to zero.
$$x^2 = 0 \quad \text{or} \quad \left(x^2 - 1 \right) = 0 \quad \text{or} \quad \left(x^2 - 2 \right) = 0$$

Solving each of these equations gives
$$x^2 = 0 \quad \text{or} \quad x^2 = 1 \quad \text{or} \quad x^2 = 2$$
$$x = 0 \qquad\qquad x = 1 \qquad\qquad x = \sqrt{2}.$$

This gives us five x-intercepts: $(0, 0) , (1, 0) , (-1, 0) , \left(\sqrt{2}, 0 \right)$, and $\left(-\sqrt{2}, 0 \right)$. See Figure 4. We can see that this is an even function.

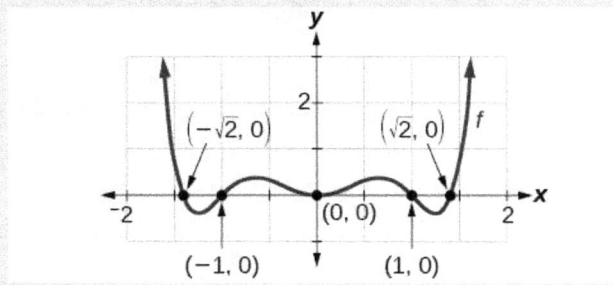

Figure 4

EXAMPLE 3: FINDING THE *X*-INTERCEPTS OF A POLYNOMIAL FUNCTION BY FACTORING

Find the x-intercepts of $f(x) = x^3 - 5x^2 - x + 5$.

Answer

Find solutions for $f(x) = 0$ by factoring.

$$x^3 - 5x^2 - x + 5 = 0$$

$$x^2(x - 5) - (x - 5) = 0 \qquad \text{Factor by grouping.}$$

$$(x^2 - 1)(x - 5) = 0 \qquad \text{Factor out the common factor.}$$

$$(x + 1)(x - 1)(x - 5) = 0 \qquad \text{Factor the difference of squares.}$$

Set each factor equal to zero and solve.

$$x + 1 = 0 \qquad \text{or} \quad x - 1 = 0 \qquad \text{or} \quad x - 5 = 0$$

$$x = -1 \qquad\qquad x = 1 \qquad\qquad x = 5$$

There are three x-intercepts: $(-1, 0), (1, 0)$, and $(5, 0)$. See Figure 5.

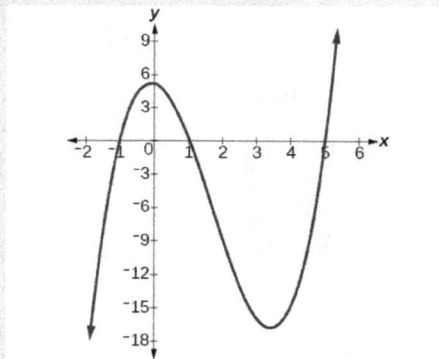

Figure 5

EXAMPLE 4: FINDING THE Y– AND X-INTERCEPTS OF A POLYNOMIAL IN FACTORED FORM

Find the y- and x-intercepts of $g(x) = (x-2)^2 (2x+3)$.

Answer

The y-intercept can be found by evaluating $g(0)$.

$$g(0) = (0-2)^2 (2(0)+3)$$
$$= 12$$

So the y-intercept is $(0, 12)$.

The x-intercepts can be found by solving $g(x) = 0$.

$$(x-2)^2 (2x+3) = 0$$
$$(x-2)^2 = 0 \qquad \text{or} \qquad (2x+3) = 0$$
$$x - 2 = 0 \qquad\qquad\qquad x = -\frac{3}{2}$$
$$x = 2$$

So the x-intercepts are $(2, 0)$ and $\left(-\frac{3}{2}, 0\right)$.

Analysis

We can always check that our answers are reasonable by using a graphing calculator to graph the polynomial as shown in Figure 6.

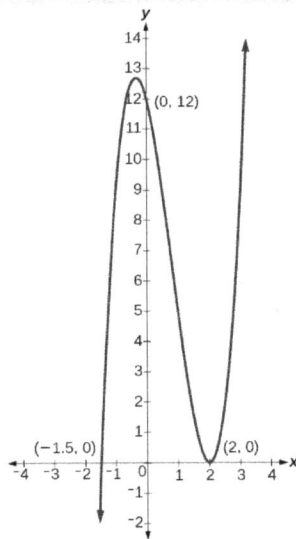

Figure 6

EXAMPLE 5: FINDING THE *X*-INTERCEPTS OF A POLYNOMIAL FUNCTION USING A GRAPH

Find the x-intercepts of $h(x) = x^3 + 4x^2 + x - 6$.

The graph of this function, is in Figure 7.

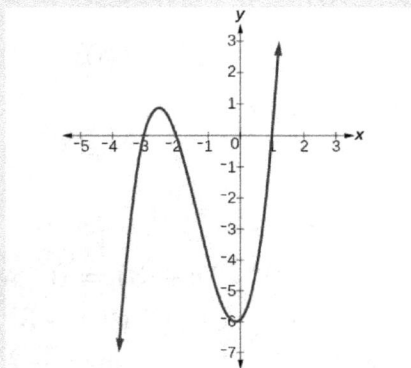

Figure 7

Answer

This polynomial is not in factored form, has no common factors, and does not appear to be factorable using techniques previously discussed. Fortunately, we can use technology to find the intercepts. Keep in mind that some values make graphing difficult by hand. In these cases, we can take advantage of graphing utilities.

Looking at the graph of this function, as shown in Figure 7, it appears that there are *x*-intercepts at $x = -3, -2,$ and 1.

We can check whether these are correct by substituting these values for x and verifying that

$$h(-3) = h(-2) = h(1) = 0.$$

Since $h(x) = x^3 + 4x^2 + x - 6$, we have:

$$h(-3) = (-3)^3 + 4(-3)^2 + (-3) - 6 = -27 + 36 - 3 - 6 = 0$$
$$h(-2) = (-2)^3 + 4(-2)^2 + (-2) - 6 = -8 + 16 - 2 - 6 = 0$$
$$h(1) = (1)^3 + 4(1)^2 + (1) - 6 = 1 + 4 + 1 - 6 = 0$$

Each x-intercept corresponds to a zero of the polynomial function and each zero yields a factor, so we can now write the polynomial in factored form.

$$h(x) = x^3 + 4x^2 + x - 6$$
$$= (x + 3)(x + 2)(x - 1)$$

Find the y- and x-intercepts of the function $f(x) = x^4 - 19x^2 + 30x$.

Answer

y-intercept $(0, 0)$; x-intercepts $(0, 0)$, $(-5, 0)$, $(2, 0)$, and $(3, 0)$

Identifying Zeros and Their Multiplicities

Graphs behave differently at various x-intercepts. Sometimes, the graph will cross over the horizontal axis at an intercept. Other times, the graph will touch the horizontal axis and bounce off.

Suppose, for example, we graph the function $f(x) = (x + 3)(x - 2)^2(x + 1)^3$.

Notice in Figure 8 that the behavior of the function at each of the x-intercepts is different.

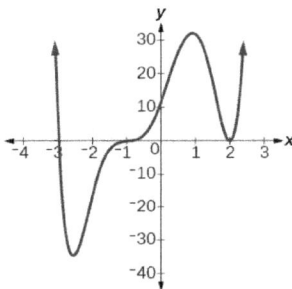

Figure 8 Identifying the behavior of the graph at an x-intercept by examining the multiplicity of the zero.

The x-intercept -3 is the solution of equation $(x + 3) = 0$. The graph passes directly through the x-intercept at $x = -3$. The factor is linear (has a degree of 1), so the behavior near the intercept is like that of a line—it passes directly through the intercept. We call this a single zero because the zero corresponds to a single factor of the function.

The x-intercept at $x = 2$ is the repeated solution of equation $(x - 2)^2 = 0$. The graph touches the axis at the intercept and changes direction. The factor is quadratic (degree 2), so the behavior near the intercept is like that of a quadratic—it bounces off of the horizontal axis at the x-intercept.

$$(x - 2)^2 = (x - 2)(x - 2)$$

The factor is repeated, that is, the factor $(x - 2)$ appears twice. The number of times a given factor appears in the factored form of the equation of a polynomial is called the **multiplicity**. The zero associated with this factor, $x = 2$, has multiplicity 2 because the factor $(x - 2)$ occurs twice.

The x-intercept at $x = -1$ is the repeated solution of factor $(x + 1)^3 = 0$. The graph passes through the axis at the intercept, but flattens out a bit first. This factor is cubic (degree 3), so the behavior near the intercept is like that of a cubic—with the same S-shape near the intercept as the toolkit function $f(x) = x^3$. We call this a triple zero, or a zero with multiplicity 3.

For zeros with even multiplicities, the graphs *touch* or are tangent to the x-axis. For zeros with odd multiplicities, the graphs *cross* or intersect the x-axis. See Figure 9 for examples of graphs of polynomial functions with multiplicity 1, 2, and 3.

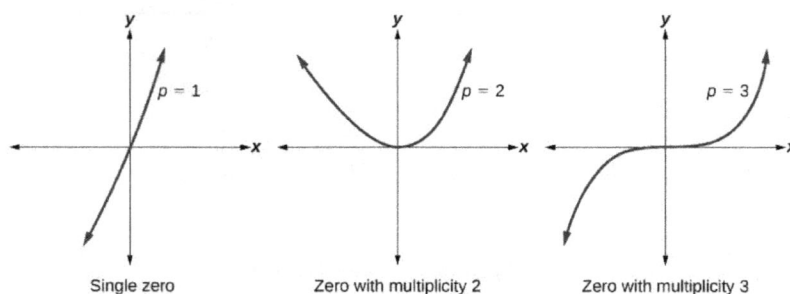

| Single zero | Zero with multiplicity 2 | Zero with multiplicity 3 |

Figure 9

For higher even powers, such as 4, 6, and 8, the graph will still touch and bounce off of the horizontal axis but, for each increasing even power, the graph will appear flatter as it approaches and leaves the x-axis.

For higher odd powers, such as 5, 7, and 9, the graph will still cross through the horizontal axis, but for each increasing odd power, the graph will appear flatter as it approaches and leaves the x-axis.

GRAPHICAL BEHAVIOR OF POLYNOMIALS AT x-INTERCEPTS

If a polynomial contains a factor of the form $(x - h)^p$, the behavior near the x-intercept $(h, 0)$ is determined by the power p. We say that $x = h$ is a zero of multiplicity p.

The graph of a polynomial function will touch the x-axis at zeros with even multiplicities. The graph will cross the x-axis at zeros with odd multiplicities.

The sum of the multiplicities is the degree of the polynomial function.

HOW TO

Given a graph of a polynomial function of degree n, identify the zeros and their multiplicities.

1. If the graph crosses the x-axis and appears almost linear at the intercept, it is a single zero.
2. If the graph touches the x-axis and bounces off of the axis, it is a zero with even multiplicity.
3. If the graph crosses the x-axis at a zero, it is a zero with odd multiplicity.
4. The sum of the multiplicities is n.

Use the graph of the function of degree 6 in Figure 10 to identify the zeros of the function and their possible multiplicities.

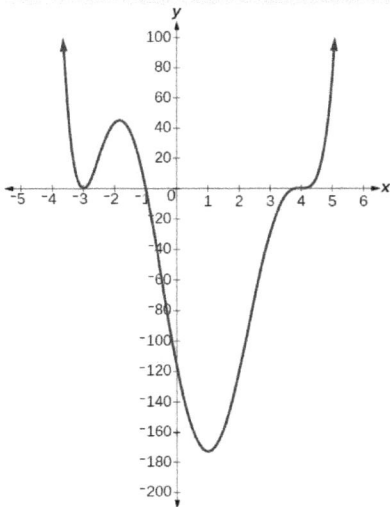

Figure 10

Answer

The polynomial function is of degree 6. The sum of the multiplicities must be 6.

Starting from the left, the first zero occurs at $x = -3$. The graph touches the x-axis, so the multiplicity of the zero must be even. The zero of -3 could have multiplicity 2.

The next zero occurs at $x = -1$. The graph looks almost linear at this point. This is a single zero of multiplicity 1.

The last zero occurs at $x = 4$. The graph crosses the x-axis, so the multiplicity of the zero must be odd. We know that the multiplicity is 3 so that the sum of the multiplicities is 6.

TRY IT #2

Use the graph of the function of degree 9 in Figure 11 to identify the zeros of the function and their multiplicities.

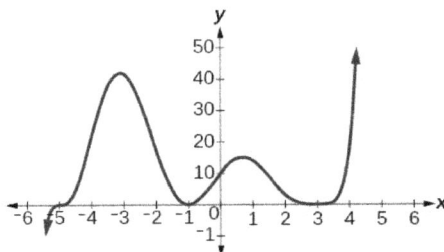

Figure 11

Answer

The graph has a zero at $x = -5$ with multiplicity 3, a zero at $x = -1$ with multiplicity 2, and a zero at $x = 3$ with multiplicity 4.

Determining End Behavior

As we have already learned, the behavior of a graph of a polynomial function of the form

$$f(x) = a_n x^n + a_{n-1} x^{n-1} + ... + a_1 x + a_0$$

will either ultimately rise or fall as x increases without bound and will either rise or fall as x decreases without bound. This is because for very large inputs, say 100 or 1,000, the leading term dominates the size of the output. The same is true for more and more negative inputs, say –100 or –1,000.

Recall that we call this behavior the **end behavior** of a function. When the leading term of a polynomial function, $a_n x^n$, is an even power function with a positive leading coefficient, as x increases or decreases without bound, $f(x)$ increases without bound. When the leading term is an odd power function with a positive leading coefficient, as x decreases without bound, $f(x)$ also decreases without bound; as x increases without bound, $f(x)$ also increases without bound. If the leading coefficient is negative, it will change the direction of the end behavior. Figure 12 summarizes all four cases.

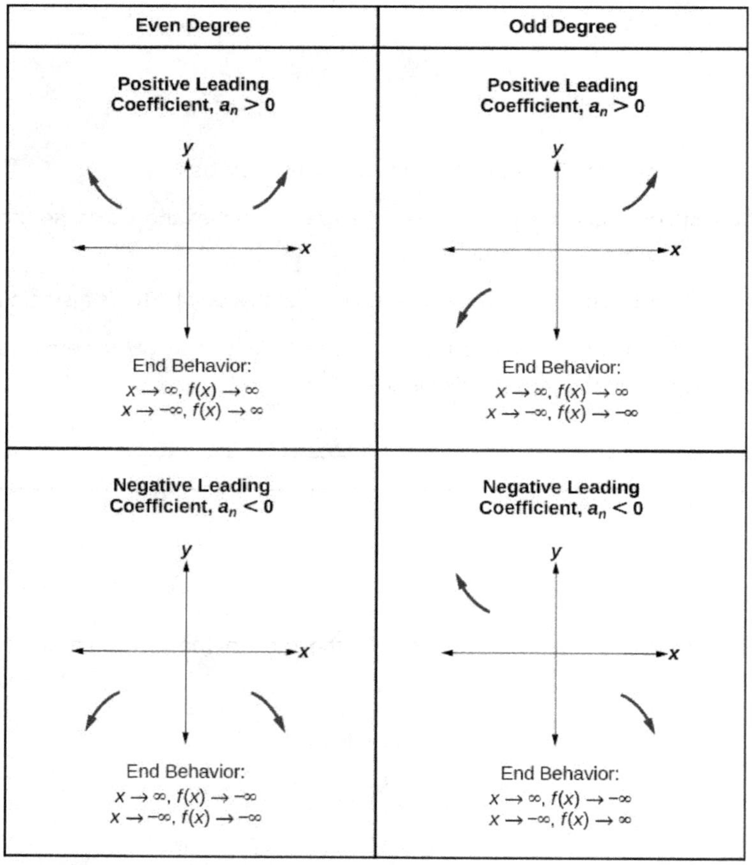

Even Degree	Odd Degree
Positive Leading Coefficient, $a_n > 0$ End Behavior: $x \to \infty, f(x) \to \infty$ $x \to -\infty, f(x) \to \infty$	**Positive Leading Coefficient, $a_n > 0$** End Behavior: $x \to \infty, f(x) \to \infty$ $x \to -\infty, f(x) \to -\infty$
Negative Leading Coefficient, $a_n < 0$ End Behavior: $x \to \infty, f(x) \to -\infty$ $x \to -\infty, f(x) \to -\infty$	**Negative Leading Coefficient, $a_n < 0$** End Behavior: $x \to \infty, f(x) \to -\infty$ $x \to -\infty, f(x) \to \infty$

Figure 12

Understanding the Relationship between Degree and Turning Points

In addition to the end behavior, recall that we can analyze a polynomial function's local behavior. It may have a turning point where the graph changes from increasing to decreasing (rising to falling or a local maximum) or decreasing to increasing (falling to rising or a local minimum). Look at the graph of the polynomial function $f(x) = x^4 - x^3 - 4x^2 + 4x$ in Figure 13. The graph has three turning points.

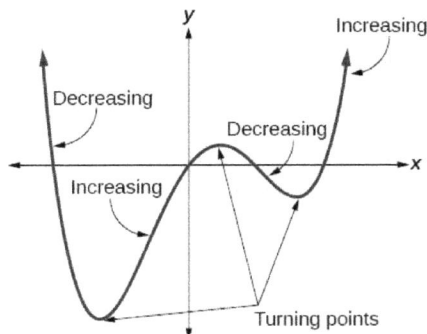

This function f is a 4th degree polynomial function and has 3 turning points. The maximum number of turning points of a polynomial function is always one less than the degree of the function.

A polynomial of degree n will have at most $n - 1$ turning points.

EXAMPLE 7: FINDING THE MAXIMUM NUMBER OF TURNING POINTS USING THE DEGREE OF A POLYNOMIAL FUNCTION

Find the maximum possible number of turning points of each polynomial function.

a. $f(x) = -x^3 + 4x^5 - 3x^2 + 1$
b. $f(x) = -(x-1)^2(1 + 2x^2)$

Answer

a. $f(x) = -x^3 + 4x^5 - 3x^2 + 1$

First, rewrite the polynomial function in descending order: $f(x) = 4x^5 - x^3 - 3x^2 + 1$

The lead term is $4x^5$. This polynomial function is of degree 5.

The maximum possible number of turning points is $5 - 1 = 4$.

b. $f(x) = -(x-1)^2(1 + 2x^2)$

First, identify the leading term of the polynomial function if the function were expanded.

$$f(x) = -(x - 1)^2(1 + 2x^2)$$

$$a_n = -(x^2)(2x^2) = -2x^4$$

Then, identify the degree of the polynomial function. This polynomial function is of degree 4.

The maximum possible number of turning points is $4 - 1 = 3$.

Graphing Polynomial Functions

We can use what we have learned about multiplicities, end behavior, and turning points to sketch graphs of polynomial functions. Let us put this all together and look at the steps required to graph polynomial functions.

HOW TO

Given a polynomial function, sketch the graph.

1. Find the intercepts.
2. Check for symmetry. If the function is an even function, its graph is symmetrical about the y-axis, that is, $f(-x) = f(x)$. If a function is an odd function, its graph is symmetrical about the origin, that is, $f(-x) = -f(x)$.
3. Use the multiplicities of the zeros to determine the behavior of the polynomial at the x-intercepts.
4. Determine the end behavior by examining the leading term.
5. Use the end behavior and the behavior at the intercepts to sketch a graph.
6. Ensure that the number of turning points does not exceed one less than the degree of the polynomial.
7. Optionally, use technology to check the graph.

EXAMPLE 8: SKETCHING THE GRAPH OF A POLYNOMIAL FUNCTION

Sketch a graph of $f(x) = -2(x + 3)^2(x - 5)$.

Answer

This graph has two x-intercepts. At $x = -3$, the factor is squared, indicating a multiplicity of 2. The graph will bounce at this x-intercept. At $x = 5$, the function has a multiplicity of one, indicating the graph will cross through the axis at this intercept.

The y-intercept is found by evaluating $f(0)$.

$$f(0) = -2(0 + 3)^2(0 - 5)$$
$$= -2 \cdot 9 \cdot (-5)$$
$$= 90$$

The y-intercept is $(0, 90)$.

Additionally, we can see the leading term, if this polynomial were multiplied out, would be $-2x^3$, so the end behavior is that of a vertically reflected cubic, with the outputs decreasing as the inputs increase without bound, and the outputs increasing as the inputs decrease without bound. See Figure 14.

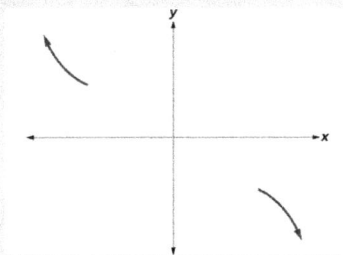

Figure 14

To sketch this, we consider that:

- As $x \rightarrow -\infty$ the function $f(x) \rightarrow \infty$, so we know the graph starts in the second quadrant and is decreasing toward the x-axis.
- Since $f(-x) = -2(-x+3)^2(-x-5)$ is not equal to $f(x)$, the graph does not display symmetry.
- At $(-3, 0)$, the graph bounces off of the x-axis, so the function must start increasing.
  At $(0, 90)$, the graph crosses the y-axis at the y-intercept. See Figure 15.

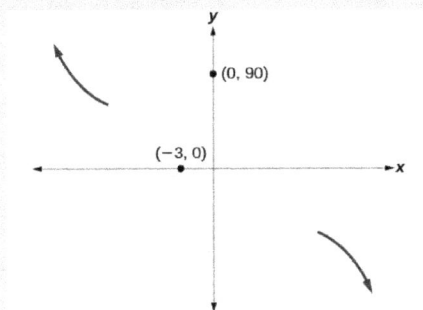

Figure 15

Somewhere after this point, the graph must turn back down or start decreasing toward the horizontal axis because the graph passes through the next intercept at $(5, 0)$. See Figure 16.

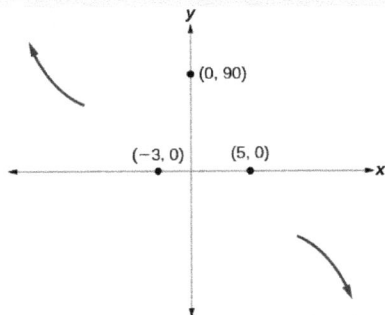

Figure 16

As $x \rightarrow \infty$ the function $f(x) \rightarrow -\infty$, so we know the graph continues to decrease, and we can stop drawing the graph in the fourth quadrant.

Using technology, we can create the graph for the polynomial function, shown in Figure 17, and verify that the resulting graph looks like our sketch in Figure 16.

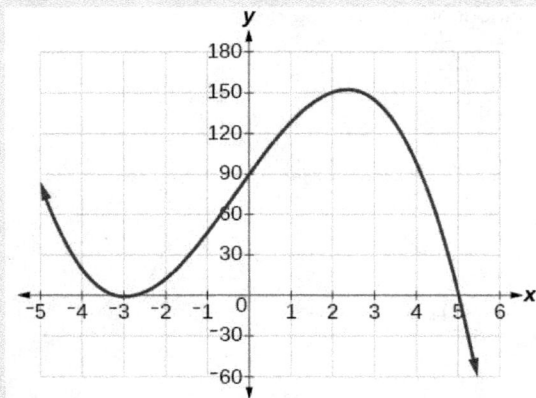

Figure 17 The complete graph of the polynomial function

$$f\left(x\right) = -2(x+3)^2\left(x-5\right).$$

TRY IT #3

Sketch a graph of $f\left(x\right) = \frac{1}{4}x(x-1)^4(x+3)^3$.

Answer

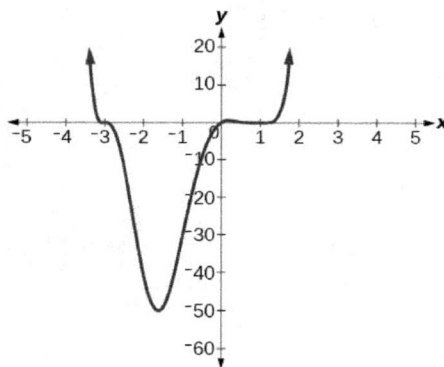

Writing Formulas for Polynomial Functions

Now that we know how to find zeros of polynomial functions, we can use them to write formulas based on graphs. Because a polynomial function written in factored form will have an x-intercept where each factor is equal to zero, we can form a function that will pass through a set of x-intercepts by introducing a corresponding set of factors.

Factored Form of Polynomials

If a polynomial of lowest degree p has horizontal intercepts at $x = x_1, x_2, \ldots, x_n$, then the polynomial can be written in the factored form: $f(x) = a(x - x_1)^{p_1}(x - x_2)^{p_2} \cdots (x - x_n)^{p_n}$ where the powers p_i on each factor can be determined by the behavior of the graph at the corresponding intercept, and the stretch factor a can be determined given a value of the function other than an x-intercept.

HOW TO

Given a graph of a polynomial function, write a formula for the function.

1. Identify the x-intercepts of the graph to find the factors of the polynomial.
2. Examine the behavior of the graph at the x-intercepts to determine the multiplicity of each factor.
3. Find the polynomial of least degree containing all the factors found in the previous step.
4. Use any other point on the graph (the y-intercept may be easiest) to determine the stretch factor.

EXAMPLE 9: WRITING A FORMULA FOR A POLYNOMIAL FUNCTION FROM THE GRAPH

Write a formula for the polynomial function shown in Figure 18.

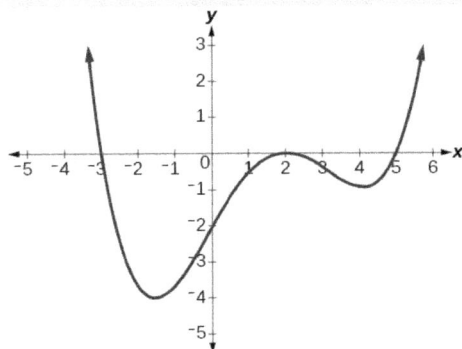

Figure 18

Answer

This graph has three x-intercepts: $x = -3, 2,$ and $5.$ The y-intercept is located at $(0, -2)$. At $x = -3$ and $x = 5$, the graph passes through the axis linearly, suggesting the corresponding factors of the polynomial will be linear. At $x = 2$, the graph bounces at the intercept, suggesting the corresponding factor of the polynomial will be second degree (quadratic). Together, this gives us

$$f(x) = a(x + 3)(x - 2)^2(x - 5).$$

To determine the stretch factor, we utilize another point on the graph. We will use the y-intercept $(0, -2)$, to solve for a.

$$f(0) = a(0+3)(0-2)^2(0-5)$$

$$-2 = a(0+3)(0-2)^2(0-5)$$

$$-2 = -60a$$

$$a = \frac{1}{30}$$

The graphed polynomial appears to represent the function

$$f(x) = \frac{1}{30}(x+3)(x-2)^2(x-5).$$

TRY IT #4

Given the graph shown in Figure 19, write a formula for the function shown.

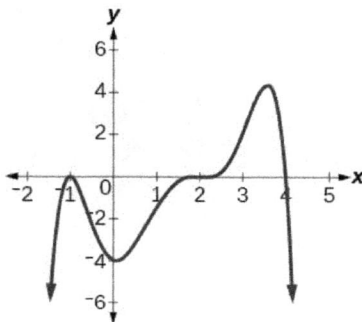

Figure 19

Answer

$$f(x) = -\frac{1}{8}(x-2)^3(x+1)^2(x-4)$$

Using Local and Global Extrema

With quadratics, we algebraically find the maximum or minimum value of the function by finding the vertex. For general polynomials, finding these turning points is not possible without more advanced techniques from calculus. Even then, finding where extrema occur can still be algebraically challenging. For now, we will estimate the locations of turning points using technology to generate a graph.

Each turning point represents a local minimum or maximum. Sometimes, a turning point is the highest or lowest point on the entire graph. In these cases, we say that the turning point is a **global maximum** or a **global minimum**. These are also referred to as the **absolute maximum** and **absolute minimum** values of the function.

We can see the difference between local and global extrema in Figure 20.

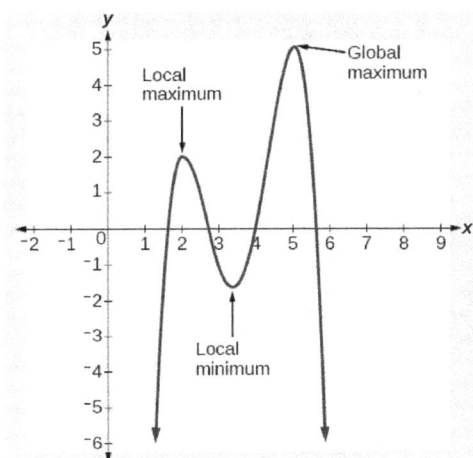

Figure 20

EXAMPLE 10: USING LOCAL EXTREMA TO SOLVE APPLICATIONS

An open-top box is to be constructed by cutting out squares from each corner of a 14 cm by 20 cm sheet of plastic then folding up the sides. Find the size of squares that should be cut out to maximize the volume enclosed by the box.

Answer

We will start this problem by drawing a picture like that in Figure 21, labeling the width of the cut-out squares with a variable, w.

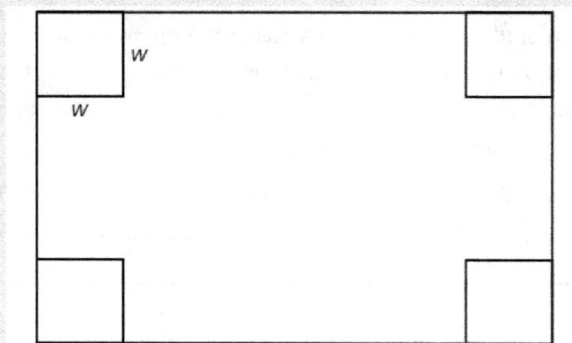

Figure 21

Notice that after a square is cut out from each end, it leaves a $(14 - 2w)$ cm by $(20 - 2w)$ cm rectangle for the base of the box, and the box will be w cm tall. This gives the volume

$$V\left(w\right) = \left(20 - 2w\right)\left(14 - 2w\right)w$$
$$= 280w - 68w^2 + 4w^3$$

Notice, since the factors are w, $20 - 2w$ and $14 - 2w$, the three zeros are 10, 7, and 0, respectively. Because a height of 0 cm is not reasonable, we consider the only the zeros 10 and 7. The shortest side is 14 and we are cutting off two squares, so values w may take on are greater than zero or less than 7. This means we will restrict the domain of this function to $0 < w < 7$. Using technology to sketch the graph of $V\left(w\right)$ on this reasonable domain, we get a graph like that in Figure 22. We can use this graph to estimate the maximum value for the volume, restricted to values for w that are reasonable for this problem—values from 0 to 7.

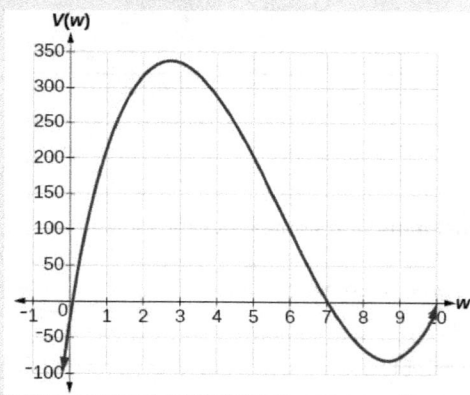

Figure 22

From this graph, we turn our focus to only the portion on the reasonable domain, $\left[0,\ 7\right]$. We can estimate the maximum value to be around 340 cubic cm, which occurs when the squares are about 2.75 cm on each side. To improve this estimate, we could use advanced features of our technology, if available, or simply change our window to zoom in on our graph to produce Figure 23.

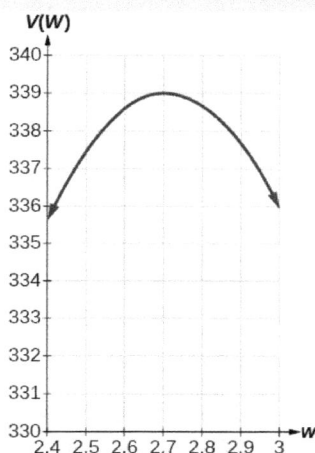

Figure 23

From this zoomed-in view, we can refine our estimate for the maximum volume to about **339** cubic cm, when the squares measure approximately 2.7 cm on each side.

TRY IT #5

Use technology to find the maximum and minimum values on the interval $[-1, 4]$ of the function $f(x) = -0.2(x-2)^3(x+1)^2(x-4)$.

Answer

The minimum occurs at approximately the point $(0, -6.5)$, and the maximum occurs at approximately the point $(3.5, 7)$.

KEY CONCEPTS

- Polynomial functions of degree 2 or more are smooth, continuous functions.
- To find the zeros of a polynomial function, if it can be factored, factor the function and set each factor equal to zero.
- Another way to find the x-intercepts of a polynomial function is to graph the function and identify the points at which the graph crosses the x-axis.
- The multiplicity of a zero determines how the graph behaves at the x-intercepts.
- The graph of a polynomial will cross the horizontal axis at a zero with odd multiplicity.
- The graph of a polynomial will touch the horizontal axis at a zero with even multiplicity.
- The end behavior of a polynomial function depends on the leading term.
- The graph of a polynomial function changes direction at its turning points.

- A polynomial function of degree n has at most $n-1$ turning points.
- To graph polynomial functions, find the zeros and their multiplicities, determine the end behavior, and ensure that the final graph has at most $n-1$ turning points.
- Graphing a polynomial function helps to estimate local and global extremas.

GLOSSARY

global maximum
highest turning point on a graph; $f(a)$ where $f(a) \geq f(x)$ for all x.

global minimum
lowest turning point on a graph; $f(a)$ where $f(a) \leq f(x)$ for all x.

multiplicity
the number of times a given factor appears in the factored form of the equation of a polynomial; if a polynomial contains a factor of the form $(x-h)^p$, $x = h$ is a zero of multiplicity p.

4.3 Rational Functions

Suppose we know that the cost of making a product is dependent on the number of items, x, produced. and that it is given by the equation $C\left(x\right) = 15,000x - 0.1x^2 + 1000$. If we want to know the average cost for producing x items, we would divide the cost function by the number of items, x.

The average cost function, which yields the average cost per item for x items produced, is

$$f\left(x\right) = \frac{15,000x - 0.1x^2 + 1000}{x}$$

Many other application problems require finding an average value in a similar way, giving us variables in the denominator. Written without a variable in the denominator, this function will contain a negative integer power.

In the last few sections, we have worked with polynomial functions, which are functions with non-negative integers for exponents. In this section, we explore rational functions, which can be thought of as the ratio of two polynomial functions.

Using Arrow Notation

We have seen the graphs of the reciprocal function and the squared reciprocal function from our study of toolkit functions. Examine these graphs, as shown in Figure 1, and notice some of their features.

Graphs of Toolkit Functions

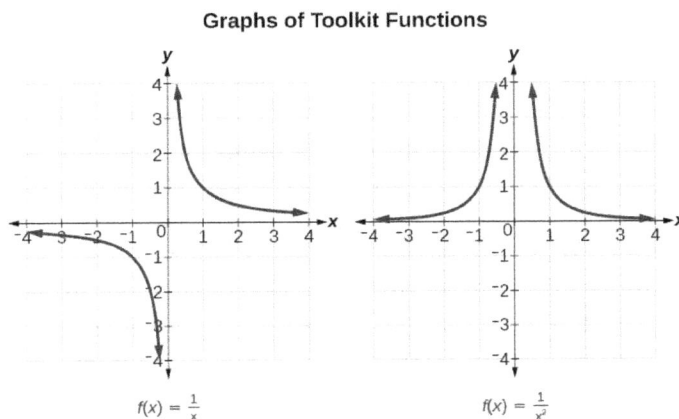

$f(x) = \frac{1}{x}$ $f(x) = \frac{1}{x^2}$

Figure 1

Several things are apparent if we examine the graph of $f(x) = \frac{1}{x}$.

1. On the left branch of the graph, the curve approaches the x-axis $(y = 0)$ as $x \to -\infty$.
2. As the graph approaches $x = 0$ from the left, the curve drops, but as we approach zero from the right, the curve rises.
3. Finally, on the right branch of the graph, the curves approaches the x-axis $(y = 0)$ as $x \to \infty$.

To summarize, we use arrow notation to show that x or $f(x)$ is approaching a particular value. See Table 1.

Table 1: Arrow Notation

Symbol	Meaning
$x \to a^-$	x approaches a from the left ($x < a$ but is increasing and getting closer and closer to a)
$x \to a^+$	x approaches a from the right ($x > a$ but is decreasing and getting closer and closer to a)
$x \to a$	the input approaches a from both sides
$x \to \infty$	x goes toward infinity (x increases without bound)
$x \to -\infty$	x goes toward negative infinity (x decreases without bound)
$f(x) \to \infty$	the output goes toward infinity (the output increases without bound)
$f(x) \to -\infty$	the output goes toward negative infinity (the output decreases without bound)
$f(x) \to a$	the output approaches a

Local Behavior of $f(x) = \frac{1}{x}$

Let's begin by looking at the reciprocal function, $f(x) = \frac{1}{x}$. We cannot divide by zero, which means the function is undefined at $x = 0$; so zero is not in the domain. As the input values approach zero from the left side (negative values very close to zero), the function values decrease without bound. We can see this behavior in Table 2.

Table 2

x	−0.1	−0.01	−0.001	−0.0001
$f(x) = \frac{1}{x}$	−10	−100	−1000	−10,000

We write in arrow notation, as $x \to 0^-$, $f(x) \to -\infty$.

As the input values approach zero from the right side (positive values very close to zero), the function values increase without bound. We can see this behavior in Table 3.

Table 3

x	0.1	0.01	0.001	0.0001
$f(x) = \frac{1}{x}$	10	100	1000	10,000

We write in arrow notation as $x \to 0^+$, $f(x) \to \infty$.

See Figure 2.

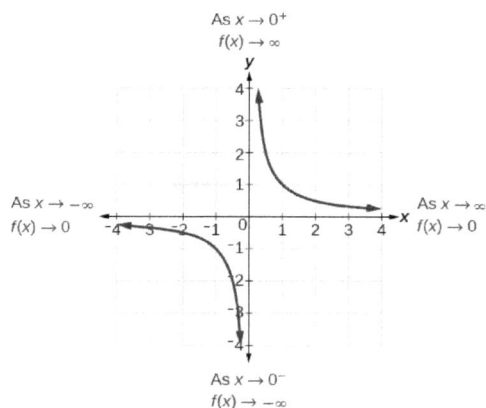

As x → 0⁺
f(x) → ∞

As x → -∞
f(x) → 0

As x → ∞
f(x) → 0

As x → 0⁻
f(x) → -∞

Figure 2

This behavior creates a **vertical asymptote**, which is a vertical line that the graph approaches but never crosses. In this case, the graph is approaching the vertical line $x = 0$ as the input becomes close to zero. See Figure 3.

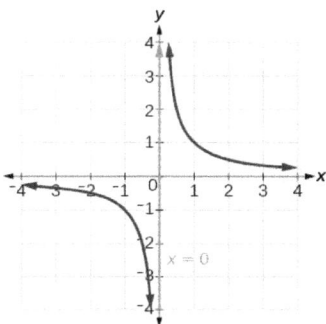

$x = 0$

Figure 3

DEFINITION

A **vertical asymptote** of a graph is a vertical line $x = a$ where the graph tends toward positive or negative infinity as the inputs approach a. We write

from the left side,

$$\text{as } x \to a^-, f(x) \to \infty, \text{ or as } x \to a^-, f(x) \to -\infty, \text{ and}$$

from the right side,

$$x \to a^+, f(x) \to \infty, \text{ or as } x \to a^+, f(x) \to -\infty.$$

End Behavior of $f(x) = \frac{1}{x}$

As the values of x increase without bound, the function values approach 0. As the values of x decrease without bound, the function values approach 0. See Figure 4. Symbolically, using arrow notation

$$\text{as } x \to \infty, f(x) \to 0, \text{ and as } x \to -\infty, f(x) \to 0.$$

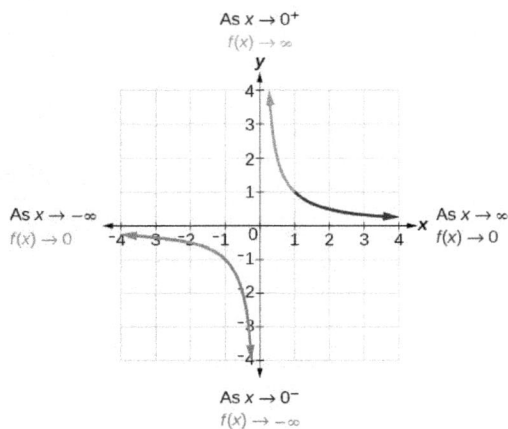

As $x \to 0^+$
$f(x) \to \infty$

As $x \to -\infty$
$f(x) \to 0$

As $x \to \infty$
$f(x) \to 0$

As $x \to 0^-$
$f(x) \to -\infty$

Figure 4

Based on this overall behavior and the graph, we can see that the function approaches 0 but never actually reaches 0; it seems to level off as the inputs become large. This behavior creates a **horizontal asymptote**, a horizontal line that the graph approaches as the input increases or decreases without bound. In this case, the graph is approaching the horizontal line $y = 0$. See Figure 5.

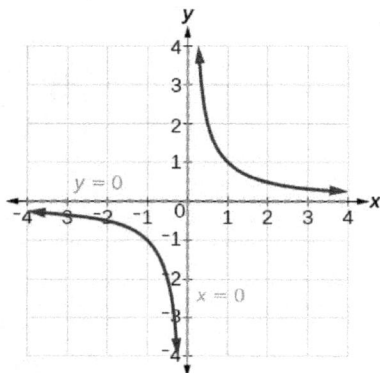

$y = 0$

$x = 0$

Figure 5

DEFINITION

A **horizontal asymptote** of a graph is a horizontal line $y = b$ where the graph approaches the line as the inputs increase or decrease without bound. We write as $x \to \infty$ or $x \to -\infty$, $f(x) \to b$.

EXAMPLE 1: USING ARROW NOTATION

Use arrow notation to describe the end behavior and local behavior of the function graphed in Figure 6.

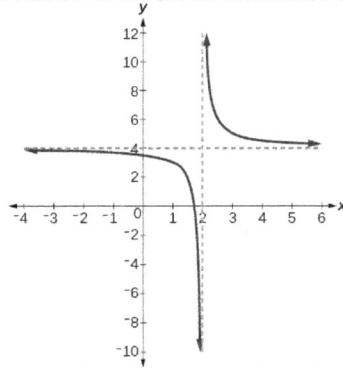

Figure 6

Answer

Notice that the graph is showing a vertical asymptote at $x = 2$, which tells us that the function is undefined at $x = 2$.

$$\text{As } x \to 2^-, f(x) \to -\infty, \text{ and as } x \to 2^+, \ f(x) \to \infty.$$

Also, as the inputs decrease without bound, the graph appears to be leveling off with output values closer and closer to 4, indicating a horizontal asymptote at $y = 4$. As the inputs increase without bound, the graph levels off approaching 4.

$$\text{As } x \to \infty, \ f(x) \to 4 \text{ and as } x \to -\infty, \ f(x) \to 4.$$

TRY IT #1

Use arrow notation to describe the end behavior and local behavior for the reciprocal squared function.

Answer

End behavior: as $x \to \infty, f(x) \to 0$; There is a horizontal asymptote at $y = 0$.

Local behavior: as $x \to 0, f(x) \to \infty$ There is a vertical asymptote at $x = 0$. There are no x– or y-intercepts.

Sketch a graph of the reciprocal function shifted two units to the left and up three units. Identify the horizontal and vertical asymptotes of the graph, if any.

Answer

Shifting the graph left 2 and up 3 would result in the function

$$f\left(x\right) = \tfrac{1}{x+2} + 3,$$

or equivalently, by adding the terms after finding a common denominator,

$$f\left(x\right) = \tfrac{3x+7}{x+2}.$$

The graph of the shifted function is displayed in Figure 7.

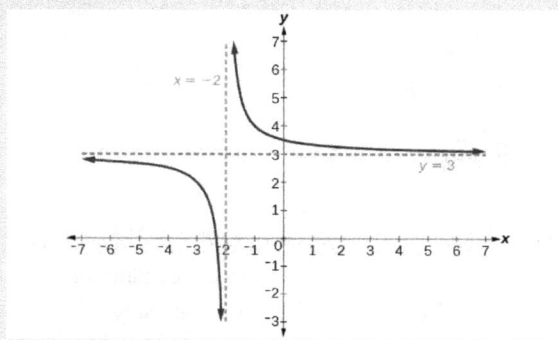

Figure 7

Notice that this function is undefined at $x = -2$, and the graph also is showing a vertical asymptote at $x = -2$.

$$\text{As } x \to -2^-, f\left(x\right) \to -\infty, \text{ and as } x \to -2^+, f\left(x\right) \to \infty.$$

As the inputs increase and decrease without bound, the graph appears to be leveling off with output values getting closer and closer to 3, indicating a horizontal asymptote of $y = 3$.

$$\text{As } x \to \infty, f\left(x\right) \to 3.$$

Analysis

Notice that horizontal and vertical asymptotes are shifted left 2 and up 3 along with the function.

TRY IT #2

Sketch the graph, and find the horizontal and vertical asymptotes of the reciprocal squared function that has been shifted right 3 units and down 4 units.

Answer

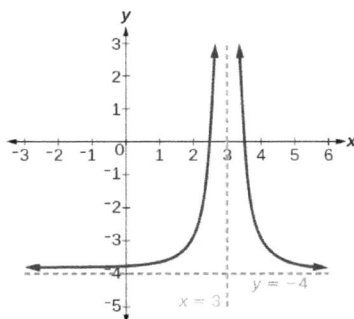

Figure 8

The function and the asymptotes are shifted 3 units right and 4 units down. As $x \to 3, f(x) \to \infty$, and as $x \to \infty, f(x) \to -4$. Therefore, the vertical asymptote is $x = 3$, and the horizontal asymptote is $y = -4$.

The function is $f(x) = \dfrac{1}{(x-3)^2} - 4$.

Solving Applied Problems Involving Rational Functions

In Example 2, we shifted a toolkit function in a way that resulted in the function $f(x) = \dfrac{3x+7}{x+2}$. This is an example of a rational function. A **rational function** is a function that can be written as the quotient of two polynomial functions. Many real-world problems require us to find the ratio of two polynomial functions. Problems involving rates and concentrations often involve rational functions.

DEFINITION

A **rational function** is a function that can be written as the quotient of two polynomial functions $P(x)$ and $Q(x)$.

$$f(x) = \frac{P(x)}{Q(x)} = \frac{a_p x^p + a_{p-1} x^{p-1} + \ldots + a_1 x + a_0}{b_q x^q + b_{q-1} x^{q-1} + \ldots + b_1 x + b_0}, Q(x) \neq 0$$

EXAMPLE 3: SOLVING AN APPLIED PROBLEM INVOLVING A RATIONAL FUNCTION

A large mixing tank currently contains 100 gallons of water into which 5 pounds of sugar have been mixed. A tap will open pouring 10 gallons per minute of water into the tank at the same time sugar is poured into the tank at a rate of 1 pound per minute. Find the concentration (pounds per gallon) of sugar in the tank after 12 minutes. Is that a greater concentration than at the beginning?

Answer

Let t be the number of minutes since the tap opened. Since the water increases at 10 gallons per minute, and the sugar increases at 1 pound per minute, these are constant rates of change. This tells us the amount of water in the tank is changing linearly, as is the amount of sugar in the tank. We can write an equation independently for each:

$$\text{water: } W(t) = 100 + 10t \text{ in gallons}$$
$$\text{sugar: } S(t) = 5 + 1t \text{ in pounds}$$

The concentration, C, will be the ratio of pounds of sugar to gallons of water.

$$C(t) = \frac{5+t}{100+10t}$$

The concentration after 12 minutes is given by evaluating $C(t)$ at $t = 12$.

$$C(12) = \frac{5+12}{100+10(12)} = \frac{17}{220}$$

This means the concentration is 17 pounds of sugar to 220 gallons of water.

At the beginning, the concentration is

$$C(0) = \frac{5+0}{100+10(0)} = \frac{1}{20}$$

Since $\frac{17}{220} \approx 0.08 > \frac{1}{20} = 0.05$, the concentration is greater after 12 minutes than at the beginning.

TRY IT #3

There are 1,200 freshmen and 1,500 sophomores at a prep rally at noon. After 12 p.m., 20 freshmen arrive at the rally every five minutes while 15 sophomores leave the rally. Find the ratio of freshmen to sophomores at 1 p.m.

Answer

$\frac{12}{11}$

Finding the Domains of Rational Functions

A vertical asymptote represents a value at which a rational function is undefined, so that value is not in the domain of the function. A rational function cannot have values in its domain that cause the denominator to equal zero. In general, to find the domain of a rational function, we need to determine which inputs would cause division by zero. The **domain of a rational function** includes all real numbers except those that cause the denominator to equal zero.

HOW TO

Given a rational function, find the domain.

1. Set the denominator equal to zero.
2. Solve to find the x-values that cause the denominator to equal zero.
3. The domain is all real numbers except those found in Step 2. Express your answer using interval notation.

EXAMPLE 4: FINDING THE DOMAIN OF A RATIONAL FUNCTION

Find the domain of $f(x) = \frac{x+3}{x^2-9}$.

Answer

Begin by setting the denominator equal to zero and solving.

$$x^2 - 9 = 0$$
$$x^2 = 9$$
$$x = 3$$

The denominator is equal to zero when $x = 3$. The domain of the function is all real numbers except $x = 3$. In interval notation we write, $(-\infty, -3) \cup (-3, 3) \cup (3, \infty)$.

Analysis

A graph of this function, as shown in Figure 9, confirms that the function is not defined when $x = 3$.

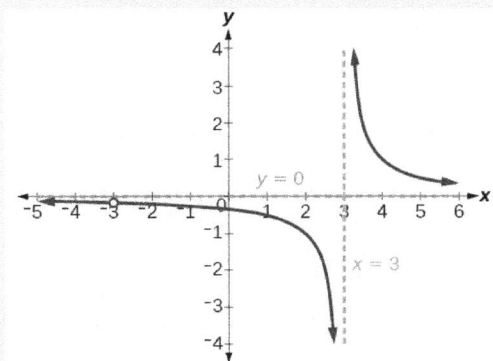

Figure 9

There is a vertical asymptote at $x = 3$ and a hole in the graph at $x = -3$. We will discuss these types of holes in greater detail later in this section.

TRY IT #4

Find the domain of $f(x) = \frac{4x}{5(x-1)(x-5)}$.

Answer

The domain is all real numbers except $x = 1$ and $x = 5$ or $(-\infty, 1) \cup (1, 5) \cup (5, \infty)$.

Identifying Vertical Asymptotes and Removable Discontinuities of Rational Functions

By looking at the graph of a rational function, we can investigate its local behavior and easily see whether there are vertical asymptotes. We may even be able to approximate their location. Even without the graph, however, we can still determine whether a given rational function has any vertical asymptotes, and calculate their location.

The **vertical asymptotes** of a rational function may be found by examining the factors of the denominator that are not common to the factors in the numerator. Vertical asymptotes occur at the zeros of such factors.

HOW TO

Given a rational function, identify any vertical asymptotes and removable discontinuties of its graph.

1. Factor the numerator and denominator.
2. List values not in the domain of the function.
3. Reduce the expression by canceling common factors in the numerator and the denominator.
4. Any values that cause the denominator to be zero in this simplified version are where the vertical asymptotes occur.
5. Any remaining values not in the domain are removable discontinuities also known as holes.

EXAMPLE 5: IDENTIFYING VERTICAL ASYMPTOTES

Find the vertical asymptotes of the graph of $k\left(x\right) = \frac{5+2x^2}{2-x-x^2}$.

Answer

First, factor the numerator and denominator.

$$k\left(x\right) = \frac{5 + 2x^2}{2 - x - x^2}$$

$$= \frac{5 + 2x^2}{\left(2 + x\right)\left(1 - x\right)}$$

To find the vertical asymptotes, we determine where this function will be undefined by setting the denominator equal to zero and solving:

$$\left(2 + x\right)\left(1 - x\right) = 0$$
$$x = -2, 1$$

Neither $x = -2$ nor $x = 1$ are zeros of the numerator, so the two values indicate two vertical asymptotes. The graph in Figure 10 confirms the location of the two vertical asymptotes.

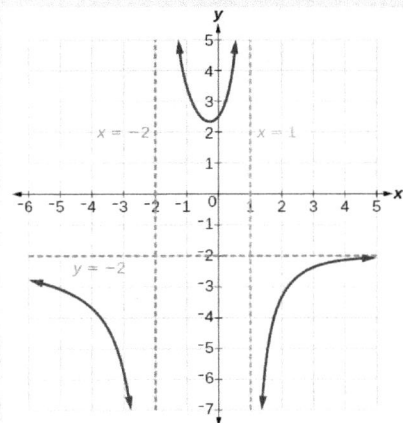

Figure 10

Occasionally, a graph will contain a hole: a single point where the graph is not defined, indicated by an open circle. We call such a hole a removable discontinuity. These often cannot be seen when technology is used to create a graph.

For example, the function $f(x) = \frac{x^2-1}{x^2-2x-3}$ may be re-written by factoring the numerator and the denominator.

$$f(x) = \frac{(x+1)(x-1)}{(x+1)(x-3)}$$

Notice that $x+1$ is a common factor to the numerator and the denominator. The zero of this factor, $x = -1$, is NOT in the domain and is the location of the removable discontinuity. Notice also that $x-3$ is not a factor in both the numerator and denominator. The zero of this factor, $x = 3$, is the vertical asymptote. See Figure 11.

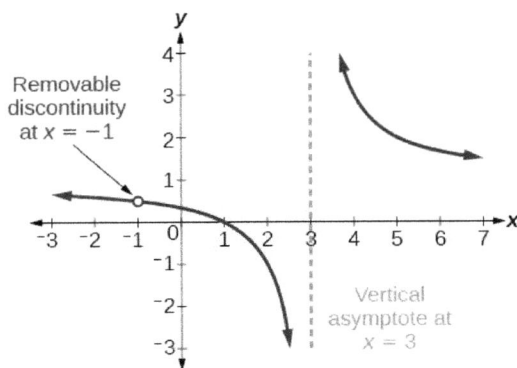

Figure 11

A **removable discontinuity** occurs in the graph of a rational function at $x = a$ if a is a zero for a factor that is both in the denominator and in the numerator. We factor the numerator and denominator and check for common factors. If we find any, we set the common factor equal to 0 and solve. This may be the location of a removable discontinuity. This is a removable discontinuity if the multiplicity of this factor in the numerator is greater than or equal to that in the denominator. If the multiplicity of this factor is greater in the denominator, then there is still a vertical asymptote at that value.

EXAMPLE 6: IDENTIFYING VERTICAL ASYMPTOTES AND REMOVABLE DISCONTINUITIES FOR A GRAPH

Find the vertical asymptotes and removable discontinuities of the graph of $k\left(x\right) = \frac{x-2}{x^2-4}$.

Answer

Factor the numerator and the denominator.

$$k\left(x\right) = \frac{x-2}{(x-2)(x+2)}$$

Notice that there is a common factor in the numerator and the denominator, $x - 2$. The zero for this factor is $x = 2$. This is the location of the removable discontinuity.

Notice that there is a factor in the denominator that is not in the numerator, $x + 2$. The zero for this factor is $x = -2$. The vertical asymptote is $x = -2$. See Figure 12.

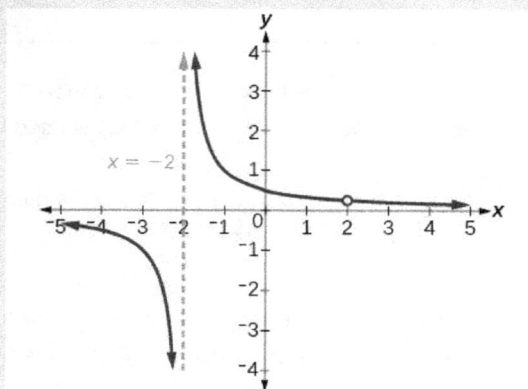

Figure 12

The graph of this function will have the vertical asymptote at $x = -2$, but at $x = 2$ the graph will have a hole.

TRY IT #5

Find the vertical asymptotes and removable discontinuities of the graph of

$$f\left(x\right) = \frac{x^2-25}{x^3-6x^2+5x}.$$

Answer

Removable discontinuity at $x = 5$. Vertical asymptotes: $x = 0, \ x = 1$.

Identifying Horizontal Asymptotes of Rational Functions

While vertical asymptotes describe the behavior of a graph as the *output* gets very large or very small, horizontal asymptotes help describe the behavior of a graph as the *input* gets very large or very small. Recall that a polynomial's end behavior will mirror that of the leading term. Likewise, a rational function's end behavior will mirror that of the ratio of the leading terms of the numerator and denominator functions.

There are three distinct outcomes when checking for horizontal asymptotes:

Case 1: If the degree of the denominator is greater than the degree of the numerator, there is a horizontal asymptote at $y = 0$.

$$\text{Example: } f(x) = \frac{4x+2}{x^2+4x-5}$$

In this case, the end behavior is $f(x) \approx \frac{4x}{x^2} = \frac{4}{x}$. This tells us that, as the inputs increase or decrease without bound, this function will behave similarly to the function $g(x) = \frac{4}{x}$, and the outputs will approach zero, resulting in a horizontal asymptote at $y = 0$. See Figure 13. Note that this graph crosses the horizontal asymptote.

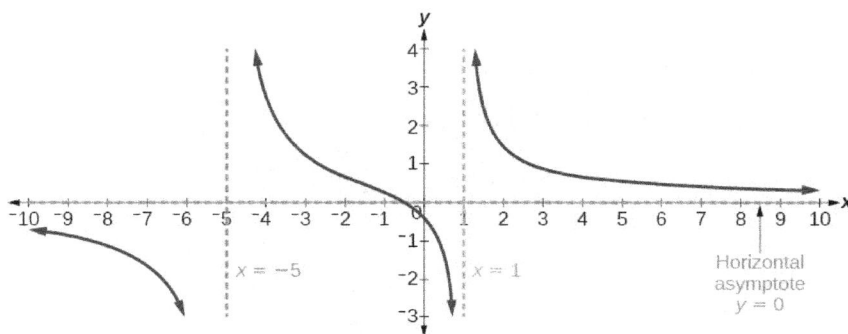

Figure 13 Horizontal Asymptote $y = 0$ when $f(x) = \frac{p(x)}{q(x)}$, $q(x) \neq 0$ where degree of $p <$ degree of q.

Case 2: If the degree of the denominator is less than the degree of the numerator by one, we get a slant asymptote.

$$\text{Example: } f(x) = \frac{3x^2-2x+1}{x-1}$$

In this case, the end behavior is $f(x) \approx \frac{3x^2}{x} = 3x$. This tells us that as the inputs increase or decrease without bound, this function will behave similarly to the function $g(x) = 3x$. As the inputs grow large, the outputs will grow and not level off, so this graph has no horizontal asymptote. However, the graph of $g(x) = 3x$ looks like a diagonal line, and since f will behave similarly to g, it will approach a line close to $y = 3x$. This line is a slant asymptote.

To find the equation of the slant asymptote, divide $\frac{3x^2-2x+1}{x-1}$. The quotient is $3x + 1$, and the remainder is 2. The slant asymptote is the graph of the line $g(x) = 3x + 1$. See Figure 14.

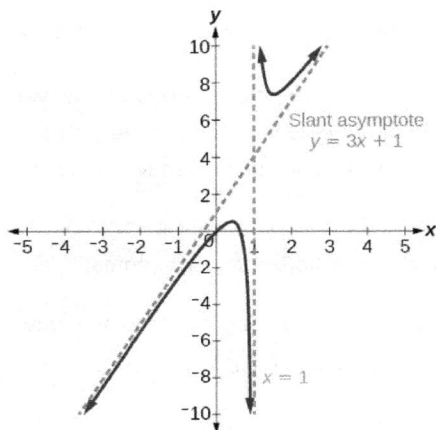

Figure 14 Slant Asymptote when
$$f\left(x\right)=\frac{p(x)}{q(x)},\ q\left(x\right)\neq 0\text{ where degree of}$$
$p>$ degree of q by 1.

Case 3: If the degree of the denominator equals the degree of the numerator, there is a horizontal asymptote at $y=\frac{a_n}{b_n}$, where a_n and b_n are the leading coefficients of $p\left(x\right)$ and $q\left(x\right)$ for $f\left(x\right)=\frac{p(x)}{q(x)}, q\left(x\right)\neq 0.$

Example: $f\left(x\right)=\frac{3x^2+2}{x^2+4x-5}$

In this case, the end behavior is $f\left(x\right)\approx\frac{3x^2}{x^2}=3.$ This tells us that as the inputs grow large, this function will behave like the function $g\left(x\right)=3$, which is a horizontal line. As $x\to\infty$, $f\left(x\right)\to 3$, resulting in a horizontal asymptote of $y=3.$ See Figure 15. Note that this graph crosses the horizontal asymptote.

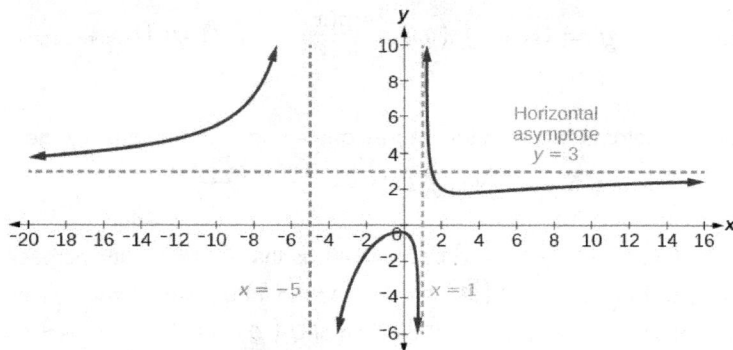

Figure 15 Horizontal Asymptote when $f\left(x\right)=\frac{p(x)}{q(x)},\ q\left(x\right)\neq 0$ where degree of $p=$ degree of $q.$

Notice that, while the graph of a rational function will never cross a vertical asymptote, the graph may or may not cross a horizontal or slant asymptote. Also, although the graph of a rational function may have many vertical asymptotes, the graph will have at most one horizontal (or slant) asymptote.

It should be noted that, if the degree of the numerator is larger than the degree of the denominator by more than one, the end behavior of the graph will mimic the behavior of the reduced end behavior fraction. For instance, if we had the function

$$f\left(x\right)=\frac{3x^5-x^2}{x+3}$$

with end behavior $f\left(x\right)\approx\frac{3x^5}{x}=3x^4$, the end behavior of the graph would look similar to that of an even polynomial with a positive leading coefficient. We write as $x\to\infty, f\left(x\right)\to\infty.$

HORIZONTAL ASYMPTOTES OF RATIONAL FUNCTIONS

The horizontal asymptote of a rational function can be determined by looking at the degrees of the numerator and denominator.

- Degree of numerator *is less than* degree of denominator: horizontal asymptote at $y = 0$.
- Degree of numerator *is greater than degree of denominator by one*: no horizontal asymptote; slant asymptote.
- Degree of numerator *is equal to* degree of denominator: horizontal asymptote at ratio of leading coefficients.

EXAMPLE 7: IDENTIFYING HORIZONTAL ASYMPTOTES

For the functions below, identify the horizontal or slant asymptote.

a. $g(x) = \frac{6x^3 - 10x}{2x^3 + 5x^2}$

b. $k(x) = \frac{x^2 + 4x}{x^3 - 8}$

Answer

For these solutions, we will use $f(x) = \frac{p(x)}{q(x)}, q(x) \neq 0$.

a. $g(x) = \frac{6x^3 - 10x}{2x^3 + 5x^2}$: The degree of p is 3 and the degree of q is 3, so we can find the horizontal asymptote by taking the ratio of the leading terms, $\frac{6x^3}{2x^3}$. There is a horizontal asymptote at $y = \frac{6}{2}$ or $y = 3$.

b. $k(x) = \frac{x^2 + 4x}{x^3 - 8}$: The degree of p is 2 which is less than the degree of q which is 3, so there is a horizontal asymptote $y = 0$.

EXAMPLE 8: IDENTIFYING HORIZONTAL ASYMPTOTES

In the sugar concentration problem earlier, we created the equation $C(t) = \frac{5 + t}{100 + 10t}$.

Find the horizontal asymptote and interpret it in context of the problem.

Answer

Both the numerator and denominator are linear (degree 1). Because the degrees are equal, there will be a horizontal asymptote at the ratio of the leading coefficients. In the numerator, the leading term is t, with coefficient 1. In the denominator, the leading term is $10t$, with coefficient 10. The horizontal asymptote will be at the ratio of these values:

$$t \to \infty, C(t) \to \frac{1}{10}$$

This function will have a horizontal asymptote at $y = \frac{1}{10}$.

This tells us that as the values of t increase, the values of C will approach $\frac{1}{10}$. In context, this means that, as more time goes by, the concentration of sugar in the tank will approach one-tenth of a pound of sugar per gallon of water.

EXAMPLE 9: IDENTIFYING HORIZONTAL AND VERTICAL ASYMPTOTES

Find the horizontal and vertical asymptotes of the function $f(x) = \frac{(x-2)(x+3)}{(x-1)(x+2)(x-5)}$

Answer

First, note that this function has no common factors, so there are no potential removable discontinuities.

The function will have vertical asymptotes when the denominator is zero, causing the function to be undefined. The denominator will be zero at $x = 1, -2,$ and 5, indicating vertical asymptotes at these values.

The numerator has degree 2, while the denominator has degree 3. Since the degree of the denominator is greater than the degree of the numerator, the denominator will grow faster than the numerator, causing the outputs to tend towards zero as the inputs get large, and so as $x \to \infty$, $f(x) \to 0$. This function will have a horizontal asymptote at $y = 0$. See Figure 16.

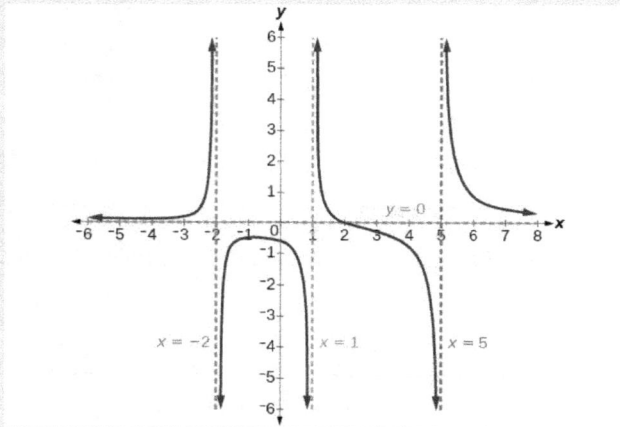

Figure 16

TRY IT #6

Find the vertical and horizontal asymptotes of the function $f(x) = \frac{(2x-1)(2x+1)}{(x-2)(x+3)}$

Answer

Vertical asymptotes at $x = 2$ and $x = -3$; horizontal asymptote at $y = 4$.

EXAMPLE 10: NUMERICAL SUPPORT FOR ASYMPTOTES AND DISCONTINUITIES

Find the horizontal and vertical asymptotes and removal discontinuities of $f\left(x\right) = \frac{\left(x+2\right)\left(x-1\right)}{\left(x+3\right)\left(x-1\right)}$. Give numerical support for your conclusions.

Answer

Setting the denominator $\left(x+3\right)\left(x-1\right)$ equal to zero and solving, we find that $x = -3$ and $x = 1$ are not in the domain of the function. Since $\left(x-1\right)$ is in the numerator and denominator an equal number of times, we conclude that there is a removable discontinuity or hole at $x = 1$. Since $\left(x+3\right)$ does not have a common factor in the numerator, we conclude that there is a vertical asymptote of $x = -3$.

To support the hole at $x = 1$, we choose input values just above and just below $x = 1$ that get closer and closer to $x = 1$ and then evaluate the function there. We will choose our x values equal to 1.1, 1.01, and 1.001 for the values slightly larger than $x = 1$ and 0.9, 0.99 and 0.999 for the values slightly smaller than $x = 1$. See Table 4 and 5.

Table 4

x	1.1	1.01	1.001
$f\left(x\right)$	0.7561	0.75062	.750062

Table 5

x	0.9	0.99	0.999
$f\left(x\right)$	0.74359	0.74937	0.749937

Notice, in Table 4, that the output has more zeros following 0.75 as we move across the table. We therefore conclude that the output is going to 0.75 as x goes to 1 from above. We write, as $x \to 1^{+}, f\left(x\right) \to 0.75$. In Table 5, we see that the output increases the number of nines that follow 0.74, indicating that the output goes to 0.75. We write, as $x \to 1^{-}, f\left(x\right) \to 0.75$. This means that there is a hole at $\left(1, 0.75\right)$.

To support the vertical asymptote at $x = -3$, we choose values just above and just below $x = -3$ that get closer and closer to $x = -3$ and then evaluate the function there. We will choose our x values equal to -2.9, -2.99, and -2.999 for the values to the right of $x = -3$ and -3.1, -3.01 and -3.001 for the values to the left of $x = -3$. See Table 6 and 7.

Table 6

x	-2.9	-2.99	-2.999
$f\left(x\right)$	-9	-99	-999

Table 7

x	-3.1	-3.01	-3.001
$f\left(x\right)$	11	101	1001

Table 6 shows that as x gets closer and closer to -3 from the right, the output is decreasing without bound or as $x \to -3^{+}$, $f(x) \to -\infty$. Table 7 shows that x gets closer and closer to -3 from the left, the output is increasing without bound or as $x \to -3^{-}$, $f(x) \to \infty$. This supports that there is a vertical asymptote at $x = -3$.

Finally, to determine the horizontal asymptote, we look at the ratio of the leading terms of the numerator and denominator:

$$\frac{x^2}{x^2} = 1.$$

There is a horizontal asymptote of $y = 1$. To support this numerically, we choose very large values for x such as 100, 1 000 and 10 000 and very negative numbers such as -100, -1 000, -10 000 and evaluate the function at these points. See Table 8 and 9.

Table 8

x	100	1000	10000
$f(x)$	0.99029	0.99900299	0.99990002999

Table 9

x	-100	-1000	-10000
$f(x)$	1.0103	1.001003	1.00010003

Look at the pattern in the output of each table. We see that as x increases or decreases without bound, the output becomes closer and closer to 1, supporting the asymptote we found algebraically. We write, as $x \to \pm\infty$, $f(x) \to 1$.

Intercepts of Rational Functions

A rational function will have a y-intercept when the input is zero, if the function is defined at zero. A rational function will not have a y-intercept if the function is not defined at zero.

Likewise, a rational function will have x-intercepts at the inputs that cause the output to be zero. Since a fraction is only equal to zero when the numerator is zero, x-intercepts can only occur when the numerator of the rational function is equal to zero. However, if the numerator is zero at a value not in the domain, then there is not an x-intercept at that point.

EXAMPLE 11: FINDING THE INTERCEPTS OF A RATIONAL FUNCTION

Find the intercepts of $f(x) = \frac{(x-2)(x+3)}{(x-1)(x+2)(x-5)}$.

Answer

We can find the y-intercept by evaluating the function at zero.

$$f(0) = \frac{(0-2)(0+3)}{(0-1)(0+2)(0-5)}$$

$$= \frac{-6}{10}$$

$$= -\frac{3}{5}$$

$$= -0.6.$$

The x-intercepts will occur when the function is equal to zero:

$$0 = \frac{(x-2)(x+3)}{(x-1)(x+2)(x-5)}$$ This is zero when the numerator is zero.

$$0 = (x-2)(x+3)$$

$$x = 2, -3$$

The y-intercept is $(0, -0.6)$, and the x-intercepts are $(2, 0)$ and $(-3, 0)$. See Figure 17.

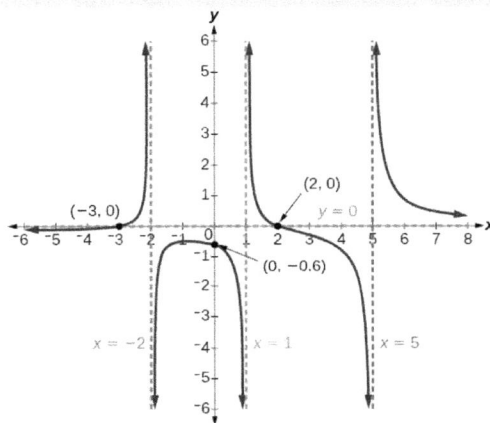

Figure 17

TRY IT #7

Given the reciprocal squared function that is shifted right 3 units and down 4 units, write this as a rational function. Then, find the x– and y-intercepts and the horizontal and vertical asymptotes.

Answer

For the transformed reciprocal squared function, we find the rational form.

$$f(x) = \frac{1}{(x-3)^2} - 4$$

$$= \frac{1 - 4(x-3)^2}{(x-3)^2}$$

$$= \frac{1 - 4\left(x^2 - 6x + 9\right)}{(x-3)(x-3)}$$

$$= \frac{-4x^2 + 24x - 35}{x^2 - 6x + 9}$$

Because the numerator is the same degree as the denominator we know that as $x \to \infty$, $f(x) \to -4$; so $y = -4$ is the horizontal asymptote. Next, we set the denominator equal to zero, and find that the vertical asymptote is $x = 3$, because as $x \to 3$, $f(x) \to \infty$. We then set the numerator equal to 0 and find the x-intercepts are at $(2.5, 0)$ and $(3.5, 0)$, since 2.5 and 3.5 are in the domain. Finally, we evaluate the function at 0 and find the y-intercept to be at $\left(0, \frac{-35}{9}\right)$.

Graphing Rational Functions

In Example 11, we see that the numerator of a rational function reveals the x-intercepts of the graph, whereas the denominator reveals the vertical asymptotes of the graph. As with polynomials, factors of the numerator may have integer powers greater than one. Fortunately, the effect on the shape of the graph at those intercepts is the same as we saw with polynomials.

The vertical asymptotes associated with the factors of the denominator will mirror one of the two toolkit reciprocal functions. When the degree of the factor in the denominator is odd, the distinguishing characteristic is that on one side of the vertical asymptote the graph heads towards positive infinity, and on the other side the graph heads towards negative infinity. See Figure 18.

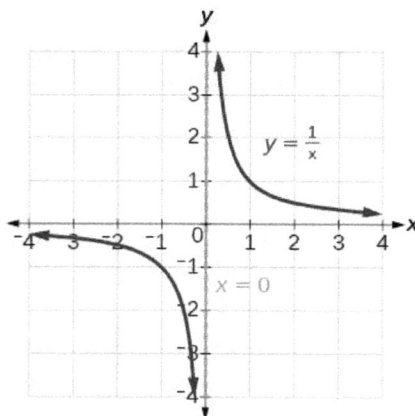

Figure 18

When the degree of the factor in the denominator is even, the distinguishing characteristic is that the graph either heads toward positive infinity on both sides of the vertical asymptote or heads toward negative infinity on both sides. See Figure 19.

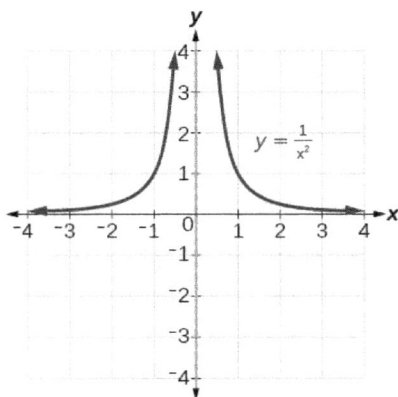

Figure 19

For example, the graph of $f\left(x\right) = \dfrac{\left(x+1\right)^2\left(x-3\right)}{\left(x+3\right)^2\left(x-2\right)}$ is shown in Figure 20.

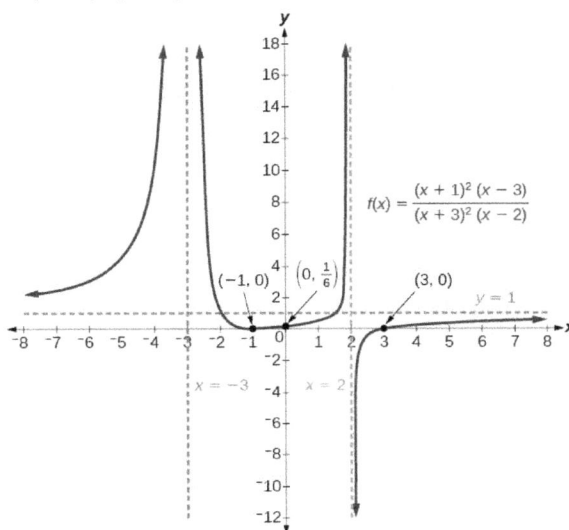

Figure 20

- At the x-intercept $x = -1$ corresponding to the $\left(x+1\right)^2$ factor of the numerator, the graph bounces, consistent with the quadratic nature of the factor.
- At the x-intercept $x = 3$ corresponding to the $\left(x-3\right)$ factor of the numerator, the graph passes through the axis as we would expect from a linear factor.
- At the vertical asymptote $x = -3$ corresponding to the $\left(x+3\right)^2$ factor of the denominator, the graph heads towards positive infinity on both sides of the asymptote, consistent with the behavior of the function $f\left(x\right) = \dfrac{1}{x^2}$.
- At the vertical asymptote $x = 2$, corresponding to the $\left(x-2\right)$ factor of the denominator, the graph heads towards positive infinity on the left side of the asymptote and towards negative infinity on the right side, consistent with the behavior of the function $f\left(x\right) = \dfrac{1}{x}$.

HOW TO

Given a rational function, sketch a graph.

---

1. Evaluate the function at 0 to find the *y*-intercept.
2. Factor the numerator and denominator.
3. For factors in the numerator that are not in the denominator, determine where each factor of the numerator is zero to find the *x*-intercepts.
4. Find the multiplicities of the *x*-intercepts to determine the behavior of the graph at those points.
5. For factors in the denominator, note the multiplicities of the zeros to determine the local behavior. For those factors not common to the numerator, find the vertical asymptotes by setting those factors equal to zero and then solve.
6. For factors that are in both the denominator AND the numerator, find the removable discontinuities by setting those factors equal to 0 and then solve.
7. Compare the degrees of the numerator and the denominator to determine the horizontal or slant asymptotes.
8. Sketch the graph.

EXAMPLE 12: GRAPHING A RATIONAL FUNCTION

Sketch a graph of $f(x) = \frac{(x+2)(x-3)}{(x+1)^2(x-2)}$.

Answer

We can start by noting that the function is already factored, saving us a step.

Next, we will find the intercepts. Evaluating the function at zero gives the *y*-intercept:

$$f(0) = \frac{(0+2)(0-3)}{(0+1)^2(0-2)}$$
$$= 3$$

To find the *x*-intercepts, we determine when the numerator of the function is zero. Setting each factor equal to zero, we find *x*-intercepts at $x = -2$ and $x = 3$. At each, the behavior will be linear (multiplicity 1), with the graph passing through the intercept.

We have a *y*-intercept at $(0, 3)$ and *x*-intercepts at $(-2, 0)$ and $(3, 0)$.

To find the vertical asymptotes, we determine when the denominator is equal to zero. This occurs when $x + 1 = 0$ and when $x - 2 = 0$, giving us vertical asymptotes at $x = -1$ and $x = 2$.

There are no common factors in the numerator and denominator. This means there are no removable discontinuities.

Finally, the degree of denominator is larger than the degree of the numerator, telling us this graph has a horizontal asymptote at $y = 0$.

To sketch the graph, we might start by plotting the three intercepts. Since the graph has no *x*-intercepts between the vertical asymptotes, and the *y*-intercept is positive, we know the function must remain positive between the asymptotes, letting us fill in the middle portion of the graph as shown in Figure 21.

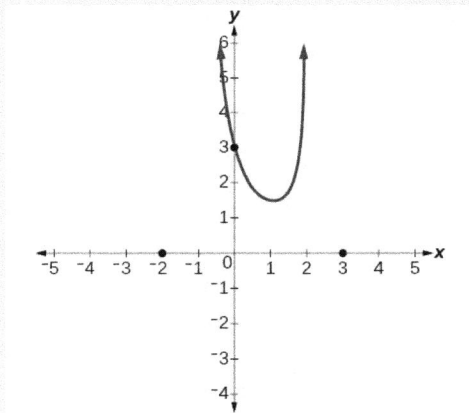

Figure 21

The factor associated with the vertical asymptote at $x = -1$ was squared, so we know the behavior will be the same on both sides of the asymptote. The graph heads toward positive infinity as the inputs approach the asymptote on the right, so the graph will head toward positive infinity on the left as well.

For the vertical asymptote at $x = 2$, the factor was not squared, so the graph will have opposite behavior on either side of the asymptote. See Figure 22. After passing through the x-intercepts, the graph will then level off toward an output of zero, as indicated by the horizontal asymptote.

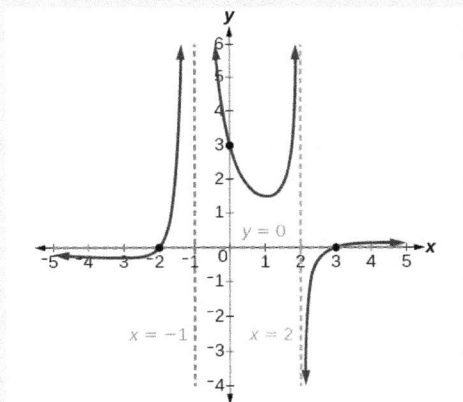

Figure 22

TRY IT #8

Given the function $f\left(x\right) = \frac{(x+2)^2(x-2)}{2(x-1)^2(x-3)}$, use the characteristics of polynomials and rational functions to describe its behavior and sketch the function.

Answer

Horizontal asymptote at $y = \frac{1}{2}$. Vertical asymptotes at $x = 1$ and $x = 3$. y-intercept at $\left(0, \frac{4}{3}\right)$.

x-intercepts at $(2, 0)$ and $(-2, 0)$. $(-2, 0)$ is a zero with multiplicity 2, and the graph bounces off the x-axis at this point. $(2, 0)$ is a single zero and the graph crosses the axis at this point.

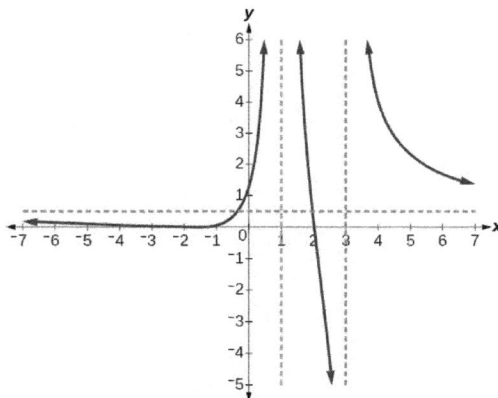

Figure 23

Writing Rational Functions

Now that we have analyzed the equations for rational functions and how they relate to a graph of the function, we can use information given by a graph to write the function. A rational function written in factored form will have an x-intercept where each factor of the numerator is equal to zero. (An exception occurs in the case of a removable discontinuity.) As a result, we can form a numerator of a function whose graph will pass through a set of x-intercepts by introducing a corresponding set of factors. Likewise, because the function will have a vertical asymptote where each factor of the denominator is equal to zero, we can form a denominator that will produce the vertical asymptotes by introducing a corresponding set of factors.

Writing Rational Functions from Intercepts and Asymptotes

If a rational function has x-intercepts at $x = x_1, x_2, \ldots, x_n$, vertical asymptotes at $x = v_1, v_2, \ldots, v_m$, and no $x_i = $ any v_j, then the function can be written in the form:

$$f(x) = a \frac{(x-x_1)^{p_1}(x-x_2)^{p_2} \cdots (x-x_n)^{p_n}}{(x-v_1)^{q_1}(x-v_2)^{q_2} \cdots (x-v_m)^{q_n}}$$

where the powers p_i or q_i on each factor can be determined by the behavior of the graph at the corresponding intercept or asymptote, and the stretch factor a can be determined given a value of the function other than the x-intercept or by the horizontal asymptote if it is nonzero.

HOW TO

Given a graph of a rational function, write the function.

1. Determine the factors of the numerator. Examine the behavior of the graph at the x-intercepts to determine the zeros and their multiplicities. (This is easy to do when finding the "simplest" function with small multiplicities—such as 1 or 3—but may be difficult for larger multiplicities—such as 5 or 7, for example.)

---

2. Determine the factors of the denominator. Examine the behavior on both sides of each vertical asymptote to determine the factors and their powers.
3. Use any clear point on the graph to find the stretch factor.

EXAMPLE 13: WRITING A RATIONAL FUNCTION FROM INTERCEPTS AND ASYMPTOTES

Write an equation for the rational function shown in Figure 24.

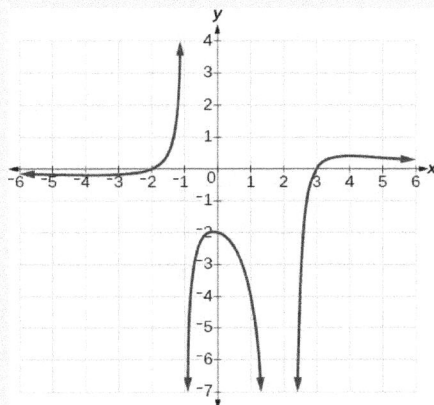

Figure 24

Answer

The graph appears to have x-intercepts at $x = -2$ and $x = 3$. At both, the graph passes through the intercept, suggesting linear factors. The graph has two vertical asymptotes. The one at $x = -1$ seems to exhibit the basic behavior similar to $\frac{1}{x}$, with the graph heading toward positive infinity on one side and heading toward negative infinity on the other. The asymptote at $x = 2$ is exhibiting a behavior similar to $\frac{1}{x^2}$, with the graph heading toward negative infinity on both sides of the asymptote. See Figure 25.

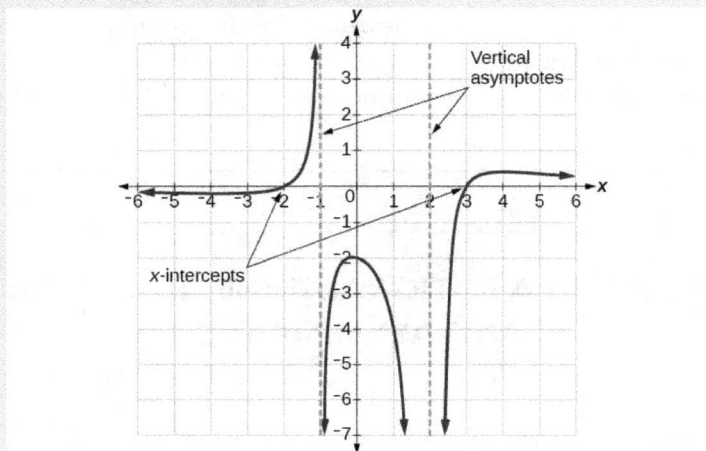

Figure 25

We can use this information to write a function of the form

$$f\left(x\right) = a\frac{(x+2)(x-3)}{(x+1)(x-2)^2}.$$

To find the stretch factor, we can use another clear point on the graph, such as the y-intercept $(0, -2)$.

$$-2 = a\frac{(0+2)(0-3)}{(0+1)(0-2)^2}$$

$$-2 = a\frac{-6}{4}$$

$$a = \frac{-8}{-6} = \frac{4}{3}$$

This gives us a final function of $f\left(x\right) = \frac{4(x+2)(x-3)}{3(x+1)(x-2)^2}.$

Access these online resources for additional instruction and practice with rational functions.

- Graphing Rational Functions
- Find the Equation of a Rational Function
- Determining Vertical and Horizontal Asymptotes
- Find the Intercepts, Asymptotes, and Hole of a Rational Function

Key Equations

Rational Function	$f\left(x\right) = \dfrac{P\left(x\right)}{Q\left(x\right)} = \dfrac{a_p x^p + a_{p-1} x^{p-1} + \ldots + a_1 x + a_0}{b_q x^q + b_{q-1} x^{q-1} + \ldots + b_1 x + b_0}, Q\left(x\right) \neq 0$

KEY CONCEPTS

- We can use arrow notation to describe local behavior and end behavior of the toolkit functions $f(x) = \frac{1}{x}$ and $f(x) = \frac{1}{x^2}$.
- A function that levels off at a horizontal value has a horizontal asymptote. A function can have more than one vertical asymptote.
- Application problems involving rates and concentrations often involve rational functions.
- The domain of a rational function includes all real numbers except those that cause the denominator to equal zero.
- The vertical asymptotes of a rational function will occur where the denominator of the function is equal to zero and the numerator is not zero.
- A removable discontinuity might occur in the graph of a rational function if an input causes both numerator and denominator to be zero.
- A rational function's end behavior will mirror that of the ratio of the leading terms of the numerator and denominator functions.
- Graph rational functions by finding the intercepts, behavior at the intercepts, asymptotes, and end behavior.
- If a rational function has x-intercepts at $x = x_1, x_2, \ldots, x_n$, vertical asymptotes at $x = v_1, v_2, \ldots, v_m$, and no $x_i = $ any v_j, then the function can be written in the form
$$f(x) = a\frac{(x - x_1)^{p_1}(x - x_2)^{p_2} \cdots (x - x_n)^{p_n}}{(x - v_1)^{q_1}(x - v_2)^{q_2} \cdots (x - v_m)^{q_n}}$$

GLOSSARY

arrow notation
a way to symbolically represent the local and end behavior of a function by using arrows to indicate that an input or output approaches a value

horizontal asymptote
a horizontal line $y = b$ where the graph approaches the line as the inputs increase or decrease without bound

rational function
a function that can be written as the ratio of two polynomials

removable discontinuity
a single point at which a function is undefined that, if filled in, would make the function continuous; it appears as a hole on the graph of a function

vertical asymptote
a vertical line $x = a$ where the graph tends toward positive or negative infinity as the inputs approach a

4.4 Root Functions and Their Transformations

A root function is a power function of the form $f\left(x\right) = x^{\frac{1}{n}}$, where n is a positive integer greater than one. For example, $f\left(x\right) = x^{\frac{1}{2}} = \sqrt{x}$ is the square-root function and $g\left(x\right) = x^{\frac{1}{3}} = \sqrt[3]{x}$ is the cube-root functions.

The root functions $f\left(x\right) = x^{\frac{1}{n}}$ have defining characteristics depending on whether n is odd or even. For all positive even integers $n \geq 2$, the domain of $f\left(x\right) = x^{\frac{1}{n}}$ is the interval $[0, \infty)$. Figure 1 shows the the functions $f\left(x\right) = x^{\frac{1}{2}} = \sqrt{x}$, $g\left(x\right) = x^{\frac{1}{4}} = \sqrt[4]{x}$ and $h\left(x\right) = x^{\frac{1}{6}} = \sqrt[6]{x}$ which are all even root functions.

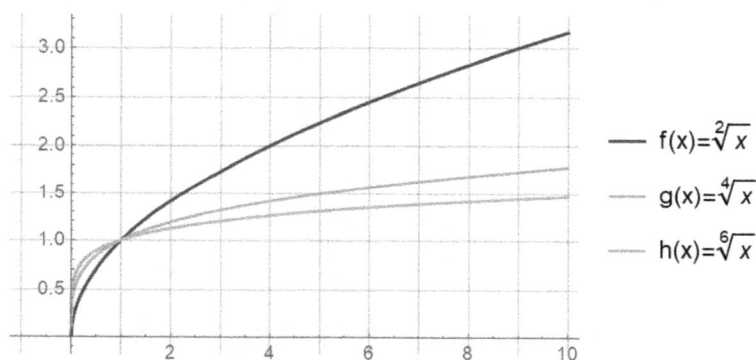

Figure 1

Notice that these graphs have similar shapes, very much like that of the square root function in the toolkit. However, as the value of n increases, the graphs steepen somewhat near the origin and become flatter away from the origin growing more slowly. The x and y intercepts of these functions are $(0, 0)$. The end behavior for the even root function only makes sense as x increases without bound since negative values are not in the domain. We observe as $x \to \infty$, $f\left(x\right) \to \infty$.

For all positive odd integers $n \geq 3$, the domain of $f\left(x\right) = x^{\frac{1}{n}}$ is the set of all real numbers. Since $x^{\frac{1}{n}} = \left(-x\right)^{\frac{1}{n}}$ for positive odd integers n, $f\left(x\right) = x^{\frac{1}{n}}$ is an odd function if n is a positive odd number. Figure 2 shows the functions $f\left(x\right) = x^{\frac{1}{3}} = \sqrt[3]{x}$, $g\left(x\right) = x^{\frac{1}{5}} = \sqrt[5]{x}$ and $h\left(x\right) = x^{\frac{1}{7}} = \sqrt[7]{x}$ which are all odd root functions.

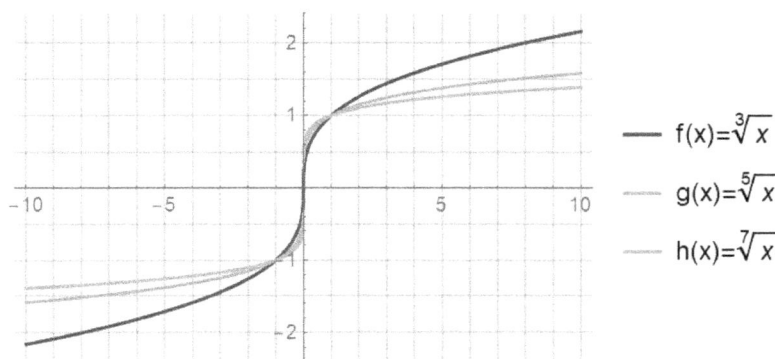

Figure 2

Notice that these graphs look similar to the cube root function in the toolkit. Again, as the value of n increases, the graphs steepens near the origin and become flatter away from the origin increasing more slowly. The x and y intercepts of these functions are $(0,0)$. The end behavior for the even root function is expressed as $x \to \infty$, $f(x) \to \infty$ for large values of x and as $x \to -\infty$, $f(x) \to -\infty$ for very negative values of x.

Transformations of Root Functions

For transformations of even root functions, the domain and range are effected by horizontal and vertical shifts, reflections and stretches. There are two methods you can use to find the domain. The first method is to use algebra and the idea that even root functions must have non-negative values under the root symbol. The expression under the root symbol is set greater than or equal to zero and the inequality is solved to find the domain. Alternatively, you can use the properties of the transformation by identifying the basic function and determining where the point (0,0) gets transformed to in the new function. The x-coordinate will be the starting or ending point for the domain. If there is not a horizontal reflection, the domain will be from that value to the right and if there is a horizontal reflection, then the domain will go from that value to the left.

The range is determined by identifying the basic function and determining what transformation is applied to get the function you are working with. After applying transformation to the point (0,0), the y-coordinate tells you where the range starts or ends. If there is not a vertical reflection the range will be from that value to infinity and if there is a vertical reflection the range will be from minus infinity to that value.

HOW TO

Given a root function, find the domain and range.

Domain Method 1: Algebraically

1. Set the expression under the root symbol greater than or equal to zero and solve.
2. Write the solution in interval notation. Remember to use the square bracket as appropriate.

Domain Method 2: Transformations

1. Identify the basic root function.
2. Describe the transformation in words and then determine where the point (0,0) gets mapped to under that transformation.

---

3. If there is not a vertical reflection, the domain is from the x-coordinate of the transformed point to infinity. If there is a vertical reflection, the domain is from minus infinity to that x-coordinate.

Range

1. Identify the basic root function.
2. Describe the transformation in words and then determine where the point (0,0) gets mapped to under that transformation.
3. The y-coordinate tells you where the range starts or ends. If there is not a vertical reflection the range will be from that value to infinity and if there is a vertical reflection the range will be from minus infinity to that value.

EXAMPLE 1: THE DOMAIN AND RANGE OF AN EVEN ROOT FUNCTION

Find the domain, range and intercepts of the square root function shifted 3 units left and 1 unit up.

Answer

First, find the equation for the function. $f(x+3) + 1 = \sqrt{x+3} + 1.$

Method 1: The domain of an even root function must have non-negative values under the root symbol, so we solve the inequality

$$x + 3 \geq 0.$$
$$x \geq -3$$

Therefore, the domain is all real numbers greater that or equal to negative 3 or in interval notation $[-3, \infty)$.

Method 2: Alternatively, the point $(0, 0)$ is shifted to the point $(-3, 1)$. The starting value for the domain is -3 and since there is no horizontal reflection the graph opens to the right like $\sqrt{x}$. Again the domain is $[-3, \infty)$.

The range of the square root will be shifted up one unit, so the range is all real numbers greater than or equal to one or in interval notation $[1, \infty)$. Notice that the starting value of 1 is reflected in the shift of the point $(0, 0)$ above and since there is no vertical reflection the interval goes in the direction of infinity.

EXAMPLE 2: DOMAIN AND INTERCEPTS OF EVEN ROOT FUNCTIONS

Find the domain and range for

a. $g(x) = \sqrt[4]{3 - 2x}.$
b. $h(x) = -3\sqrt[4]{x}.$

Answer

a. *Method 1:* Even root functions have non-negative input, so the domain of $g\left(x\right) = \sqrt[4]{3 - 2x}$ is found by solving the inequality

$$3 - 2x \geq 0.$$

The solution is

$$x \leq \tfrac{3}{2}.$$

Therefore the domain is $\left(-\infty, 1.5\right]$.

Method 2: The function is a horizontal shift of $\sqrt[4]{x}$ left by 3 units followed by a horizontal compression by a factor of 1/2 and then a horizontal reflection. The point (0,0) is mapped as follows:

$$(0, 0) \rightarrow (-3, 0) \rightarrow (-1.5, 0) \rightarrow (1.5, 0)$$

It is the horizontal reflection that makes the graph open to the left rather than the right.

The domain is $\left(-\infty, 1.5\right]$.

The range remains $\left(0, \infty\right)$ since there are no vertical transformations.

The domain and range can be seen in the graph below.

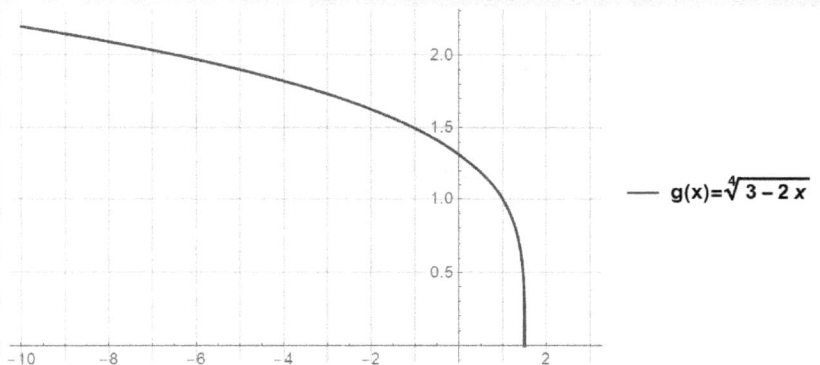

$g(x) = \sqrt[4]{3 - 2x}$

b. Notice that $h\left(x\right) = -3\sqrt[4]{x}$ is a vertical reflection and stretch by a factor of 3 of the function $\sqrt[4]{x}$. This tells us that the properties associated with the output value will change so we need to consider the range carefully.

The domain will be $\left[0, \infty\right)$ since there are no horizontal transformations.

The vertical reflection does effect the range. Since the fourth root function outputs positive numbers or zero, when we multiply by a negative number will will have negative values or zero. Therefore, the range is $\left(-\infty, 0\right]$.

The vertical reflection and range is shown in the graph below.

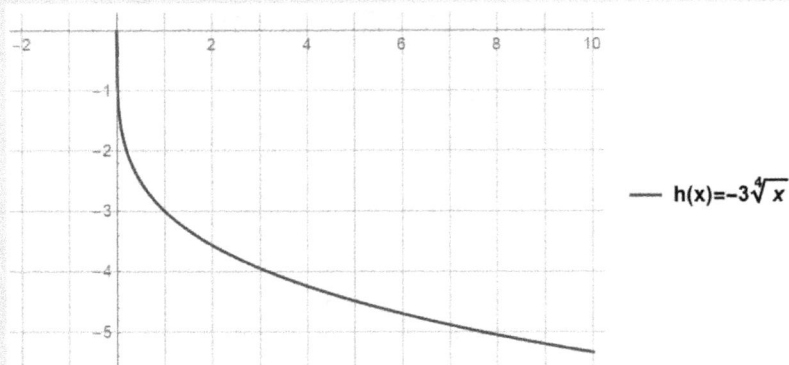

$$h(x) = -3\sqrt[4]{x}$$

Intercepts of Even Root Functions

Transformations of even root functions may or may not have x or y intercepts. If $x = 0$ is in the domain of the transformed function then there will be a y-intercept found by evaluating $f(0)$. If $y = 0$ is in the range, then there will be an x-intercept and we solve $f(x) = 0$.

EXAMPLE 3: INTERCEPTS OF TRANSFORMATIONS OF EVEN ROOT FUNCTIONS

Find x-intercepts and y-intercepts for

a. $g(x) = \sqrt[4]{3 - 2x}$.

b. $h(x) = -3\sqrt[4]{x}$.

c. $f(x) = \sqrt{x + 3} + 1$.

Answer

a. The x-intercept has $y = 0$. We solve the equation $\sqrt[4]{3 - 2x} = 0$ by raising both sides to the 4th power to get $3 - 2x = 0$. Finally, $x = 1.5$. The x-intercept is $(1.5, 0)$.

For the y-intercept, $f(0)$ is evaluated. We get $f(0) = \sqrt[4]{3 - 2(0)} = \sqrt[4]{3} \approx 1.316$. Therefore, the y-intercept is $(0, 1.316)$.

b. Since there is no horizontal or vertical shift both the x and y intercepts are $(0, 0)$.

c. Evaluate $f(0)$ to get $f(0) = \sqrt{0 + 3} + 1 \approx 2.73$. The y-intercept is approximately $(0, 2.73)$.

Since $y = 0$ is not in the range because the graph was shifted up one unit, there is no x-intercept.

TRY IT #1

Find the domain, range and intercepts of the fourth root function shifted 2 units right and 1 unit down.

Answer

The domain is $[2, \infty)$. The range is $[-1, \infty)$. There is no y-intercept because $x = 0$ is not in the domain. The x-intercept is $[3, 0)$.

TRY IT #2

Find the domain, x-intercepts and y-intercepts for $g(x) = \sqrt{7 - 0.5x}$.

Answer

The domain is $(-\infty, 14]$. The x-intercept is $(14, 0)$ and the y-intercept is $(0, 2.646)$. These values can be confirmed in the graph below.

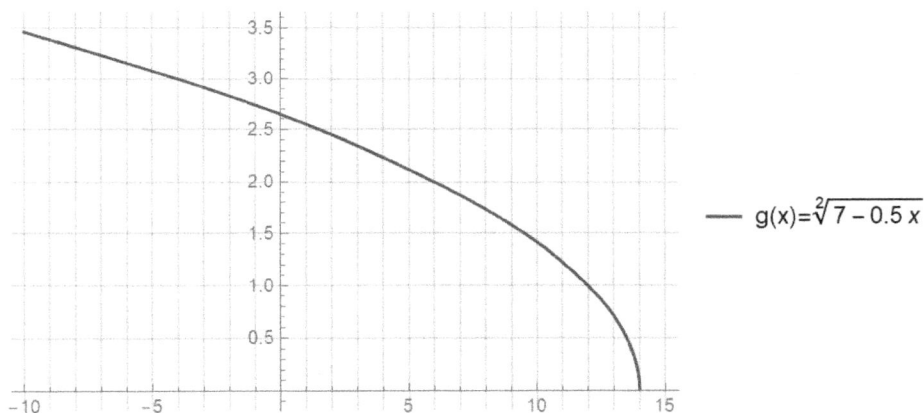

$$g(x) = \sqrt[2]{7 - 0.5\,x}$$

End Behavior of Even Root Functions

The final property to examine for even root functions and their transformations is the end or long term behavior. Since the domain is only part of the real numbers only behavior to the left or right needs to be determined depending on whether the domain goes toward minus infinity or plus infinity.

EXAMPLE 4: END BEHAVIOR OF A HORIZONTALLY REFLECTED EVEN ROOT FUNCTION

Determine the end behavior of $k\left(x\right)=\sqrt[6]{2-x}$.

Answer

$k\left(x\right)$ is a shift left by 2 units of $\sqrt[6]{x}$ followed by a horizontal reflection. The shift by 2 units will not effect the end behavior but the reflection will make it so the graph opens to the left. Since (0,0) transforms as follows

$$\left(0,0\right)\to\left(-2,0\right)\to\left(2,0\right)$$

the domain is $\left(-\infty,2\right)$. Right end behavior does not make sense for this function.

As x goes further and further to the left the output will become larger and larger. We write this as $x\to-\infty,f\left(x\right)\to\infty.$

We can confirm this in the table and graph below. For the table, even when we chose extremely large values for x, the output is not really large but we see that it continues to increase and does not level off.

x	$f\left(x\right)$
100	2.16
1,000	3.16
10,000	4.64
1,000,000	10
1,000,000,000	31.6

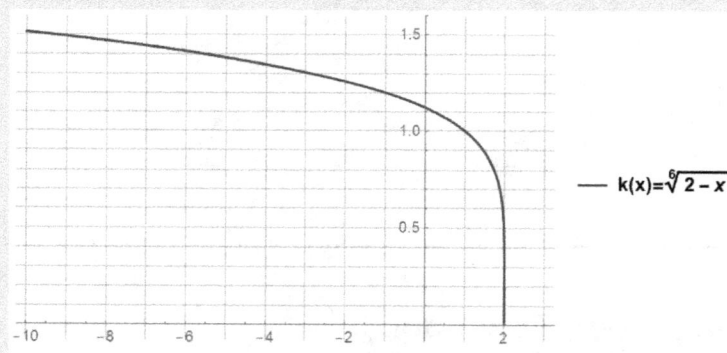

$$— \quad k(x)=\sqrt[6]{2-x}$$

Properties of Odd Root Functions

Odd root functions do not have their domains and ranges change under transformations since they are defined on $\left(-\infty,\infty\right)$. However with horizontal and vertical shifts, the intercepts are expected to change and if there is a horizontal or vertical reflection, the end behavior may be effected.

EXAMPLE 5: PROPERTIES OF A REFLECTED ODD ROOT FUNCTION

Determine the domain, range, x-intercept, y-intercept and end behavior of the function $f(x) = \sqrt[3]{3-x} + 1$.

Answer

This equation is a shift of $\sqrt[3]{x}$ left by 3 units and then a horizontal reflection. Finally the graph is shifted up 1 unit.

Since the domain and range of $\sqrt[3]{x}$ is all real numbers, these transformations will not effect these properties. Therefore the domain and range of $k(x)$ are $(-\infty, \infty)$.

To find the x-intercept we solve the equation $\sqrt[3]{3-x} + 1 = 0$. Begin by subtracting one from each side to get $\sqrt[3]{3-x} = -1$. Next, cube both sides to get $3 - x = -1$. Finally $x = 4$. The x-intercept is $(4, 0)$.

To find the y-intercept we evaluate $f(0)$. We get $f(0) = \sqrt[3]{3-0} + 1 = \sqrt[3]{3} + 1 \approx 2.442$. The y-intercept is $(0, 2.442)$.

The end behavior will be similar to $\sqrt[3]{-x}$ or $-\sqrt[3]{x}$ since odd root functions are odd. Therefore, as x becomes more and more negative, $f(x)$ increases without bound and as x becomes larger and larger $f(x)$ decreases without bound. We write, as $x \to -\infty$, $f(x) \to \infty$ and $x \to \infty$, $f(x) \to -\infty$.

These properties can be confirmed in the graph below.

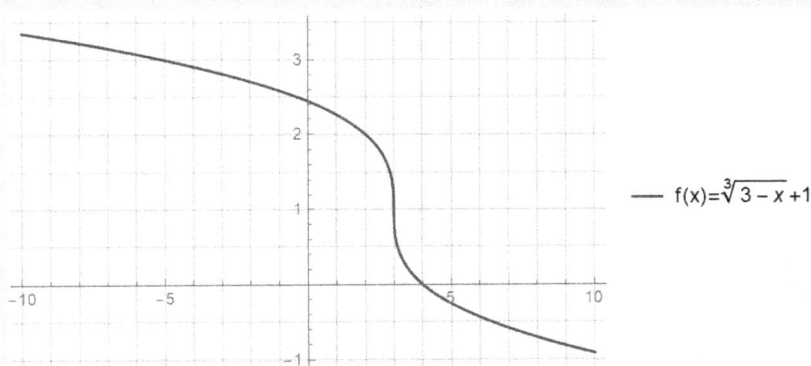

$$f(x) = \sqrt[3]{3-x} + 1$$

KEY CONCEPTS

- Root functions are steep near the origin and then grow slowly.
- The domain, range, intercepts and end behavior may change when even root functions are transformed.

  ◦ Intercepts may not exist for all transformed even root functions.
  ◦ Only one side of end behavior makes sense for transformed even root functions.

- The domain and range for transformed odd root functions remains $(-\infty, \infty)$
- The intercepts and end behavior may change when odd root functions are transformed. However, there will and x and y intercept and the end behavior must be considered on both the right and left.

Basic Functions and Identities

Graphs of the Parent Functions

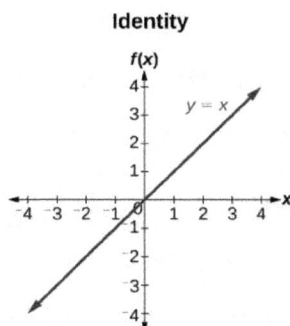

Identity

$$f(x)$$

$$y = x$$

Domain: $(-\infty, \infty)$
Range: $(-\infty, \infty)$

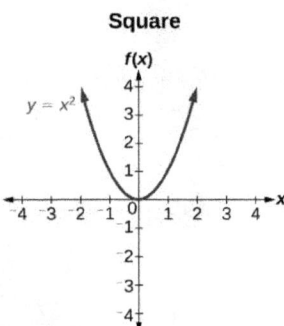

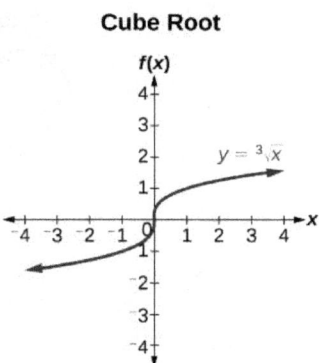

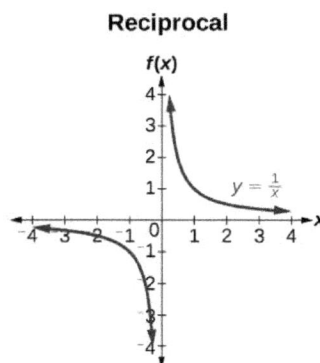

Square

$$f(x)$$

$$y = x^2$$

Domain: $(-\infty, \infty)$
Range: $[0, \infty)$

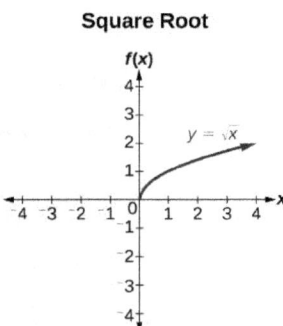

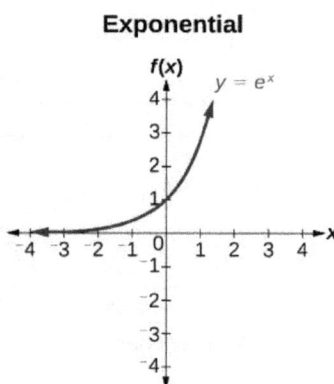

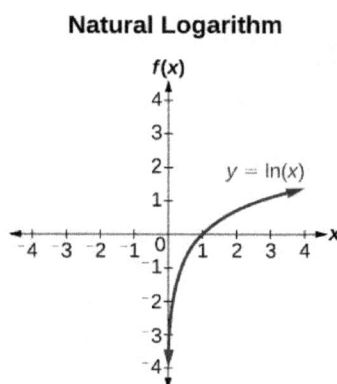

Square Root

$$f(x)$$

$$y = \sqrt{x}$$

Domain: $[0, \infty)$
Range: $[0, \infty)$

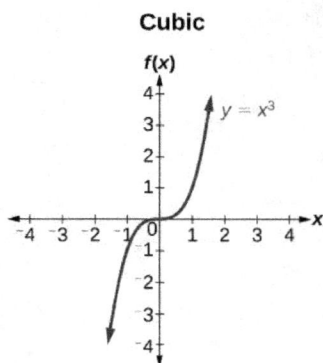

Cubic

$$f(x)$$

$$y = x^3$$

Domain: $(-\infty, \infty)$
Range: $(-\infty, \infty)$

Cube Root

$$f(x)$$

$$y = \sqrt[3]{x}$$

Domain: $(-\infty, \infty)$
Range: $(-\infty, \infty)$

Reciprocal

$$f(x)$$

$$y = \frac{1}{x}$$

Domain: $(-\infty, 0) \cup (0, \infty)$
Range: $(-\infty, 0) \cup (0, \infty)$

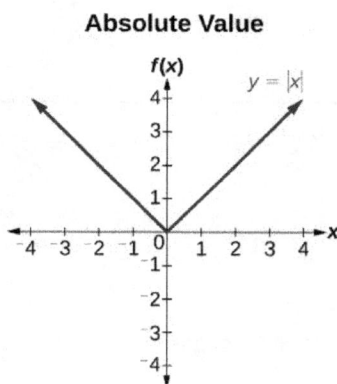

Absolute Value

$$f(x)$$

$$y = |x|$$

Domain: $(-\infty, \infty)$
Range: $[0, \infty)$

Exponential

$$f(x)$$

$$y = e^x$$

Domain: $(-\infty, \infty)$
Range: $(0, \infty)$

Natural Logarithm

$$f(x)$$

$$y = \ln(x)$$

Domain: $(0, \infty)$
Range: $(-\infty, \infty)$

Graphs of the Trigonometric Functions

Sine

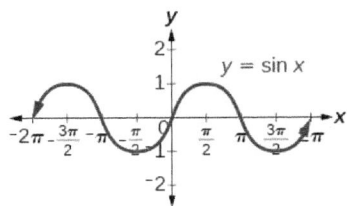

Domain: $(-\infty, \infty)$
Range: $(-1, 1)$

Cosine

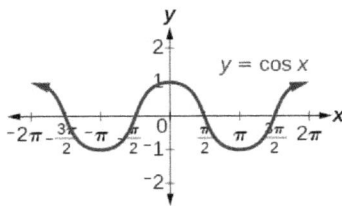

Domain: $(-\infty, \infty)$
Range: $(-1, 1)$

Tangent

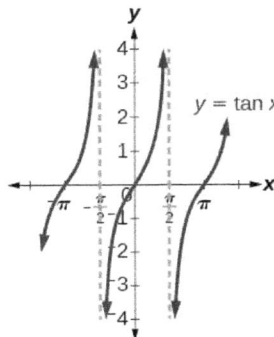

Domain: $x \neq \frac{\pi}{2}k$,
where k is an odd integer
Range: $(-\infty, \infty)$

Cosecant

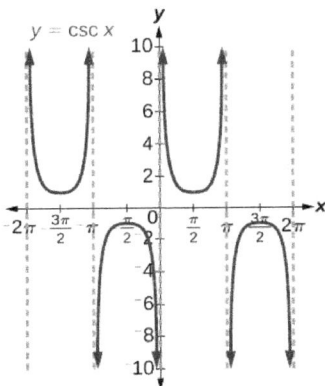

Domain: $x \neq \pi k$,
where k is an integer
Range: $(-\infty, -1] \cup [1, \infty)$

Secant

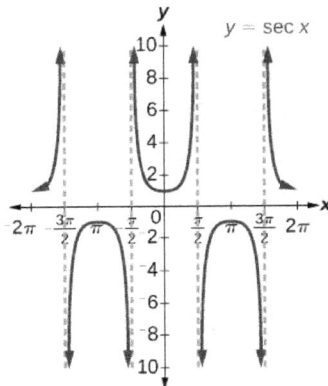

Domain: $x \neq \frac{\pi}{2}k$,
where k is an odd integer
Range: $(-\infty, -1] \cup [1, \infty)$

Cotangent

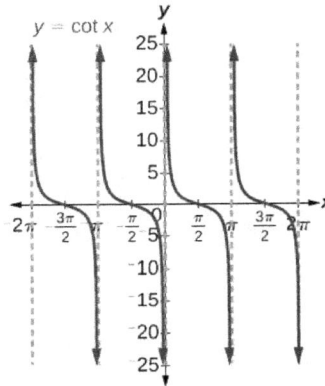

Domain: $x \neq \pi k$,
where k is an integer
Range: $(-\infty, \infty)$

Inverse Sine

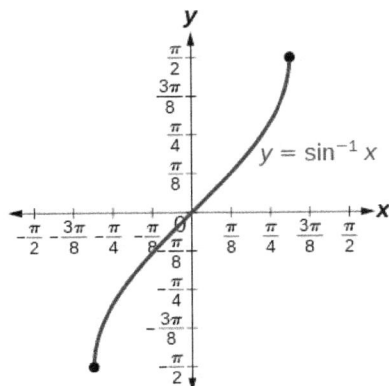

Domain: $[-1, 1]$
Range: $\left[-\frac{\pi}{2}, \frac{\pi}{2}\right]$

Inverse Cosine

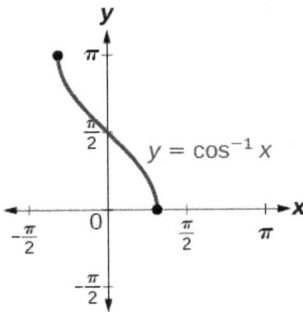

Domain: $[-1, 1]$
Range: $[0, \pi]$

Inverse Tangent

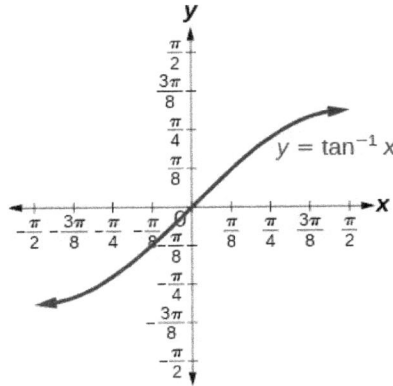

Domain: $(-\infty, \infty)$
Range: $\left(-\frac{\pi}{2}, \frac{\pi}{2}\right)$

Inverse Cosecant

$y = \csc^{-1} x$

Domain: $(-\infty, -1] \cup [1, \infty)$
Range: $\left[-\frac{\pi}{2}, 0\right) \cup \left(0, \frac{\pi}{2}\right]$

Inverse Secant

$y = \sec^{-1} x$

Domain: $(-\infty, -1] \cup [1, \infty)$
Range: $\left[0, \frac{\pi}{2}\right) \cup \left(\frac{\pi}{2}, \pi\right]$

Inverse Cotangent

$y = \cot^{-1} x$

Domain: $(-\infty, \infty)$
Range: $\left[-\frac{\pi}{2}, 0\right) \cup \left(0, \frac{\pi}{2}\right]$

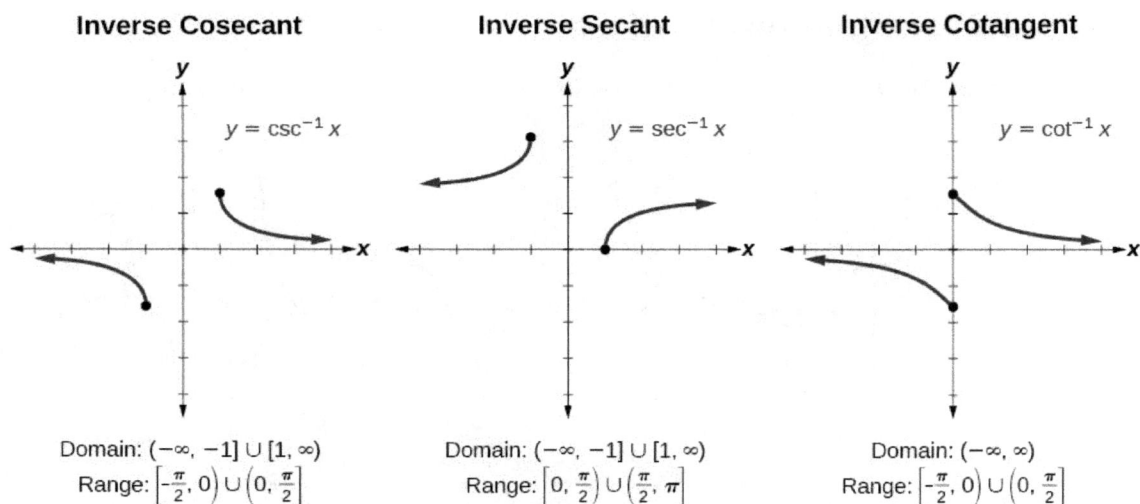

Trigonometric Identities

Pythagorean Identities	$\cos^2 t + \sin^2 t = 1$ $1 + \tan^2 t = \sec^2 t$ $1 + \cot^2 t = \csc^2 t$
Even-Odd Identities	$\cos(-t) = \cos t$ $\sec(-t) = \sec t$ $\sin(-t) = -\sin t$ $\tan(-t) = -\tan t$ $\csc(-t) = -\csc t$ $\cot(-t) = -\cot t$
Cofunction Identities	$\cos t = \sin\left(\frac{\pi}{2} - t\right)$ $\sin t = \cos\left(\frac{\pi}{2} - t\right)$ $\tan t = \cot\left(\frac{\pi}{2} - t\right)$ $\cot t = \tan\left(\frac{\pi}{2} - t\right)$ $\sec t = \csc\left(\frac{\pi}{2} - t\right)$ $\csc t = \sec\left(\frac{\pi}{2} - t\right)$
Fundamental Identities	$\tan t = \frac{\sin t}{\cos t}$ $\sec t = \frac{1}{\cos t}$ $\csc t = \frac{1}{\sin t}$ $\cot t = \frac{1}{\tan t} = \frac{\cos t}{\sin t}$

Sum and Difference Identities	$\cos(\alpha + \beta) = \cos\alpha\cos\beta - \sin\alpha\sin\beta$ $\cos(\alpha - \beta) = \cos\alpha\cos\beta + \sin\alpha\sin\beta$ $\sin(\alpha + \beta) = \sin\alpha\cos\beta + \cos\alpha\sin\beta$ $\sin(\alpha - \beta) = \sin\alpha\cos\beta - \cos\alpha\sin\beta$ $\tan(\alpha + \beta) = \frac{\tan\alpha + \tan\beta}{1 - \tan\alpha\tan\beta}$ $\tan(\alpha - \beta) = \frac{\tan\alpha - \tan\beta}{1 + \tan\alpha\tan\beta}$
Double-Angle Formulas	$\sin(2\theta) = 2\sin\theta\cos\theta$ $\cos(2\theta) = \cos^2\theta - \sin^2\theta$ $\cos(2\theta) = 1 - 2\sin^2\theta$ $\cos(2\theta) = 2\cos^2\theta - 1$ $\tan(2\theta) = \frac{2\tan\theta}{1 - \tan^2\theta}$
Half-Angle Formulas	$\sin\frac{\alpha}{2} = \sqrt{\frac{1 - \cos\alpha}{2}}$ $\cos\frac{\alpha}{2} = \sqrt{\frac{1 + \cos\alpha}{2}}$ $\tan\frac{\alpha}{2} = \sqrt{\frac{1 - \cos\alpha}{1 + \cos\alpha}}$ $\tan\frac{\alpha}{2} = \frac{\sin\alpha}{1 + \cos\alpha}$ $\tan\frac{\alpha}{2} = \frac{1 - \cos\alpha}{\sin\alpha}$
Reduction Formulas	$\sin^2\theta = \frac{1 - \cos(2\theta)}{2}$ $\cos^2\theta = \frac{1 + \cos(2\theta)}{2}$ $\tan^2\theta = \frac{1 - \cos(2\theta)}{1 + \cos(2\theta)}$
Product-to-Sum Formulas	$\cos\alpha\cos\beta = \frac{1}{2}\left[\cos(\alpha - \beta) + \cos(\alpha + \beta)\right]$ $\sin\alpha\cos\beta = \frac{1}{2}\left[\sin(\alpha + \beta) + \sin(\alpha - \beta)\right]$ $\sin\alpha\sin\beta = \frac{1}{2}\left[\cos(\alpha - \beta) - \cos(\alpha + \beta)\right]$ $\cos\alpha\sin\beta = \frac{1}{2}\left[\sin(\alpha + \beta) - \sin(\alpha - \beta)\right]$
Sum-to-Product Formulas	$\sin\alpha + \sin\beta = 2\sin\left(\frac{\alpha + \beta}{2}\right)\cos\left(\frac{\alpha - \beta}{2}\right)$ $\sin\alpha - \sin\beta = 2\sin\left(\frac{\alpha - \beta}{2}\right)\cos\left(\frac{\alpha + \beta}{2}\right)$ $\cos\alpha - \cos\beta = -2\sin\left(\frac{\alpha + \beta}{2}\right)\sin\left(\frac{\alpha - \beta}{2}\right)$ $\cos\alpha + \cos\beta = 2\cos\left(\frac{\alpha + \beta}{2}\right)\cos\left(\frac{\alpha - \beta}{2}\right)$
Law of Sines	$\frac{\sin\alpha}{a} = \frac{\sin\beta}{b} = \frac{\sin\gamma}{c}$ $\frac{a}{\sin\alpha} = \frac{b}{\sin\beta} = \frac{c}{\sin\gamma}$
Law of Cosines	$a^2 = b^2 + c^2 - 2bc\cos\alpha$ $b^2 = a^2 + c^2 - 2ac\cos\beta$ $c^2 = a^2 + b^2 - 2ab\cos\gamma$